Princ... ...lth
and ... ork

seventh ed...

Allan

Chartered

Hascom Ne...

Published by IOSH Services Limited

Publishing history

First edition   June 1991
Supplement   March 1992
Second edition   January 1993
Third edition   February 1995
Fourth edition   July 1997
Fifth edition   August 1999
(revised)   October 2000
Sixth edition   September 2002
(revised)   July 2003
Seventh edition   September 2005

ISBN-10 0 901357 38 3
ISBN-13 978 0 901357 38 0

Publisher's note

Published by IOSH Services Limited, The Grange, Highfield Drive, Wigston, Leicestershire LE18 1NN, UK

Printed in England by the Lavenham Press Limited

Contents

Part 1: Safety management techniques

Part 2: Workplaces and work equipment

Part 3: Occupational health and hygiene

Part 4: Law

Self-assessment questions and answers

Foreword

A successful business cannot run without healthy employees and a healthy reputation. The public is now more informed than ever about health and safety issues as a result of the saturation of news in the media, forcing businesses to consider more than just the bottom line of production and profit. It is not enough for a business to produce something a consumer wants to buy – the product must be made responsibly. Companies have a clear ethical and legal duty to protect their employees and the general public.

But, quite simply, investing in health and safety should also save your company money by reducing costs, increasing efficiency and improving the morale of your employees. In some cases, it can even avert an incident which can threaten your business's future survival.

Principles is essential reading for NEBOSH Certificate students, 'Managing safely' students and safety representatives for good reason. It is a clear and concise introductory text, providing all the legal, practical and referential tools required to ensure best practice is followed. I highly recommend it to both students and managers who need to understand the basics of health and safety.

Adam Crozier
Chief executive, Royal Mail

About the author

Allan St John Holt is one of Britain's best-known safety experts. He has also lectured widely in the USA, Canada, the Asia–Pacific region and Australia on health and safety matters. Twice President of IOSH and a Fellow of the Institution, he is a Chartered Safety and Health Practitioner. Allan founded the profession's examining body, NEBOSH, in 1979.

Now a non-executive chairman of the consultancy Hascom Network Ltd, he has been global director of environment, health and safety for Bovis Lend Lease and is currently head of health and safety management at Royal Mail Group.

An Ambassador of Veterans of Safety International, he was inducted into the Safety and Health Hall of Fame International in 1997 and in 2000 received the Distinguished Service to Safety Award of the US National Safety Council. A magistrate in Southampton since 1987, Allan was elected honorary president of the Southampton Occupational Safety Association in 1991.

Introduction

It is hard to believe that fourteen years have passed since the first edition of *Principles of health and safety at work* was published: so much has changed since then in the health and safety world. The seventh edition has been further expanded to include recent changes in the law, and developments in professional practice. Although the contents are not set out to follow the syllabus for the NEBOSH General Certificate, all the significant points within the revised April 2005 syllabus can be found here. Students should not assume that the self-assessment questions which follow most of the sections, or their answers, reflect the degree of difficulty of any particular examination's likely questions and the standard of answer looked for by examiners.

Rather than being 'just an exam textbook', *Principles* is designed for use before, during and after the examination as a desk reference which aims to provide basic information in a readable and helpful way.

It does not cover product and machinery safety requirements, or delve deeply into the environment and the management of waste, which are the subject of detailed laws and standards. Readers are strongly advised to acquaint themselves with these, since there are links between them and the regulations and topics discussed in this book.

As always, the comments of many readers have been used to improve the book, and my thanks go to all those who have suggested improvements. In particular, I would like to thank my colleagues Jim Allen, Dr Nicholas Batten and Geoff Morris. At Royal Mail I am grateful for the continued support of Dr Steve Boorman, director of corporate social responsibility, and my many professional colleagues whose support has made this seventh edition possible.

Allan St John Holt
Southampton
September 2005

Part 1
Safety management techniques

1 Accident prevention

Introduction

Few of the things we do in life are free from risk or uncertainty. As Benjamin Franklin observed as long ago as 1789: "Nothing in this world can be said to be certain, except death and taxes." Nevertheless, we order our affairs in the reasonable expectation that things usually turn out as planned. Nobody goes to work prepared to die that day, but sadly some people do. Accident prevention is aimed at spotting what could go wrong and preventing it from doing so, or at least minimising the consequences. Since the subject first received significant attention – from lawmakers about 200 years ago and from research over the past 100 years – we have learned a good deal about the 'accident phenomenon'.

Accidents are the direct results of unsafe activities and conditions, both of which can be controlled by management. Management is responsible for the creation and maintenance of the working environment, into which workers must fit and interact. Control of this environment is discussed elsewhere in this book. Control of workers and their behaviour is more difficult. They have to be given information, and taught to understand that accidents are not inevitable but are caused.

Workers need training to develop skills, to recognise the need to develop and comply with safe systems of work, and to report and correct unsafe conditions and practices. Their health and safety awareness and attitudes require constant improvement, and the social environment of the workplace must be one which fosters good health and safety practices and conditions, not one which discourages them. Such an environment is known as a **positive safety culture**.

A primary requirement of management is to appreciate the need to concentrate on the nature of the accident phenomenon, rather than its outcome, the injury or damage/loss. Also, there is a need for awareness that the primary cause of an accident is not necessarily the most important feature; secondary causes, usually system failures, will persist unless action is taken. Thus, a 'simple' fall from a ladder may be dismissed as 'carelessness', but this label may hide other significant factors, such as lack of training, maintenance, adequate job planning and instruction, and no safe system of work. These topics will be discussed in this Part.

Three statements must be adopted by management in order to achieve success by planning rather than by chance:
1   accidents have causes
2   steps must be taken to prevent them
3   accidents will continue to happen if these steps are not taken.

Accident prevention objectives

Moral objectives derive from the concept that a duty of reasonable care is owed to others. Greater awareness of the quality of life at, and as affected by, work has focused popular attention on the ability of employers to handle a wide variety of issues, previously seen only as marginally relevant to the business enterprise. Environmental affairs, pollution, product safety and other matters are now commonly discussed, and there is a growing belief that it is simply morally unacceptable to put the safety and health of others, inside or outside the workplace, at risk, for profit or otherwise. Physical pain and hardship resulting from death and disability is impossible to quantify. Moral obligations are now more in the minds of employers and business stakeholders than ever before.

A dimension of the moral objective is morale, which also interlinks with the following two objectives. Workers' morale is strengthened by active participation in accident prevention programmes, and weakened following accidents. Adverse publicity affects the fortunes of an organisation both internally in this way and externally, as reduced public confidence may weaken local community ties, market position, stakeholder confidence, market share and reputation generally.

Legal objectives are given in statute law, which details steps to be taken and carries the threat of prosecution or other enforcement action as a consequence of failure to comply. Civil law enables injured workers and others to gain compensation as a result of a breach of statutory duties or because a reasonable standard of care was not provided under the circumstances.

Economic objectives are to ensure the continuing financial health of a business and avoid the costs associated with accidents. These include monetary loss to employers, to the community and society from injuries to workers, damage to property and work interruptions. Some, but not all, of these costs are insurable and are known as direct costs. Increased premiums will be a consequence of claims, so an increase in overheads is predictable following accidents. Indirect costs include uninsured property damage, delays, overtime costs, management time spent on investigations, and decreased output from the replacement worker(s).

Basic terms

An **accident** is an **incident** plus its consequences; the end product of a sequence of events or actions resulting in an undesired consequence (injury, property damage, interruption and/or delay). The incident is that sequence of events or actions. An incident does not necessarily have a definable start or finish. (Think about a road bulk tanker overturning and spilling its contents onto the road and down drains. Can you say when the incident started or finished?) Thus, an **injury** is a consequence of failure – but not the only possible one.

An **accident** can be defined more fully as 'an undesired event, which results in physical harm and/or property damage, usually resulting from contact with a source of energy above the ability of the body or structure to withstand it'. The idea of energy transfer as part of the definition of an accident is a relatively recent one, and one which helps us to understand the accident process.

Hazard means the inherent property or ability of something to cause harm – the potential to interrupt or interfere with a process or person – which is or may be causally related to an accident, by itself or with other variables. Hazards may arise from components which interact with or influence each other, eg two chemicals reacting to produce a third.

Risk is the chance or probability of loss, an evaluation of the potential for failure. It is easy to confuse the terms 'hazard' and 'risk', and many writers have done so. The terms are often incorrectly used, sometimes interchanged. A simple way to remember the difference is that 'hazard' describes potential for harm, while 'risk' is the likelihood that harm will result in the particular situation or circumstances.

Another way of defining risk is as the probability that a hazard will result in an accident with definable consequences. In a wider sense, we can look at 'risk' as the product of the severity of the consequences of any failure and the likelihood of that failure occurring. Thus, an event with a low probability of occurrence but a high severity can be compared against an event likely to happen relatively often but with a comparatively trivial consequence. Comparisons between risks can be made using simple numerical formulae (see Part 1 Section 3).

Accident ratio studies

There have been several studies on the frequency of the possible outcomes of incidents, and an interest in this was one of the influences on H W Heinrich, an American safety specialist whose *Industrial accident prevention* was first published in 1931 and became the first widely available textbook for health and safety practitioners. In it, he developed the Domino Theory of accident causation and distinguished between unsafe acts and unsafe conditions as major causative agents (see below). Seventy years ago, he found that less than 10 per cent of all 'accidents' result in personal injury. From a study of 75,000 reported accidents, he saw that, on average, for every disabling injury there were 29 less serious injuries and 300 accidents involving no personal injury at all. Much depends on the definitions chosen, but the principle was new then and holds good today.

Heinrich's work on the outcomes of incidents was repeated by Frank Bird (1969) and in the UK by Tye & Pearson (1974/75) and, most recently, by the Health and Safety Executive (HSE) Accident Prevention Advisory Unit. The results of frequency outcomes are usually represented in a triangle:

The frequencies shown are taken from recent HSE work in five industries, including construction, food and transport. They will vary between employers according to the nature of the hazards and risk levels. 'Non-injury accidents' need to be measured, and conventionally this is done by recording incidents resulting in property damage rather than people damage.

The main conclusions that can be drawn from this work are:
* injury incidents are less common than non-injury incidents
* there is a consistent relationship between all loss producing incidents
* non-injury incidents could have been injury incidents, and the outcome in each case is mostly determined by chance.

Overall, because the consequences of incidents are usually distributed randomly throughout a range of outcomes (with death at one end of the range through to property damage and 'near miss' at the other), we cannot distinguish usefully between those which result in injury and those that do not. Therefore, it is important to look at all incidents as sources of information on what is going wrong; relying on injury records only allows a look at a minority of incidents which happened to result in a serious consequence. Each incident can offer information as to its cause, and limiting study to those resulting in physical injury wastes the opportunity to gain more information about system and control failures. Furthermore, this is why prevention efforts need to be aimed at loss prevention as a whole.

Accident causes

Immediate or primary causes of accidents are often grouped into unsafe acts and unsafe conditions. This is convenient, but can be misleading as accidents typically include both groups of causes at some stage in the chain of causation.

Unsafe acts can include:
* working without authority
* failure to warn others of danger
* leaving equipment in a dangerous condition
* using equipment at the wrong speed
* disconnecting safety devices such as guards
* using defective equipment
* using equipment the wrong way or for wrong tasks
* failure to use or wear personal protective equipment
* bad loading of vehicles
* failure to lift loads correctly
* being in an unauthorised place
* unauthorised servicing and maintaining of moving equipment
* horseplay
* smoking in areas where this is not allowed
* drinking alcohol or taking drugs.

Unsafe conditions can include:
* inadequate or missing guards to moving machine parts

- defective tools or equipment
- inadequate fire warning systems
- fire hazards
- hazardous atmospheric conditions
- excessive noise
- exposure to radiation
- inadequate illumination or ventilation.

These are all deviations from required safe practice, but they must be seen as the symptoms of more basic underlying indirect or secondary causes which allow them to exist and persist.

Indirect causes include:
- management system pressures:
  - financial restrictions
  - lack of commitment
  - lack of policy
  - lack of standards
  - lack of knowledge and information
  - restricted training and selection for tasks
- social pressures:
  - group attitudes
  - trade customs
  - tradition
  - societal attitudes to risk-taking
  - 'acceptable' behaviour in the workplace.

Direct and indirect costs of accidents

It is often assumed, wrongly, that the compulsory employers' liability insurance covers the costs of accidents at work. Research by the HSE and other international authorities has demonstrated that the 'below the line' or indirect and uninsurable costs far exceed the insurable or recoverable costs. This is known as the 'iceberg effect', since the material below the water is greater than that which is easily visible. There are more than 20 recognised headings for indirect costs, which include legal fees and fines, supervisory time spent on investigations, clean-up costs, loss of experience and expertise, reputation damage and rehabilitation.

Indirect costs were shown in a short HSE study to be at best eight times the direct costs and at worst 34 times, so the difference is considerable. It can be seen that the indirect

costs are much harder to calculate, and will always be different for each organisation. Despite the effort required to do the analysis, this will be repaid when it brings the true loss to the organisation to the attention of management.

Priorities in prevention

Basic principles should be observed in setting up strategies for control and management of health and safety at work. These are:
- if possible, avoid a risk altogether by eliminating the hazard
- tackle risks at source – avoid the quick temporary fix, or putting up a sign where a better physical control could be used (eg quieten machines rather than provide personal protective equipment or erect warning signs)
- adapt work to the individual when designing work areas and selecting methods of work
- use technology to improve conditions
- where options are available, give priority to protection for the whole workplace rather than to individuals (eg protect a roof edge rather than supply safety harnesses)
- ensure that everyone understands what they have to do to be safe and healthy at work
- make sure that health and safety management is accepted by everyone, and that it applies to all aspects of an organisation's activities.

More information on strategies is given in Part 1 Section 2.

Principles of accident prevention

In summary, these are: the systematic use of techniques to identify and remove hazards, the control of risks which remain, and the use of techniques to influence behaviour and encourage safe attitudes. They are the primary responsibility of management, and are discussed in the next Sections.

Section 2 covers techniques of health and safety management, including the central role of the management organisation and policy needed to put them into practice, Section 3 explores the assessment of risk, and Section 4 looks at successful written health and safety policies. Detailed treatments of practical aspects are contained in Parts 2 and 3 of this book.

Revision

3 accident prevention objectives:
- moral/morale
- legal
- economic

Definitions: accident, incident, hazard, risk

Accident causes are immediate or indirect, and involve unsafe acts and/or conditions

3 principles of accident prevention:
- use techniques to identify and remove hazards
- use techniques to assess and control remaining risks
- use techniques to influence behaviour and attitudes

Selected references

Guidance (HSE)

HSG65   *Successful health and safety management*
INDG355   *Reduce risks – cut costs: the real cost of accidents and ill-health at work*

Self-assessment questions

1 Explain the difference between hazard and risk, using examples from your workplace.

2 Write a short summary of an accident with which you are familiar, and list the immediate and indirect cause(s).

3 Add another layer to the bottom of the 'accident triangle'. How could you classify the incidents that could be included there? How could they be counted?

2 Techniques of health and safety management

Introduction

Health and safety management is concerned with, and achieved by, all the techniques which promote the subject. Some are described in other Parts of this book; some will be considered in Sections later in this Part. Safety management is also concerned with influencing human behaviour, and with limiting the opportunities for mistakes to be made which could result in harm or loss. To do this, safety management must take into account the ways in which people fail (ie fail to do what is expected of them and/or what is safe). The techniques are aimed at the recognition and elimination of hazards, and the assessment and control of those risks which remain.

Objectives

The practical objectives of safety management are:

- gaining support from all concerned for the health and safety effort and fostering a positive safety culture
- motivation, education and training so that everyone can recognise and correct hazards
- achieving hazard and risk control by design and purchasing policies
- operation of a suitable inspection programme to provide feedback (see Section 10)
- ensuring hazard control principles form part of supervisory training
- devising and introducing controls based on risk assessments
- complying with regulations and standards.

To achieve these objectives, a safety policy statement is required. The design of this policy and other relevant considerations are discussed in Part 1 Section 4.

Benefits

Successful safety management can lead to substantial cost savings, as well as a good accident record. Some companies have become well known for the success of their safety management system. Du Pont claims that several of its plants, each with more than 1,000 employees, have run for more than 10 years without recording a single lost-time injury accident. Du Pont has 10 principles of safety management, which are worthy of study:

1 all injuries and occupational illnesses are preventable

2 management is directly responsible for doing this, with each level accountable to the one above and responsible for the level below

3 safety is a condition of employment, and is as important to the company as production, quality or cost control

4 training is required in order to sustain safety knowledge, and includes establishing procedures and safety performance standards for each job

5 safety audits and inspections must be carried out

6 deficiencies must be corrected promptly, by modifications, changing procedures, improved training and/or consistent and constructive disciplining

7 all unsafe practices, incidents and injury accidents will be investigated

8 safety away from work is as important as safety at work

9 accident prevention is cost-effective; the highest cost is human suffering

10 people are the most critical element in the health and safety programme. Employees must be actively involved and complement management responsibility by making suggestions for improvements.

Key elements

The five key elements of successful health and safety management are:

- policy
- organisation
- planning and implementation
- measuring performance
- reviewing performance and auditing.

The reader interested in a complete treatment of this topic is referred to the HSE publication HSG65 – *Successful health and safety management*. In general, the principles it presents are also to be found in internationally recognised safety management standards.

It is important to appreciate that the five elements exist independently, not just in a cycle to be reviewed at intervals. Thus, the policy needs to be kept under continuous review for relevance, applicability, and other factors which will change over time.

Policy

Successful health and safety management demands comprehensive health and safety policies which are effectively implemented and which are considered in all business practice and decision-making. Effective policies contain commitments to continuous improvement and avoid references to 'compliance' alone, because that can foster a culture of 'doing just enough' to satisfy minimum legal requirements. At the heart of the policy must be an acceptance of successful health and safety management as a key business aim of the organisation, equal in importance to more obvious aims such as profit and financial viability.

Some jurisdictions require written safety policy statements to be created by all employers, except for the smallest organisations. This has been the case in the United Kingdom since April 1975, and is simply a reflection in the law of what has been known for many years – written policies are the centrepieces of good health and safety management. They insist, persuade, explain and assign responsibilities. An essential requirement for management involvement at all levels is to define health and safety responsibility in detail within the written document, and then to check at intervals that the responsibility has been adequately discharged. This process leads to ownership of the health and safety programme, and it is based on the principle of accountability discussed in Section 4.

Organising

To make the health and safety policy effective, both management and employees must be actively involved and committed. Organisations which achieve high standards in health and safety create and sustain a 'safety culture' which motivates and involves all members of the organisation in the control of risks. They establish, operate and maintain structures and systems which are intended to:

* secure **control** – by ensuring managers lead by example, by the setting of performance standards and measuring against them, and by the appointment of a key senior person at the highest level to oversee and be responsible for the implementation of aspects of the policy. The visible accountability of individuals is an important aspect of control – performance reviews and appraisals should contain measures on individual safety performance
* encourage **co-operation** – both of employees and their safety representatives. The reinforcement of shared objectives and a heightened awareness of health and safety goals is an important aspect of the maintenance of a good safety culture
* secure effective **communication** – by providing information about hazards, risks and preventive measures. Visible behaviour of key senior people is an important tool and motivator in its own right – regular safety tours demonstrate interest and active involvement
* achieve **co-ordination** of activities – internally between departments and other operating areas, and externally with other organisations which interface with them
* ensure **competence** – by assessing the skills needed to carry out all tasks safely, and then providing the means to ensure that all employees (including temporary ones) are adequately instructed and trained. The Management of Health and Safety at Work Regulations 1999 (MHSWR) encourage this by requiring employers to recruit, select, place, transfer and train on the basis of assessments and capabilities, and to ensure that appropriate channels are open for access to information and specialist advice

when required. Competence is not simply a question of training – it also involves the application of knowledge, skills and appearance.

Planning and implementation

Planning ensures that health and safety efforts really work. Success in health and safety relies on the establishment, operation and maintenance of planning systems which:

- **identify** objectives and targets that are attainable and relevant
- **set performance standards** for management, and for the control of risks, which are based on hazard identification and risk assessment and which take legal requirements as the accepted minimum standard
- **consider and control** risks both to employees and to others who may be affected by the organisation's activities, products and services
- ensure **documentation** of all performance standards
- **react** to changing internal and external demands
- **sustain** a positive safety culture.

Organisations which plan and control in this way can expect fewer injuries (and claims resulting from them), reduced insurance costs, less absenteeism, higher productivity, improved quality and lower operating costs.

Monitoring

Just like finance, production or sales, health and safety has to be monitored to establish the degree of success. For this to happen, two types of monitoring system need to be operated. These are:

- **active** monitoring systems – intended to measure the achievement of objectives and specified standards before things go wrong. This involves regular inspection and checking to ensure that standards are being implemented and that management controls are working properly
- **reactive** monitoring systems – intended to collect and analyse information about failures in health and safety performance

when things do go wrong. This will involve learning from mistakes, whether they result in accidents, ill health, property damage incidents or 'near misses'.

Information from both active and reactive monitoring systems should be used to identify situations that create risks and enable something to be done about them.

Priority should be given to the greatest risks. The information should then be referred to people within the organisation who have the authority to take any necessary remedial action, and also to effect any organisational and policy changes which may be necessary.

Reviewing and auditing performance

Auditing enables management to ensure that its policy is being carried out and that it is having the desired effect. Auditing complements the monitoring programme. Economic auditing of a company is well established as a tool to ensure economic stability, and it has been shown that similar systematic evaluation of performance in health and safety has equal benefits. It is a feature of good governance.

An audit is not the same as an inspection. Essentially, the audit assesses the organisation's ability to meet its own standards on a wide front, rather than providing a 'snapshot' of a particular site or premises.

The two main objectives of an audit are:

- to ensure that standards achieved conform as closely as possible to the objectives set out in the organisation's safety policy
- to provide information to justify continuation of the same strategy, or a change of course.

The best health and safety audit systems are capable of identifying deviations from agreed standards, analysing events leading to these deviations and highlighting good practice. They look especially at the 'software' elements of health and safety, such as systems of work, management practices, instruction, training and supervision, as well as the more traditional 'hardware' elements, which include machinery guarding and the use of personal protective equipment.

In organisations with advanced positive safety cultures, audits are welcomed as coaching opportunities and for their ability to allow high-performing work areas to demonstrate their success. Where they are seen as exercises in 'bayoneting the wounded', audits can actually worsen the safety culture. World-class audits go well beyond 'yes/no' scoring and evaluate the deployment of control systems as well as their technical presence. Experience shows that audit systems which include questions on environmental and health management, and even diversity and social policy issues, remove the need for independent audits in these areas, which is welcomed by local management often pressed for time. Quantified targets should be agreed in advance of auditing.

The process of **benchmarking** with other enterprises is very useful as a comparative measure, even though it is often difficult to ensure that the same topics are being accurately compared. Benchmarking partners can often be found through local safety groups and trade associations, with appropriate arrangements for confidentiality.

Key performance indicators are a matter for individual companies to adopt, but because accidents are essentially lagging measures which describe failures in controls, it is important to identify factors other than accidents which can demonstrate whether the control measures are working. For example, measures and targets can be set for management inspections, induction and refresher training achieved; for environmental management, illustrated by carbon dioxide production and water and energy use; and for the control of contractors on premises by interview and performance management.

People problems
In all control measures, reliance is placed on **human behaviour** to carry out the solutions, so a major task of health and safety

management is to assure safe behaviour by motivation, education, training and the creation of work patterns and structures which enable safe behaviour to be practised. In a major study in 1977, it was found that supervisors in the construction industry gave a variety of reasons for their inactivity on health and safety matters. In order of frequency, the most common responses were:
* resource limitations
* seen as outside the boundaries of their duties
* acceptance of hazards as inevitable
* influences of the social climate
* industry tradition
* lack of technical competence
* incompatible demands on their time
* reliance on the worker to take care
* lack of authority
* lack of information.

It can be seen that much of the inactivity of **supervision** can be corrected by establishing a favourable environment, with clear responsibilities given and accountability practised, together with necessary training in the complex nature of the accident phenomenon and in solutions to health and safety problems. Supervisors' and workers' attitudes to safety generally reflect their perception of the attitudes of their employer.

Attempts to motivate the individual meet with greatest success when persuasion rather than compulsion is used to achieve agreed health and safety objectives. Consultation with workers, through representatives such as trade unions, and locally through the formation of **safety committees**, is generally a successful strategy, provided that an adequate role is given to those being consulted. The importance of consulting employees on matters affecting their health and safety has been recognised in UK law for some time (see Part 4 Section 28), but is widely ignored.

The benefits of safety committees include the involvement of the workforce, encouragement to accept safety standards and rules, help in arriving at practicable solutions to problems, and recognition of hazards which may not be apparent to management.

At an individual level, an appreciation of some of the more common reasons why people fail to carry out tasks safely is useful for management. People may actually decide to act wrongly – sometimes in the belief that non-observance of safety rules is expected of them by managers, supervisors or work colleagues. Few deliberately decide to injure themselves, but a deliberate decision to err may be made, for example after a poor estimate of the risk of injury in an activity. Others err because of traps – where the correct action cannot be taken because it is physically impossible, for example as a result of the design of equipment. Valves placed out of reach, dials too far away to be properly read, poor ergonomic design of workstations – all contribute to the likelihood that mistakes will be made. Under these circumstances, it is not surprising that motivational aids such as posters are found to be relatively ineffective in producing safe behaviour. More information on human factors can be found in Part 1 Section 12, and on ergonomics in Part 3 Section 9.

Recent work in the USA, confirmed by studies in the UK at the University of Manchester Institute of Science and Technology (UMIST), has shown that a behaviour-based approach to health and safety management can be an effective tool for increasing safety on construction sites and elsewhere, despite some practical problems of implementation. The technique involves sampling, recording and publicising the percentage of safe (versus unsafe) behaviours, as noted by specially trained observers drawn from workforce and management. This gives more data on potential system and individual failures than could be obtained from a study of accident records. The attraction of the technique is that it offers **measurement** of potential for harm, independent of the accident record. Disadvantages may include the need to achieve an altered safety climate inside management and workforce to adopt the techniques, and employee suspicion of hidden motives for the observations.

Safety culture

Producing and maintaining a positive safety culture requires recognition of the significance of human factors elements (see Part 1 Section 12). Strategies need to be planned to identify and resolve these issues. Key points are:

- good communications between employees and management
- ensuring a real and viable commitment to high standards by senior management, including adequate budgetary provision
- maintaining a personal example by all levels of management
- maintaining good training standards to achieve and raise competence
- achievement of good working conditions.

Without a positive safety culture, the health and safety management effort is likely to consist of a 'command and control' atmosphere, heavy on rules and procedures, and heavy on discipline for offenders. Indicators which can measure the safety culture include absenteeism, average or mean length of absence following injury, staff turnover and complaints about working conditions.

In order to change the safety culture, there must be recognition from the most senior levels that behaviours and attitudes can be influenced directly or indirectly by management actions and activities and, for example, the way in which policy change is introduced. Often a change in culture is initiated and levered by expectations of society generally, and stakeholders particularly – such as the confidence of customers.

The acid test of safety culture is when unsafe behaviour is not accepted by anyone, and is challenged by all employees whenever it is seen. At this point, a 'can do' spirit is pervasive, and replaces previous mindsets which generate responses on the pattern of 'Here's why that won't work.'

Revision

6 **objectives of health and safety management:**
- to gain support from all concerned
- motivation, education and training
- achieving hazard control
- operating inspection programme
- devising and introducing risk controls
- compliance with regulations and standards

2 **monitoring systems:**
- active
- reactive

5 **key elements of health and safety management:**
- policy
- organisation
- planning and implementation
- measuring performance
- review and auditing

4 **safety culture factors:**
- communications
- commitment
- competence
- conditions

Self-assessment questions

1 What are the advantages of a works safety committee? Are there any disadvantages?

2 How is the success of your management programme evaluated? Are positive measurement methods used as well as negative ones?

3 Risk assessment

Introduction

There are at least two senses in which risk assessment has been carried out subconsciously over a long period. Firstly, we all make assessments many times each day of the relative likelihood of undesirable consequences arising from our actions in particular circumstances. Whether to cross a road by the lights or take a chance in the traffic is one such example. In making a judgement, we evaluate the chance of injury and also its likely severity, and in doing so we identify a hazard and evaluate the risk.

The second sense of risk assessment is based on the requirement for the employer under the Health and Safety at Work etc Act 1974, in many of its sections, to take 'reasonably practicable' precautions in various areas to safeguard employees. In doing this, we make a balanced judgement about the extent of the risk and its consequences against the time, trouble and cost of the steps needed to remove or reduce it. If the cost is 'grossly disproportionate', we are able to say that the steps are not reasonably practicable. Thus, in a very real sense, risk assessments have been carried out in the UK since at least 1974.

The difference between these assessments and those required by the Management of Health and Safety at Work Regulations 1999 (MHSWR) is that the significant results must be recorded by most employers, and information based on them is to be given to employees in a much more specific way than before. Those carrying out the statutory assessments found that some activities were being viewed afresh for the first time, and that the exercise was useful as it forced challenges to long-held assumptions about the safety of traditional work practices – which often did not stand up to the scrutiny. 'We've always done it that way' is neither an assessment nor a defence!

Benefits

Risk assessments are carried out to enable control measures to be devised. We need to have an idea of the relative importance of risks and to know as much about them as we can in order to take decisions on controls which are both appropriate and cost-effective.

It is not possible to produce a safety policy, together with arrangements and organisation for carrying it out, until risk assessments have been done – without them there will be no sure knowledge of the significant issues to be controlled.

Types of risk assessment

There are two major types of risk assessment, which are not mutually exclusive. The first type produces an objective probability estimate based on known risk information applied to the circumstances being considered – this is a **quantitative** risk assessment. The second type is subjective, based on personal judgement backed by generalised data on risk – the **qualitative** assessment. Except in cases of specially high risk, public concern, or where the severity consequences of an accident are felt to be large and widespread, qualitative risk assessments are adequate for the purpose, and much simpler to make. The legal requirements refer to this type of assessment unless the circumstances require more rigorous methods to be used.

Special cases

In addition to the general requirement for risk assessments, several sets of regulations contain their own more specific requirements, often specifying the parameters to be used. These include regulations governing substances hazardous to health, manual handling, work equipment and display screen equipment.

Some categories of workers require more detailed assessment because of their condition, including expectant and nursing mothers, people with disabilities, lone workers and young persons. Information on the more detailed assessment needs can be found in Part 4, where they appear in the commentary to the various regulations, particularly the MHSWR.

What to assess

Items of work equipment will require appropriate and probably individual risk assessment. So will tasks of various kinds. Where there is similarity of activities, and of the hazards and risks associated with them, although carried out in different physical areas or workplaces, a general risk assessment can be made which covers their common and basic features. This is known as a 'generic' or 'model' assessment, and may be included or referred to in the safety policy document.

There are likely to be situations in specific areas or on specific occasions when such an assessment will not be sufficiently detailed, and those circumstances should be indicated in the policy to alert to the need to take further action. There may also be work situations where hazards associated with particular situations will be unique, so that a special assessment must be made every time that the work is done. A method statement may be required (see Part 1 Section 5). Examples where this holds true include demolition work, the erection of steel structures and asbestos removal.

Contents of risk assessments

Significant findings are to be recorded where five or more people are employed. According to the MHSWR Approved Code of Practice (ACoP), the record should contain a statement of the significant hazards identified, the control measures in place and the extent to which they control the risk(s) (cross-references can be made to manuals and other documents), and the population exposed to the risk(s).

As the Regulations require review of assessments in specified circumstances, and as it will be necessary to review for changed circumstances over time, it would be prudent to include in the documentation a note of the date the assessment was made and the date for the next regular review. Similarly, it would be wise to include a note to employees reminding them of their duties under Regulation 14 of the MHSWR to inform the employer of any circumstances which might indicate a shortcoming in the assessment.

Risk assessments are living documents, and need review over time – usually annually. They must also be reviewed following changes in the process or pattern of work, or any other circumstances which could vary the risks to be addressed, or more appropriate control measures as technology advances. Accidents, cases of ill health and dangerous occurrences or near misses should always prompt a review of the relevant assessment(s).

In 1994 the HSE published guidance entitled *Five steps to risk assessment*, a leaflet setting out the most basic steps in the process and providing a specimen format for recording the findings of assessments. The leaflet has since been updated and reissued. The simplistic approach presented has much to commend it. Comments contained in the text give useful pointers to the approach to risk assessment which will satisfy minimum requirements in the view of the enforcing authority, for example:

"An assessment of risk is nothing more than a careful examination of what, in your work, could cause harm to people so that you can weigh up whether you have taken enough precautions or should do more..."

A more sophisticated approach will often be required where hazards and risks are changing frequently, as in the construction industry, and also where clients, work partners and other external organisations seek evidence of more detailed analysis than *Five steps to risk assessment* is likely to provide.

The test of assessments is that they shall be both 'suitable' and 'sufficient'. Plainly, the test is a simple one – does the assessment and the method chosen to carry it out pay enough attention to detail (ie neither too much nor too little)? It should record the hazards, the control measures in place, what more needs to be done to control the hazards, and details of those exposed to the hazard. Following the assessment, a re-evaluation of the remaining, or **residual**, risk should be made. Advice on acceptable levels of residual risk can be obtained from a variety of sources, including official guidance on what constitute 'reasonably practicable' controls.

Estimates of occupational health risks and decisions on acceptable controls for them are more difficult, not least because the hazards are less easy to detect and measure, and because their effects can take a considerable time to become apparent.

There is no shortage of more complicated commercial systems, often computer-based so as to enable rapid review and quick updating in changing conditions.

Hazard evaluation

The hazards to be identified are those associated with machinery, equipment, tools, procedures, tasks, processes and the physical aspects of the plant and premises – in other words, everything. Evaluation of the hazards is achieved by assembling information from those familiar with the hazards, such as insurance companies, professional societies, government departments and agencies, manufacturers, consultants, trade unions, old inspection reports (both internally and externally produced), accident reports and standards.

Some hazards may not be readily identifiable and there are techniques which can be applied to assist in this respect. These include inductive analysis techniques, which predict failures – Failure Modes and Effects Analysis (FMEA) is one of these; Job Safety Analysis (JSA) is another. Inductive analysis assumes failure has occurred and then examines ways in which this could have happened by using logic diagrams.

This is time-consuming and, therefore, expensive, but it is extremely thorough. MORT (Management Oversight and Risk Tree) Analysis is an example of an analytical technique which is not difficult to use.

Ranking hazards by risk

There may be occasions when a priority list for action will be called for. Techniques for producing a ranking vary from the very simple to the very complex. What is looked for is a priority list of hazards to be controlled, on a 'worst first' basis. There is no legal need to introduce numbers into a risk assessment, but many organisations find it useful to do so in order to show a 'before' and 'after' risk reduction within the assessment.

The following system offers a simple way to determine the relative importance of risks. It takes account of the consequence (likely severity) and the probability of the event occurring. Estimation of the first is easier than the second, as data may not be available for all hazards. Estimates derived from experience can be used. It is possible to carry out ranking using a simple formula, where risk = severity estimate x probability estimate. These estimates can be given any values, as long as they are consistently used. The simplest set of values provides a 16-point scale as illustrated overleaf.

The categories are capable of much further refinement. Words like 'time' can be defined, increasing if necessary the number of categories. Many organisations increase the categories to take account of numbers exposed to the hazard as well as the duration of exposure. However, the more precise the definitions, the more necessary it will be to possess accurate predictive data.

It should also be noted that ranking systems of this kind introduce their own problems, which must be addressed or at least be known to the user. One is that long-term health hazards may receive inadequate evaluation because of lack of data. Another is that hazards of low severity but high frequency can produce the same risk

Severity rating of hazard	Value
Catastrophic – Imminent danger exists, hazard capable of causing death and illness on a wide scale	1
Critical – Hazard can result in serious illness, severe injury, property and equipment damage	2
Marginal – Hazard can cause illness, injury or equipment damage, but the results would not be expected to be serious	3
Negligible – Hazard will not result in serious injury or illness, remote possibility of damage or injury beyond minor first aid case	4

Probability rating of hazard	Value
Probable – Likely to occur immediately or shortly	1
Reasonably probable – Probably will occur in time	2
Remote – May occur in time	3
Extremely remote – Unlikely to occur	4

score through multiplication as high severity, low frequency ones. Although the scores may be the same, the response to them in terms of priority for correction may be very different. Access to good data and evaluation and categorisation by experts are possible cures.

Decision-making

This process requires information to be available on alternatives to the hazard, as well as other methods of controlling the risk. Factors which will influence decision-making are training, possibility of replacement of equipment or plant, modification possibilities and the cost of the solutions proposed. Cost–benefit analysis will be used formally or informally at this point.

Cost–benefit analysis requires a value to be placed on the costs of improvements suggested or decided on. These will include the cost of reducing the risk, eliminating the hazard, any capital expenditure needed, and any ongoing costs applicable. An estimate of the payback period will be needed. Decisions on action to be taken are often based on this – a three to five-year payback period is often

associated with health and safety improvements. The use of these techniques will direct an organisation's resources to where they will be most effective.

Introduction of protective and preventive measures

The reader's attention is drawn to the principles contained in Schedule 1 of the ACoP to MHSWR, which are summarised at the end of Section 1 of this Part. These are complemented by the further concept of **progressive risk reduction**, which encourages the setting of improving objectives of risk reduction year by year.

Readers need to be aware that some corrective measures are better able to produce the desired results than others, and that some are very ineffective indeed as controls. The safety precedence sequence shows the order of effectiveness of measures:

1. **hazard elimination** (eg use of alternative work methods, design improvements, change of process)
2. **substitution** (eg replacement of a chemical with another with less risk)
3. **use of barriers**:
   - isolation (removes hazard from the worker, puts hazard in a box)
   - segregation (removes worker from the hazard, puts worker in a box)
4. **use of procedures**:
   - limiting exposure time, dilution of exposure
   - safe systems of work, which depend on human response
5. **use of warning systems** (signs, instructions, labels, which depend on human response)
6. **use of personal protective equipment (PPE)** (depends on human response, used as a sole measure only when all other options have been exhausted). PPE is the last resort.

PPE is discussed in Part 3 Section 6. The MHSWR require that it should only be used if there is no immediately feasible way to control the risk by more effective means, and as a temporary measure pending installation of more effective solutions.

Disadvantages of PPE include:
* interference with ability to carry out the task
* PPE may fail and expose the worker to the full effect of the hazard
* continued use may mask the presence of the hazard and may result in no further preventive action being taken.

Training is a control measure, but at second hand. When the controls have been identified, training is given in what needs to be done to put the controls into action, or to keep them effective.

Design issues

Wherever design can have an influence on health, safety or the environment, knowledge of the relative effectiveness of some measures compared with others is very important for the designer. This is emphasised in guidance issued by the HSE to accompany the Construction (Design and Management) Regulations 1994, in which the major choices open to designers are listed, in order of preference and effectiveness, as:

1 avoid the hazard by design alone
2 combat the hazard at source
3 protect the entire population affected by the hazard
4 only then rely solely on PPE.

The principles of design safety are, of course, not limited to construction work. To address the potential for the collapse of a structure during erection, for example, the designer should **avoid** the use of designs which would or could involve temporary instability during the erection process. **Combating problems at source** is done by specifying an erection sequence so as to avoid instability, and **communal protection** by including the principles of any necessary temporary support into the design itself.

Communal protection by incorporating fixed barriers and guardrails into designs reduces the need for the use of PPE by contractors and maintenance staff. PPE always requires at least one positive human act before it can be effective – that of making use of it.

Specifying non-hazardous chemicals or finishes and producing designs which do not require noisy or dusty work processes are further examples of the influence which designers can have on the wellbeing of those who construct workplaces.

Monitoring

Risk assessments must be checked to ensure their validity, or when reports indicate that they may no longer be valid. It is important to remember that fresh assessments will be required when the risks change as conditions change, and also when new situations and conditions are encountered for the first time. Other information relevant to risk assessments will come from monitoring by way of inspection (see Section 2), air quality monitoring and other measurements (see Part 3 Section 3), and health surveillance.

Health surveillance

Risk assessments will identify circumstances where health surveillance will be appropriate. Requirements for such surveillance now extend beyond exposure to substances hazardous to health. Generally, there will be a need if:
* there is an identifiable disease or health condition related to the work *and*
* there is a valid technique for its identification *and*
* there is a likelihood that the disease or condition may occur as a result of the work *and*
* the surveillance will protect further the health of employees.

Examples where these conditions may apply are hand–arm vibration, including vibration white finger, and other forms of work-related upper limb disorders.

Information to others

Regulations 11 and 12 of the MHSWR require co-operation and sharing of information between employers sharing or acting as host at a workplace. Risk assessments will form the basis of that information. In the formal circumstances of contracts, there may be a contractual or legal requirement to exchange risk assessment data. The information given to others about risks must include the health and safety measures in place to address them, and be sufficient to allow the other employers to identify anyone nominated to help with emergency evacuation.

Revision

2 **types of risk assessment:**
- qualitative
- quantitative

Individual risk = severity x probability

6 **contents of risk assessments:**
- hazard details
- applicable standards
- evaluation of risks
- existing preventive measures
- what, if anything, needs to be done to reduce risks further to an acceptable level
- review dates/feedback details

Selected references

Guidance (HSE)

INDG163	*Five steps to risk assessment*
INDG213	*Five steps to information, instruction and training: meeting risk assessment requirements*
INDG218	*A guide to risk assessment requirements: common provisions in health and safety law*

Self-assessment questions

1 What information will be needed before a risk assessment is carried out?

2 Thinking about qualitative and quantitative risk assessments, explain the difference with examples, showing where it would be appropriate for each to be used.

4 The health and safety policy

Introduction

Without active support, any attempt at organised accident prevention will be useless – or even worse than useless, since there may be an illusion that health and safety matters are under control, resulting in complacency. Avoidance of accidents requires a sustained, integrated effort from all departments, managers, supervisors and workers in an organisation. Only management can provide the authority to ensure this activity is co-ordinated, directed and funded. Its influence will be seen in the policy made, the amount of scrutiny given to it, and the ways in which violations are handled.

The most effective means of demonstrating management commitment and support is by issuing a safety policy statement, signed and dated by the most senior member of the management team, and then ensuring that the detailed requirements of the policy are actively carried out by managers, supervisors and workers. Lack of firm management direction of this kind encourages the belief that 'safety is someone else's business'.

The safety policy

Section 2(3) of the Health and Safety at Work etc Act 1974 requires written safety policy statements to be created by all employers, except for the smallest organisations (those with fewer than five employees). As stated in Section 2 of this Part, this is simply a reflection in the law of what has been known for many years – written policies are the centrepieces of safety management. They insist, persuade, explain and assign responsibilities.

An essential requirement for management involvement at all levels is to define health and safety responsibility in detail within the written document, and then to check at intervals that the responsibility has been adequately discharged. This is **accountability** – the primary key to management action. It is not the same thing as responsibility; accountability is responsibility that is evaluated and measured, possibly during appraisal sessions.

Safety policy contents

What is usually referred to as 'the safety policy' will contain three separate parts: a statement of the employer's policy, the details of the organisation set up to make it happen (responsibilities at each level within the operation), and the arrangements (how, in detail, health and safety will be managed). These contents are prescribed by the Act. The statement itself is an expression of management

intention, and as such does not need to contain detail.

It is important to distinguish between a health and safety policy and a safety manual – these are not the same, but are often found combined. The safety policy will refer to the manual for details on technical points. The main problem is that the likelihood of a document being read and used is often inversely proportional to its length and complexity. Current opinion is that safety policies should be shorter rather than longer, and accompanied by explanatory manuals. The test of the health and safety policy lies in the actual conditions achieved as a result of it, not in the wording or the length.

The safety policy should provide all concerned with concise details of the organisation's health and safety objectives and the means of achieving them, including the assignment of responsibility and detailed arrangements for each workplace. Organisations with several workplaces in different locations find it convenient to express management philosophy in an overall statement, leaving a detailed statement and policy to be written and issued at local level – by department, branch or other organisational unit.

Risk assessments as required by the Management of Health and Safety at Work Regulations 1999 will be part of the organisation's documentation, as they form the basis for deciding on the control measures – the arrangements – and detailing the responsibilities within the organisation.

The Regulations require that the significant findings of risk assessments be recorded where there are five or more employees. The practicality of attaching the risk assessments in full to the safety policy document is such that it may be better to include the findings of generic assessments (see Section 3) in detail, and refer to circumstances when these will need modification and where specific assessments will be required on each occasion the risk is faced. The policy arrangements should state who carries them out, when and how.

The employer's statement

It is sadly still possible to find employer's statements that are impossible to achieve – 'We will use every means possible to prevent accidents' – or employee-dependent – 'It is the responsibility of the workforce to take care of their own health and safety; those failing to do so will be subject to disciplinary procedures'.

In the employer's opening statement, there should be commitment to the following seven aims as a minimum:

1 to identify the hazards to employees and third parties affected by the work, and control the attendant risks adequately
2 to maintain healthy and safe working conditions, including provision of safe plant and equipment
3 to work, through continuous improvement, to prevent accidents and work-related ill health
4 to ensure that employees are competent to do their work and provide them with appropriate and adequate training
5 to ensure the safe transport, storage, handling and use of hazardous substances
6 to consult with employees on health and safety issues affecting their wellbeing, giving them necessary information, instruction and supervision
7 to review and revise the safety policy at regular intervals as necessary.

The statement should be signed by the most senior person with corporate responsibility for health and safety, and dated. Nothing less than this will convey the necessary authority to the contents.

Organisation and arrangements

Responsibilities of management and supervisors at all levels amount to the details of the 'organisation for health and safety' required by the Act. At every level managers and workforce alike should have set out for them a precise summary of their duties in ensuring their own and others' health and safety at work. For managers, non-compliance with the employer's own requirements may be viewed by courts as amounting to negligence. Similarly, the duties of employees set out in the safety policy will define what amounts to taking 'reasonable care' of themselves and others. Duties laid down in the policy may include summaries of relevant statutory and organisational rules.

Opinion varies about the amount of detail which the **arrangements** within a health and safety policy should contain. One local authority uses a checklist of more than 150 matters which it expects to find set out and adequately covered within a policy. Figure 1 is a suggested checklist for evaluation of policies, allowing space for a review of adequacy of the contents as well as their presence or otherwise.

Other considerations

Safety policies, as written statements of the intentions of management, acquire a quasi-legal status. Among other things, they serve as a record of the intended standard of care to be provided by the employer. This offers a useful method of evaluating an organisation in terms of health and safety, especially because its standards, beliefs and commitments are on view and potentially measurable. Therefore, the document will be useful when evaluating contractors and their competence (see Part 1 Section 8). Others may wish to use it for different purposes – to gauge the record of a possible business partner, and, importantly, for the purpose of establishing an employer's self-assessed standard of care as a prelude to

Figure 1: Safety policy adequacy checklist

Topic	Adequate	Less than adequate	Not present
General statement of policy			
Organisation for health and safety: Organisation chart			
Clear responsibilities given for all management levels and employees			
Role and functions of health and safety professional staff			
Appointment of a competent person as required by the Management Regulations			
Arrangements			
Allocation of finance for health and safety			
Systems used to monitor safety performance (not just injury recording)			
Identification of main hazards likely to be encountered by the workforce, including eg: process noise asbestos manual handling hand–arm vibration			
Arrangements for making and reviewing risk assessments			
Details of any circumstances when specific risk assessments will be required			
Arrangements (or cross-references) for dealing with specific risk assessments			
Safety training policy, with details of arrangements including induction training and young persons training and risk assessment			
Design safety			
Fire arrangements and fire risk assessments			
Emergency evacuation summary and practice arrangements			
Arrangements for maintaining mechanical and electrical work equipment and systems			
COSHH implementation, assessments, review, information to employees			
Occupational health facilities, including first aid			
Environmental monitoring policy and arrangements			
Purchasing policy (eg on safety, noise, chemicals)			
Methods of reporting accidents and incidents			
Methods used to investigate accidents and incidents			
Arrangements for the procurement and management of contractors			
Personal protective equipment policy, requirements, availability			
Employee consultation arrangements (eg safety committees)			
Listed health and safety rules, eg: wearing of PPE drugs and alcohol horseplay			

making a civil claim for damages. Claimants may be able to use any extravagant wording or undertakings to further claims, as they can demonstrate what should have been available or shown in their case.

Revision of safety policies should be done at regular intervals to ensure that the organisation and arrangements are still applicable and appropriate to the needs. After changes in structure, senior personnel, work arrangements, processes or premises, the hazards and their risks may change. After incidents, one of the objectives of the investigation (see Part 1 Section 9) will be to check that the arrangements in force had anticipated the circumstances and foreseen the causes of the accident. If they had not, then a change to the policy will be required. The revision mechanism should be written into the policy.

Circulation/distribution of the policy is important – it must be read and understood by all those affected by it. How this is best achieved will be a matter for discussion. In some organisations, a complete copy is given to each employee. In others, a shortened version is given out, with a full copy available for inspection at each workplace. Members of the management team should be familiar with the complete document. If it is likely to be revised frequently, a loose-leaf format is advisable. This is especially true if the names and contact phone numbers of staff are printed in it, as these may change frequently.

Getting a third party to write a policy, or using 'boilerplate' material borrowed or bought, can save an organisation time, but care must be taken to make very sure that the statements and the controls relate specifically to the user organisation. Legal compliance is only the first of several uses for the policy – it can be used in civil claims as it sets out 'reasonable' measures the employer is on record as taking, and also in the evaluation of contract bids. Successful policies meet all the needs of the organisation.

The function of the safety policy

Safety policy statements serve as the bridge between health and safety activities of management and the legal system, because a requirement exists in the statute law for a policy to be written, and because the policy can be used in civil claims. Increasingly, the same is true of risk assessments, especially for manual handling and display screen equipment.

Generally, legal compliance for its own sake is the poorest of reasons for accident prevention – apart from the moral considerations, the financial losses associated with poor safety performance outweigh likely penalties for breaches of statutory duty.

Revision

5 **contents of a safety policy:**
- a general statement of management commitment
- details of the organisation (responsibilities)
- the arrangements in force to control the risks
- the signature of the most senior member of management
- the date of the last revision

Selected references

Legal
Health and Safety at Work etc Act 1974
Management of Health and Safety at Work Regulations 1999

Guidance (HSE)
INDG259   *An introduction to health and safety – health and safety in small businesses*

Self-assessment questions

1 Why are safety policies useful in the management of health and safety? Why would one consisting of only one page be of no value?

2 Obtain a copy of your organisation's safety policy and identify your personal responsibilities for health and safety at work. Can you think of any improvements which might usefully be made to the policy?

5 Safe systems of work

Introduction

It has been estimated that at least a quarter of all fatal accidents at work involve failures in systems of work – the way things are done. A safe system of work is a formal procedure which results from a systematic examination of a task in order to identify all the hazards and assess the risks, and which identifies safe methods of work to ensure that the hazards are eliminated or the remaining risks are minimised.

Many hazards are clearly recognisable and can be overcome by separating people from them physically, eg using guarding on machinery (see Part 2 Section 2). There will often be circumstances where hazards cannot be eliminated in this way, and elements of risk remain associated with the task. **Where the risk assessment indicates this is the case, a safe system of work will be required.** Some examples where safe systems will be required as part of the controls are:

- cleaning and maintenance operations
- changes to normal procedures, including layout, materials and methods
- working alone or away from the workplace and its facilities
- breakdowns and emergencies
- control of the activities of contractors in the workplace
- vehicle loading, unloading and movements
- lone working
- entry into and working in confined spaces.

Developing safe systems

Some safe systems can be verbal only – where instructions are given on the hazards and the means of overcoming them – for short duration tasks. These instructions must be given by supervisors or managers: leaving workers to devise their own method of work is not a safe system of work. The law requires that a suitable and sufficient risk assessment be made of all the risks to which employees and others who may be affected by them are exposed. Although some of these assessments can be carried out using a relatively unstructured approach, a more formal analysis can be used to develop a safe system of work. Sometimes these may be carried out as a matter of policy, with the task broken down into stages and the precautions associated with each stage written into the final document. This can be used for training new workers in the required method of work. The technique is known as **job safety analysis**.

There are five basic steps necessary in producing all safe systems of work:

- assessment of the task
- hazard identification and risk assessment
- identification of safe methods

- implementing the system
- monitoring the system.

Task assessment

All aspects of the task must be looked at, and should be put in writing to ensure nothing is overlooked. This should be done by supervisory staff in conjunction with the workers involved, to ensure that assumptions of supervisors about methods of work do not differ from reality. Account must be taken of:

- **what is used** – the plant and equipment, potential failures of machinery, substances used, electrical needs of the task
- **sources of errors** – possible human failures, short cuts, emergency work
- **where the task is carried out** – the working environment and its demands for protection
- **how the task is carried out** – procedures, potential failures in work methods, frequency of the task, training needs.

Hazard identification and risk assessment

Against a list of the elements of the task, associated hazards can be clearly identified, and a risk assessment can be made (a review of risk assessment can be found in Part 1 Section 3). Where hazards cannot be eliminated and risks remain, procedures to ensure a safe method of work should be devised.

Definition of safe methods

The chosen method can be explained orally as already mentioned. Simple written methods can be established, or a more formal method known as a permit-to-work system can be used. All of these involve:

- setting up the task and any authorisation necessary
- planning of job sequences
- specification of the approved safe working methods including the means of getting to and from the task area if appropriate
- conditions which must be verified before work starts, eg atmospheric tests, machinery lockout
- dismantling/disposal of equipment or waste at the end of the task.

Implementing the system

There must be adequate communication if the safe system of work is to be successful. The details should be understood by everyone who has to work with it, and it must be carried out on each occasion. It is important that everyone appreciates the need for the system and its place in the accident prevention programme.

Supervisors must know that their duties include devising and maintaining safe systems of work, and making sure they are put into operation and revised where necessary to take account of changed conditions or accident experience. Training is required for all concerned, to include the necessary skills, and awareness of the system and the hazards which it is aimed to eliminate by the use of safe procedures. Part of every safe system should be the requirement to stop work when a problem appears which is not covered by the system, and not to resume until a safe solution has been found.

Monitoring the system

Effective monitoring requires that regular checks be made to ensure that the system is still appropriate for the needs of the task, and that it is being fully complied with. Checking only after accidents is not an acceptable form of monitoring. Simple questions are required – Do workers continue to find the system workable? Are procedures laid down being carried out? Are the procedures still effective? Have there been any changes which require a revision of the system? A system devised as above which is not followed is **not** a safe system of work – the reasons must be found and rectified. Safe systems of work are associates of, not substitutes for, the stronger prevention techniques of design, guarding and other methods which aim to eliminate the possibility of human failure.

Permit-to-work systems

Written permit-to-work systems are normally reserved for occasions when the potential risk is high, and where at the same time the precautions needed are complicated and require written reinforcement. These systems will often be found where the activities of groups of workers or multiple employers have to be co-ordinated to ensure safety.

When permits to work are used in the workplace, their use should be covered specifically during induction training of all employees and contractors, and during preliminary discussions with contractors.

Permit-to-work systems normally use preprinted forms, listing specific checks and/or actions required at specific stages of the task. These may include isolation of supply systems and the fitting of locking devices to controls. Most permits are only designed to cover work lasting up to 24 hours, and require an authorisation signature for any time extension. The sequence of operation for a permit-to-work system is shown in Figure 1 overleaf.

An experienced, trained and authorised person will make an advance assessment of the hazards and risks involved in the work to be done, and will then complete and sign a certificate giving authority for the work to proceed under controlled conditions specified on the permit. No-one should be in a position

Figure 1: Permit-to-work system operation

Task sequence	Commentary
Request for permit to work is made	
Hazards are assessed	
Type of permit is selected	
Permit is raised, and precautions are stated	Possible precautions include: isolation of equipment and areas use of protective devices and clothing firefighting equipment and fire watchers removal of waste generated
Initial precautions taken	
Precautions are verified by the issuer and recipient	
Sampling and testing carried out	Testing for: toxic concentrations flammable concentrations lack of oxygen
Permit issued to competent person in charge of the work	
Work carried out in accordance with permit and precautions detailed in it	
Work conditions monitored	Continued monitoring of tested conditions: sampling spot checks dose monitoring
Work is completed and area handed back	
Permit is cancelled	
Record is kept of permit issue	

to authorise a permit for themselves to do work. A permit will include details of the work to be done and what is involved, including all precautions required and emergency procedures, who is to do it and when, and any limits on the work area or equipment. The permit-to-work system will usually require written acknowledgement by the person who will do the work, or who is in charge, and will also allow for signed confirmation that the workplace or the equipment has been restored to safety, for any time extension which may be permitted, and for the cancellation of the permit. There will also usually be some system for keeping a record that a permit has been issued.

A crucial point of principle is that no permit should be issued unless the persons issuing and receiving it have personally satisfied themselves that the permit is properly made out, and have visited the scene of the work and verified that the specified precautions are in place. Safe systems of work depend on people to carry them out and are, therefore, fallible. It is wise to remember that proceeding on assumptions is dangerous: 'assume' makes an ass out of you and me.

An example of a general-purpose permit is shown in Figure 2 opposite. This permit can be used for circumstances where a more specialised version is not required. For example,

Figure 2: General permit to work

To:	Company:
Location:	
Every item must be completed or deleted as appropriate	

A Job details

1 Area or equipment to which this permit applies   2 Work to be done

B Isolations (specify where necessary)   Initials and comments

1 Circuit breaker locked out/fuses withdrawn/isolator locked off	YES/NO
2 Circuit tested and confirmed to be dead	YES/NO
3 Mechanical or physical isolation	YES/NO
4 Valves closed/locked off/spades inserted	YES/NO
5 Pipelines drained/purged/disconnected/vented to atmosphere	YES/NO
6 Documented isolation procedure attached	YES/NO
7 Other	YES/NO

C Precautions (to be taken as indicated, additional to those specified on other permits)

1 Protective clothing	YES/NO Type:
2 Respiratory equipment	YES/NO Type:
3 Protected electrical equipment	YES/NO Type:

4 State additional precautions required (if none, state none):

5 Atmosphere tests are not/are required at intervals of   and results must be recorded overleaf

D Additional permits and signatures required before work starts

1 Confined space entry	YES/NO	6 In my opinion, the engineering precautions are adequate
2 HV electrical	YES/NO	
3 Hot work	YES/NO	Signed (Engineer) Date
4 Excavation	YES/NO	7 In my opinion, the precautions against special hazards within my
5 Other	YES/NO	knowledge are adequate
		Signed (Specialist) Date

E Issue and receipt before work starts

1 Issue	2 Receipt
I have examined the area specified and permission is given for the work to start, subject to the conditions hereon, under the control of	I have read and understood the conditions of this permit
	Signed Date
	3 This permit is valid from hours to hours
Signed Date	(maximum 24 hours)

F Clearance and cancellation after work

1 Clearance	2 Cancellation		
All workers under my control have been withdrawn. The permitted work is/is not complete	Work complete	YES/NO	I have notified those affected. This permit is cancelled
	Isolations removed	YES/NO	
	Area/equipment is		
Signed	safe to use	YES/NO	
Time Date			
	Signed		
	Time		Date

it could be used for work on or near overhead crane installations, or for work on pipelines with hazardous contents. There are many different types of specialised permit. Some common examples are:

- electrical permits to work – a useful example of this type of permit is contained in Appendix 1 of the HSE booklet HSG85 – *Electricity at work: safe working practices*
- hot work permits
- permits to enter premises or confined spaces
- permits to work on pressurised systems
- permits to excavate in contaminated ground, or where there are congested or buried services.

Method statements

The key feature of method statements is that they provide a sequence for carrying out an identified task – some work activities must be done in sequence to ensure safety. In such cases, it is necessary not only to know what the control measures are, but also to carry the work out in a particular order. Examples of activities where the HSE expects method statements to be provided include demolition work, asbestos removal and structural steel erection.

Method statements usually contain more detail than risk assessments, and normally include the following information:

- originator and date
- identification of individual(s) who will be responsible for the whole operation and for compliance with the method statement. Key personnel responsible for particular operations may also be named
- training requirements for personnel carrying out tasks which have a competency requirement (eg crane and fork-lift drivers, testing and commissioning staff)
- details of access equipment which will be used, safe access routes and maintenance of emergency routes
- equipment required to carry out the work, including its size, weight, power rating, necessary certification
- locations and means of fixing the stability of any lifting equipment to be used

- material storage, transportation, handling and security details
- detailed work sequence including hazard identification and risk control measures, including co-operation between trades which may be required, limitations for part completion of works and any temporary supports or supplies required
- details of all personal protective equipment and other measures such as barriers, signs, local exhaust ventilation, rescue equipment, fire extinguishers, gas detection equipment
- any environmental limitations which may be applicable, such as wind speed, rain, temperature
- details of measures to protect third parties who may be affected
- the means by which any variations to the method statement will be authorised.

In some industries there is a tendency to demand method statements for the most trivial of tasks. Provided that competence has been verified, it can be acceptable for work to proceed on the basis of risk assessment alone. This is a matter for judgement, and also the workplace safety culture.

Legal requirements

Section 2(2)(a) of the Health and Safety at Work etc Act 1974 requires the provision and maintenance of systems of work that are, so far as is reasonably practicable, safe and without risks to health. Under the employer's general duty of care at common law, a failure to do so gives rise to a claim based on an allegation of the employer's negligence. Specific legislation may require the use of formal permits to work, either directly or by implication as a means of compliance.

Further requirements for safe systems of work following on risk assessments are contained in the Management of Health and Safety at Work Regulations 1999, which also place duties on employees to follow the systems and procedures set up for their protection following risk assessments.

Revision

5 **steps in devising a safe system:**
- assessment of the task
- identification of the hazards and assessment of the risks
- definition of safe method
- implementation of the safe system
- monitoring the system

8 **areas where formal safe systems may be required:**
- cleaning tasks
- maintenance work
- changes to routine
- working alone
- working away from normal environment
- breakdowns
- emergencies
- contractors' vehicle operations
- lone working
- entry into and working in confined spaces

Selected references

Legal
Construction (Design and Management) Regulations 1994
Electricity at Work Regulations 1989

Guidance (HSE)
L113	*Safe use of lifting equipment*
L122	*Safety of pressure systems: Pressure Systems Safety Regulations 2000*
HSG85	*Electricity at work: safe working practices*
HSG150	*Health and safety in construction*
HSG224	*Managing health and safety in construction: Construction (Design and Management) Regulations 1994* ACoP, ISBN 0 7176 2139 1

Designing for health and safety in construction, ISBN 0 7176 0807 7
A guide to managing health and safety in construction, ISBN 0 7176 0755 0

Self-assessment questions

1 Can you think of tasks in your workplace which require correct steps to be taken while carrying them out to ensure worker safety? Are they covered by written documentation? Can you justify those which are only given orally?

2 Write out a safe system of work for moving filing cabinets between offices.

6 Health and safety training

Introduction

Training for health and safety is not an end in itself, it is a means to an end. Talking in general terms to employees about the need to be safe is not training; workers and management alike need to be told what to do for their own health and safety and that of others, as well as what is required by statute. A knowledge of what constitutes safe behaviour in a variety of different occupational situations is not inherited but must be acquired, either by trial and error or from a reputable source of expertise. Trial and error methods are likely to exact too high a price in modern industry, where the consequences of forced and unforced errors may be very serious, even catastrophic.

Experience and research also shows that even if individuals gain knowledge of safe behaviour patterns through instruction, films, videos, posters and booklets, there is no guarantee that they will behave safely. Training is, therefore, never a substitute for healthy and safe working conditions and good design of plant and equipment. Because humans are fallible, the priority is to lessen the opportunities for mistakes and unsafe behaviour to occur, and to minimise the consequences when they do. Safety training may be a part of other training in work or organisational procedures for reasons of time or cost, and there is merit in combined training as it serves to emphasise the need to regard health and safety generally as an integral part of good business management. Training in any subject requires the presence of three necessary conditions before it commences: the active **commitment**, support and interest of management, necessary finance and organisation to provide the **opportunity** for learning to take place, and the availability of suitable **expertise** in the subject. The support of management demonstrates the presence of an environment into which the trained person can return and exercise new skills and knowledge. The management team also demonstrates support by setting good examples; it is pointless to train workers to obey safety rules if supervisors are known to ignore them.

Trainers must not only be knowledgeable in their subject, but also be qualified to answer questions on the practical application of their knowledge in the working environment, which will include a familiarity with organisational work practices, procedures and rules.

Training needs

Organisational

New employees are known to be more likely to have accidents than those who have had time to recognise the hazards of the workplace. Formal health and safety training is now required by law as part of the induction programme. Training must also take place when job conditions change and result in exposure to new or increased risks. It must be repeated periodically, where appropriate, and be adapted to any new circumstances. By law, no health and safety training can take place outside working hours.

There may also be opportunities for self-instruction, perhaps using modern technology to assist, eg computer-based interactive learning programmes. The key points which should be covered in induction training are:

- review and discussion of the organisation's overall safety programme or policy, and the policy relating to the work activities of the newcomers
- safety philosophy: safety is as important as production or any other organisational activity; accidents have causes and can be prevented; prevention is the primary responsibility of management; each employee has a personal responsibility for his or her own safety and that of others
- local, national and organisational health and safety rules or regulations will be enforced, and those violating them may be subject to some form of disciplinary action
- the health and safety role of supervisors and other members of the management team includes taking action on, and giving advice

about, potential problems, and they are to be consulted if there are any questions about the health and safety aspects of work

- where required, the wearing or use of PPE is not a matter for individual choice or decision – its use is a condition of employment
- in the event of any injury, no matter how trivial it may appear, workers must seek first aid or medical treatment and notify their supervisor immediately. For any work involving repetitive, awkward, heavy physical or timed movements, workers should be specifically instructed to report any adverse physical symptoms immediately. These will need to be recorded and investigated without delay
- fire and emergency procedure(s)
- welfare and amenity provision
- arrangements for joint consultation with workers and their representatives should be made known to all newcomers. A joint approach to health and safety problems, as well as the regular reviewing of work practices, procedures, systems and written documentation, is an essential part of a good health and safety programme. (But joint or balanced participation should not be used as a method of removing or passing off the prime responsibility of management at all levels to manage health and safety at least as efficiently as other aspects of the organisation.)

Job-specific

Job and task training should include skills training, explanations of applicable safety regulations and organisational rules and procedures, a demonstration of any PPE which may be required and provided for the work (including demonstration of correct fit, method and circumstances of use, and cleaning procedures), and the handing over of any documentation required, such as permit-to-work documents, safety booklets and chemical information sheets. There should also be a review of applicable aspects of emergency and evacuation procedures. Use of risk assessment findings is proving popular and worthwhile as a training aid, and by this management can also fulfil the requirement to bring risk assessment

findings to the notice of those affected by them. This training may be carried out by a supervisor, but it should be properly planned and organised by the use of checklists.

The need for training following job and process changes must not be overlooked. Introduction of new legislation may require additions to be made to the training programme, as will introduction of new technology. The programme should allow for refresher training as necessary.

Supervisory and general management training at all levels is necessary to ensure that responsibilities are known and the organisation's policy is carried out. Management failures which have come to light following investigations into disasters, plant accidents and other health and safety incidents have been concentrated in the following areas:

- lack of awareness of the safety systems, including their own job requirements for health and safety
- failure to enforce health and safety rules adequately or at all
- failure to inspect and correct unhealthy or unsafe conditions
- failure to inform or train workers adequately
- failure to promote health and safety awareness by participation in discussions, motivating workers and setting an example.

It is not sufficient simply to tell supervisors they are responsible and accountable for health and safety; they must be told the extent of the responsibilities and how they can discharge them.

Key points to cover in the training of supervisors and managers are:

- the organisation's safety programme and policy
- legal framework and duties of the organisation, its management and the workforce
- specific laws and rules applicable to the work area
- safety inspection techniques and requirements
- causation and consequences of accidents
- basic accident prevention techniques

- disciplinary procedures and their application
- control of hazards likely to be present in the work area, including:
  - machinery safety
  - fire
  - materials handling
  - hazards of special equipment related to the industry
  - use of PPE
- techniques for motivating employees to recognise and respond to organisational goals in health and safety.

Senior managers should be given essentially the same information, as this gives them a full appreciation of the tasks of subordinates, makes them more aware of standards of success and failure, and equips them to make cost–benefit decisions on health and safety budgeting. External assessment of the training given to management at all levels is desirable. This can be done by training to the appropriate syllabuses of national or international professional organisations, and encouraging those trained to take the relevant examination. In the UK, these are the 'Managing safely' and 'Working safely' courses moderated by the Institution of Occupational Safety and Health (IOSH), and professional training courses leading to the two-part diploma which is administered by the National Examinations Board in Occupational Safety and Health (NEBOSH). A further IOSH course, 'Directing safely', is particularly significant as it aims to get the commitment of the most senior management to the organisation's safety programme.

Individual

Training needs of individuals at all levels are likely to emerge over time, usually through performance appraisal. They can arise from a variety of inputs, eg because of observed inappropriate behaviour, from requests from the individual, following promotion or relocation, or following an accident. The main objective is to achieve competence in the individual.

Examples of specialised training needs

First aid training has been proved to be a significant factor in accident prevention, as well as obviously beneficial in aiding accident victims. People trained in first aid are more safety conscious and are less likely to have accidents. Project FACTS – First Aid Community Training for Safety – was launched in 1970 in Ontario, Canada, and arranged for the training of 5,500 people in industry, schools and among the public in the community of Orillia, Ontario. Results showed that first aid training throughout the workforce can cut accident rates by up to 30 per cent. A second project based on industry alone found that employees not trained in first aid had accident rates double those of trained employees of similar age, sex, job and employer.

Apart from the above, and legal requirements which may exist, the need for first aid training in the workplace depends on a number of factors. These include the nature of the work and the hazards, what medical services are available in the workplace, the number of employees, and the location of the workplace relative to external medical assistance. Shift working may also be taken into account, as well as the ratio of trained persons present to the total number of workers.

Driver training and certification may be a requirement for particular classes of vehicle, according to national or local regulations in force, and in these cases the detail of the training programme may be defined in law. A common cause of death at work (and away from work) is the road traffic accident. As the loss of a key worker can have a severe impact on the viability of an organisation, training for all workers who drive should be considered. Defensive driver training has been found to be effective and cost-beneficial in reducing numbers of traffic accidents, and has been extended to members of workers' families, particularly those entitled to drive the employer's vehicles.

Fire training – to the extent that all workers should know the action to be taken when fire alarms sound – should be given to all employees and included in induction training.

Knowledge of particular emergency plans and how to tackle fires with equipment available may be given in specific training at the workplace. At whatever point the training is given, the following key points should be covered:

- evacuation plan for the building in case of fire, including assembly point(s)
- how to use firefighting appliances provided
- how to use other protective equipment, including sprinkler and other protection systems, and the need for fire doors to be unobstructed
- how to raise the alarm and operate the alarm system from call points
- workplace smoking rules
- housekeeping practices which could permit fires to start and spread if not carried out, eg waste disposal, flammable liquid handling rules
- any special fire hazards peculiar to the workplace.

Fire training should be accompanied by practices, including regular fire drills and rehearsals of evacuation procedures. No employees should be exempted from these exercises.

Reinforcement training will be required at appropriate intervals, which will depend on observation of the workforce (training needs assessment), on the complexity of the information needed to be held by the worker, the amount of practice required, and the opportunity for practice in the normal working environment. Assessment will also be needed of the likely severity of the consequences of behaviour which does not correspond with training objectives. If it is absolutely vital that only certain actions be taken in response to plant emergencies, then more frequent refresher training will be needed to ensure that routines are always familiar to those required to follow them.

Other occasions when training will be required include the appearance and introduction of new legal requirements which may alter or add to the organisation's control arrangements, and the introduction of new technology in the form of new processes, plant, equipment or substances for use at work.

Legal requirements

Many specific pieces of health and safety legislation contain requirements to provide training for employees engaged in certain tasks, and the most commonly applicable of these have been selected for discussion in Part 4. A general duty for all employees to be trained (and, in addition, provided with information, instruction and supervision) as necessary to ensure their health and safety so far as is reasonably practicable is contained in Section 2(2)(c) of the Health and Safety at Work etc Act 1974. More specific requirements are contained in the Management of Health and Safety at Work Regulations 1999 (see Part 4 Section 6).

Revision

6 **types of health and safety training needs:**
- new employee induction
- job-specific for new starters
- supervision and management
- specialised
- reinforcement
- individual

Selected references

Legal
Management of Health and Safety at Work Regulations 1999
Health and Safety (First Aid) Regulations 1981
Health and Safety (Training for Employment) Regulations 1990
Health and Safety (Safety Signs and Signals) Regulations 1996

Guidance (HSE)
CoP20	*Standards of training in gas safety installations*
L114	*Safe use of woodworking machinery*
L117	*Rider-operated lift trucks: operator training*
L74	*First aid at work*

British Standard
BS 7121	Code of practice for the safe use of cranes: Part 1: 1989 – General

Self-assessment questions

1 What topics should be covered in induction safety training to provide necessary knowledge and skills for fire prevention?

2 'Managers don't need to know about safety rules – they are not at risk.' Discuss.

7 Maintenance

Introduction

'Maintenance' can be defined as work carried out in order to keep or restore every facility (part of a workplace, building and contents) to an acceptable standard. It is not simply a matter of repair; some mechanical problems can be avoided if preventive action is taken in good time. Health and safety issues are important in maintenance because statistics show the maintenance worker to be at greater risk of accidents and injury. This is partly because these workers are exposed to more hazards than others, and partly because there are pressures of time and money on the completion of the maintenance task being achieved as quickly as possible. Often, less thought is given to the special needs of health and safety in the maintenance task than is given to routine tasks which are easier to identify, plan and control.

Maintenance policy

Whether for equipment or for premises and structures, maintenance can be:

- planned preventive – at predetermined intervals
- condition-based – centred on monitoring safety-critical parts and areas, maintaining them as necessary to avoid hazards
- breakdown – fix when broken only. This is not appropriate if a failure presents an immediate risk without correction. Breakdown maintenance needs an effective fault-reporting system.

Maintenance standards are a matter for each organisation to determine; there must be a cost balance between intervention with normal operations by planned maintenance and the acceptance of losses because of breakdowns or other failures. From the health and safety aspect, however, defects requiring maintenance attention which have led or could lead to increased risk for the workforce should receive a high priority. Linking inspections with maintenance can be useful, so that work areas and equipment are checked regularly for present and possible future defects. Some work equipment items may be subject to statutory maintenance requirements, eg portable electrical equipment. The manufacturer's instructions in this respect should be complied with as well.

Maintenance accidents

These are caused by one or more of the following factors:

- lack of perception of risk by managers/ supervisors, often because of lack of necessary training
- unsafe or no system of work devised, eg no permit-to-work system in operation, no facility to lock off machinery and electrics before work starts and until work has finished
- no co-ordination between workers, or communication with other supervisors or managers
- lack of perception of risk by workers, including failure to wear protective clothing or equipment
- inadequacy of design, installation or siting of plant and equipment
- use of contractors who are inadequately briefed on health and safety aspects, or who are selected on cost grounds and not for competence in health and safety.

Maintenance control to minimise hazards

The safe operation of maintenance systems requires steps to be taken to control the above factors. These steps can be divided into the following phases:

Planning – identification of the need for the class of maintenance appropriate, and arranging a schedule for this to meet any statutory requirements. A partial list of items for equipment and structural issues for consideration includes air receivers and all pressure vessels, boilers, lifting equipment, electrical tools and machinery, fire and other emergency equipment, and structural items subject to wear, such as floor coverings.

Evaluation – the hazards associated with each maintenance task must be listed and the risks of each considered (frequency of the task and possible consequences of failure to carry it out correctly). The tasks can then be graded and the appropriate type and degree of management control applied to each.

Control – the control(s) for each task will take the above factors into account, and will include any necessary review of design and installation, training, introduction of written safe working procedures to minimise risk (which is known as a safe system of work), allocation of supervisory responsibilities, and necessary allocation of finances. Review of the activities of contractors engaged in maintenance work will be required in addition.

Monitoring – random checks, safety audits and inspections, and the analysis of any reported accidents for causes which might trigger a review of procedures constitute necessary monitoring to ensure the control system is fully up to date. The introduction of any change in the workplace may have maintenance implications and should, therefore, be included in the monitoring process.

Planned preventive maintenance

The need for maintenance of any piece of equipment should have been anticipated in its design. Lubrication and cleaning will still be required for machinery, but the tasks can be made safer by consideration of maintenance requirements at the design stage and, later, at installation. A written plan for preventive maintenance is required, which documents the actions to be taken, how often this needs to be done, all associated health and safety matters, the training required (if any) before maintenance work can be done, and any special

operational procedures such as permits to work and locking off.

Condition-based maintenance

Safety-critical features will have been identified during the risk assessment process, and appropriate arrangements are needed to ensure that inspection at agreed intervals actually takes place, and is recorded. This type of maintenance can easily become breakdown maintenance unless the system is followed scrupulously, which gives a reason for the regular audit of maintenance practice and review of risk assessments justifying it.

Breakdown maintenance

The number of failures which require this will be reduced by planned maintenance, but many circumstances (such as severe weather conditions) can arise which require workers to carry out tasks beyond their normal work experience, and/or which are more than usually hazardous by their nature. Records of all breakdowns should be kept to influence future planned maintenance policy revisions, safety training and design.

Legal requirements

Many pieces of health and safety legislation contain both general and specific requirements to maintain premises, plant and equipment. The general duty is contained in Section 2(2)(d) of the Health and Safety at Work etc Act 1974, and more specific requirements are contained in the Management of Health and Safety at Work Regulations 1999 (see Part 4). Other requirements are contained in the Workplace (Health, Safety and Welfare) Regulations 1992, the Personal Protective Equipment at Work Regulations 1992, and the Provision and Use of Work Equipment Regulations 1998, all of which are summarised in Part 4.

Revision

5 causes of maintenance accidents:
- inadequate design
- lack of perception of risk
- lack of a safe system of work
- communication failures
- failure to brief and supervise contractors

4 ways of preventing maintenance accidents:
- planning
- evaluation
- controls
- monitoring

3 types of maintenance:
- planned preventive
- condition-based
- breakdown

Selected references

Legal
Electricity at Work Regulations 1989
Control of Substances Hazardous to Health Regulations 2002 (as amended)

Guidance (HSE)

L8	*Legionnaires' disease: the control of legionella bacteria in water systems*
L113	*Safe use of lifting equipment*
L122	*Safety of pressure systems: Pressure Systems Safety Regulations 2000*
HSG54	*The maintenance, examination and testing of local exhaust ventilation*

British Standards

BS 4430	Recommendations for the safety of industrial trucks. Part 2: Operation and maintenance
BS 6867	Code of practice for maintenance of electrical switchgear (see also BS 6423 and BS 6626)
BS 6913	Operation and maintenance of earthmoving machinery (not all parts are current)

Self-assessment questions

1 Outline the maintenance system which would lead to the rapid repair of a faulty machine guard.

2 How are repairs made to the roof of your workplace? Can the maintenance system be made safer?

8 Management and control of contractors

Introduction

Anyone entering premises for the purposes of carrying out specialised work for the client, owner or occupier must be regarded as a 'contractor' – to whom duties are owed and, indeed, who owes duties with regard to health and safety matters. Because of this, the same kinds of control measures must be applied to all who work on premises who are not in the direct employment of the occupier – caterers, window cleaners, agency staff, equipment repairers and servicers. The list of potential 'contractors' is long and must be written down as a first control task.

Analysis of investigations into accidents involving contractors shows that financial pressures, whether real or perceived, are nearly always present. The making and acceptance of the low bid in competitive tendering is often at the expense of health and safety standards. Other major factors include:

- a transient labour force which never gets properly or fully trained
- the small size of most contracting companies, which claim not to be aware of legislation or safe practices
- the inherent danger of the work and work conditions
- pressure of work
- poor management awareness of the need for safety management.

Research has shown that the effective control of any activity requires ownership of responsibility for the activity and where it occurs, and acknowledgement of legitimate decision-making authority and compliance with its decisions. In relation to the presence of contractors, it is, therefore, necessary that health and safety decisions are made by, or have received the advice of, those people who are competent to make them.

The legal requirements for co-operation, co-ordination of activities and sharing of relevant information can only be achieved if all parties contribute to the risk assessment process for a particular activity. It is not sufficient to rely on generic assessments, as conditions are likely to vary significantly away from a standard or baseline condition. It follows that the assessment process can take account of standard arrangements but must be capable of reflecting actual conditions as well. Problems can arise when contracts are placed by and awarded to those not familiar with local working conditions and requirements for health and safety, which, in turn, may not be met through ignorance and lack of adequate specification before a price is agreed.

The arrival of the Construction (Design and Management) Regulations 1994 has resulted, within the construction industry at any rate, in a formal observance of all the practices to be outlined in this Section. (A detailed summary of these Regulations can be found in Part 4 Section 25.) It will be appreciated that the principles apply to a wider range of 'contractors' than those engaged simply in the construction process as defined by the Regulations. The practice of selecting contractors on the basis of their competence in health and safety terms as well as on their ability to do the work adequately is one deserving wider recognition. Furthermore, it is the actual competence of those individuals selected to do the work (and supervise it) on the client's site which is critical, and so the dangers which can result from subcontracting below a contractor originally approved as competent should not be overlooked.

In recent times, outsourcing has extended to areas that were traditionally the sole province of the employer/client, including equipment maintenance, catering and facilities management. This Section discusses the general principles of managing the contracting relationship; however, long-term arrangements for managing contractors may stretch to the secondment of staff to contractors and other semi-merged positions, and these are not discussed here.

A control strategy

There are six parts to a successful control strategy. The extent to which each part is relevant will depend on the degree of risk and the nature of the work to be contracted. The parts are:

- identification of suitable bidders
- identification of hazards within the specification
- checking of (health and safety aspects of) bids and selection of contractor
- contractor acceptance of the client's health and safety rules
- control of the contractor on site
- checking after completion of contract.

Identification of suitable contractors

It is clearly necessary to work out a system aimed at ensuring that a contractor with knowledge of safety standards and a record of putting them into practice is selected for the work. This process has three stages:

1 each contractor wishing to enter an 'approved list' should be asked to provide his/her safety policy and sample risk assessments. Arrangements will be required for vetting these for adequacy. It is important to recognise that smaller contractors may well have much less sophisticated systems than the client, and that this is not necessarily 'bad'. Requests for policy and assessments can place a significant burden on contractors, which may add significantly to the workload and the pressure faced normally. Therefore, the depth of enquiry and the adequacy of the response should be tempered against the known hazards and risks to be controlled; a certain amount of discretion should be exercised relative to the significance of the contract and the risks likely to be associated with the work. In some jurisdictions, the law may not require small businesses to produce complex (or any) health and safety documentation. In such cases, the observation of actual work practices and conditions on other contracts is likely to give a better understanding of competence than relying purely on paperwork

2 subject to the foregoing, a prequalification questionnaire should be completed by each contractor, providing necessary information about his/her policy on health and safety,

including details of responsibility, experience, safe systems of work and training standards

3 at this stage, it should be possible to identify potentially 'competent' contractors for approval, but feedback will be required to identify any who do not in practice conform to their own stated standards. This means that the list will require regular scrutiny and updating, especially in relation to contractors used by the employer regularly. The 'approved list' should be screened regularly.

Specification

A checklist should be followed which will give a pointer to most, if not all, of the common health and safety problems which may arise during the work. These should be communicated to the contractor in the specification before the bid is made, and the received bid checked against them to ensure that proper provision is being made for the control of risk and that the contractor has identified the hazards. Suitable headings for a checklist for construction work include:

- special hazards and applicable national or local regulations and codes of practice (eg asbestos, noise, permits to work)
- training required for the contractor's employees, which the client may have to provide
- safe access/egress to, from and on the site, and to places of work within the site
- electrical and artificial lighting requirements
- manual/mechanical lifting
- buried and overhead services
- fire protection
- occupational health risks, including noise
- confined space entry
- first aid/emergency rescue
- welfare amenities
- safe storage of chemicals
- PPE
- documentation and notifications
- insurance and special terms and conditions of the contract.

The process of drawing up the specification should include appropriate consultations with the workforce to ensure that they are fully aware of the proposed work and have a chance to comment on those aspects of it which may affect their own health and safety.

Checking the bids

When the bids are returned, it should be possible to distinguish the potentially competent at this stage. An approved list of contractors, scrutinised at intervals, can save the need for carrying out a complete selection process as described on every occasion.

Safety rules

A basic principle of control is that as much detail as possible should be set down in the contract. An important condition should be that the contractor agrees to abide by all the provisions of the client's safety policy which may affect his/her employees or the work, including compliance with any local health and safety rules. Exchange of information and summaries of risk assessments are requirements of Regulations 11 and 12 of the Management of Health and Safety at Work Regulations 1999 (MHSWR), where construction or any contract workers interact with the employer's undertaking.

Often, the contractor may delegate the performance of all or part of the contract to other subcontractors. In these cases, it is essential to ensure that the subcontractors are as aware as the original contractor of the site rules and safety policy. A condition which can be attached to the contract is that the contractor undertakes to:

- check the competence, in turn, of any subcontractors he/she intends to use
- inform them, in turn, of all safety requirements
- incorporate observance of them as a requirement of any future subcontract and require the subcontractor to do likewise if he/she, in turn, subcontracts any work
- limit the number of subcontractors in the chain for any particular task – no more than three is a workable number used by some major organisations.

Written orders containing detailed terms and conditions such as the above should be the basis of the contract and should be

acknowledged by the contractor before work starts. The loan of tools and equipment by the client should be avoided unless part of the original contractual arrangement.

Areas of concern should be covered by general site rules and by the client's safety policy. They should be communicated to the contractor in the form of site rules. For construction work on a client's premises, they should cover at least the following topics:

- materials storage, handling, disposal
- use of equipment which could cause fires
- noise and vibration
- scaffolding and ladders, access
- cartridge-powered fixing tools
- welding equipment – and use of client's electricity supply
- lifting equipment – certificated, adequate
- competence of all plant operators
- vehicles on site – speed, condition, parking restrictions
- use of lasers or ionising radiations
- power tools – voltage requirements
- machinery brought onto the site
- site huts – location, ventilation, gas appliances
- use of site mains services
- electricity – specialised equipment required?
- firefighting rules
- waste disposal procedures
- use of client's equipment
- permit-to-work systems in force
- hazardous substances in use on site by client
- basic site arrangements, times, reporting, first aid, fire
- site boundaries and restricted areas.

Practical control of contractors

A reminder of the key questions at the heart of the issue:

- does the level of risk from contractor activities match the extent of contractor management in the organisation?
- is there an appropriate balance between the level of supervisory activity by the client and the level of self-management by the contractor?
- are contractors adequately assessed for performance and is that information fed back to those placing work orders?

The majority of the following measures are either explicit or implicit requirements of legislation, including the MHSWR. They are essential for all contractor operations, however large or small the contract.

1   **appointment/nomination** of a person or team to co-ordinate all aspects of the contract, including health and safety matters
2   **a pre-contract commencement meeting** held with the contractor and subcontractors as necessary to review all safety aspects of the work. The contractor should also be asked to appoint a liaison person to ease later communication problems which may arise. Also, communication paths should be developed to pass on all relevant safety information to those doing the work. Any permitted borrowing of equipment should be formally discussed at this time
3   **arrangement of regular progress meetings** between all parties, where health and safety is the first agenda item
4   **regular (at least weekly) inspections of the contractor's operations** by the client
5   **participation in any safety committees** on site by contractors should be a condition of the contract
6   **provision by the contractor of written method statements** in advance of undertaking particular work, as agreed. This applies particularly to construction work, where activities may include:
    * demolition
    * asbestos operations
    * work which involves disruption or alteration to mains services or other facilities which cause interruption to the client's activities
    * erection of falsework or temporary support structures
    * steel erection.
    An essential feature, but one often missing, is the stipulation that, in the event of the need for a deviation from the method statement, no further work will be done

until agreement has been reached and recorded in writing between the client and the contractor on the method of work to be followed in the new circumstances
7   the **formal reporting** to the client by the contractor of all lost-time accidents and dangerous occurrences, including those involving subcontractors
8   the client must **set a good example** by the following of all site rules
9   provide adequate **safety propaganda** material, including posters and handbooks
10 allow no machinery on site until **documentation on statutory inspections** has been provided, including details of driver training and experience
11 monitor the contractor's **safety training programme**.

Contract completion

The contractor should leave the work site clean and tidy at the completion of the contract, removing all waste, materials, tools and equipment. This should be checked. Checking on the quality of the contracted work is seldom overlooked, but the actual performance and behaviours of contractors should be reviewed at the same time.

Legal requirements

Failure to manage contractors has wide implications under the Health and Safety at Work etc Act 1974, where Sections 2, 3 and 4 can be applied to occupiers and contractors, depending on the circumstances. Similarly, civil claims for damages can be made against occupiers as well as contractors. The MHSWR also apply, particularly in respect of the provision of information required to be in the possession of all employees. The application of the Construction (Design and Management) Regulations 1994 should be checked by clients with contractors to ensure that all work within their scope is conducted in accordance with the requirements of the Regulations.

Revision

6 parts to the control plan for contractors:

- questionnaire to identify potentially safe contractors
- hazard identification within the specification
- check the bid and select contractor
- put health and safety rules in the contract
- control the contractor on site
- check safe completion of work

Selected references

Legal

Construction (Design and Management) Regulations 1994
Health and Safety (Information for Employees) Regulations 1989
Occupiers Liability Acts 1957 and 1984
Management of Health and Safety at Work Regulations 1999

Guidance (HSE)

HSG65	*Successful health and safety management*
HSG150	*Health and safety in construction*
HSG224	*Managing health and safety in construction: Construction (Design and Management) Regulations 1994* ACoP, ISBN 0 7176 2139 1

Self-assessment questions

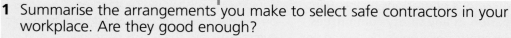

1 Summarise the arrangements you make to select safe contractors in your workplace. Are they good enough?

2 Develop a prequalification questionnaire suitable for use by contractors such as window cleaners, caterers and repair firms, which, on completion of the contracted work, would help you evaluate their safety performance.

9 Accident investigation, recording and analysis

Introduction

"We ... are still unable to see a worker safely through a day's work. Why? Because we have not thoroughly analyzed our accident causes. As a result of inadequate accident reporting we have had insufficient data to target in on unsafe tools, machines, equipment or facilities. We also continue to place all of our safety bets on 'human performance' to avoid hazards that are not being identified. We cannot expect to reduce our accident experience by a solitary approach of attempting always to change human behavior to cope with hazards. If there is a hole in the floor, we cannot reliably expect to avoid an accident by training all of the people to walk around the hole. It is far simpler to cover the hole." (David V MacCollum, past president, American Society of Safety Engineers)

The hardest lessons to be learned in accident prevention come from the investigation of accidents and incidents which could have caused injury or loss. Facing up to those lessons can be traumatic for all concerned, which is one reason why investigations are often incomplete and simplistic. Nevertheless, the depth required of an investigation must be a function of the value it has for the organisation and other bodies which may make use of the results, such as enforcement agencies. Conducting one can be expensive in time. After the investigation, a standard system of recording and analysing the results should be used.

Investigation of accidents

Purpose

The number of purposes is large; the amount of detail necessary in the report depends on the uses to be made of it. Enforcement agencies look for evidence of blame, claims specialists look for evidence of liability, trainers look for enough material for a case study. From the viewpoint of prevention, the purpose of the investigation and report is to establish whether a recurrence can be prevented, or its effects lessened, by the introduction of safeguards, procedures, training and information, or any combination of these.

The procedure

There should be a defined procedure for investigating all accidents, however serious or trivial they may appear to be. The presence of a form and checklist will help to concentrate attention on the important details. Supervisors at the workplace where the accident occurred will be involved; for less serious accidents, they may be the only people who take part in the investigation and reporting procedure. Workers' representatives may also be involved as part of the investigating team.

The equipment

The following are considered as essential tools in the competent investigation of accidents and damage/loss incidents:

- report form, possibly a checklist as a routine prompt for basic questions
- notebook or pad of paper
- tape recorder for on-site comments or to assist in interviews
- camera – instant-picture cameras are useful (but further reproduction of them may be difficult, expensive or of poor quality). Digital cameras are becoming very popular, but the evidential value of their images has yet to be tested in court
- measuring tape, which should be long and robust, like a surveyor's tape
- special equipment in relation to the particular investigation, eg meters, plans, video recorder.

The investigation

Information obtained during investigations is given verbally or provided in writing. Written documentation should be gathered to provide evidence of policy or practice followed in the workplace, and witnesses should be talked to as soon as possible after the accident. The injured person should also be seen promptly.

Key points to note about investigations are:
- events and issues under examination should not be prejudged by the investigator
- total reliance should not be placed on any one sole source of evidence
- the value of witness statements is inversely proportional to the amount of time which passes between the events or circumstances described and the date of a statement or written record (theorising by witnesses increases as memory decreases)
- the first focus of the investigation should be on when, where, to whom and the outcome of the incident
- the second focus should be on how and why, giving the immediate cause of the injury or loss, and then the secondary or contributory causes
- the amount of detail required from the investigation will depend on a) the severity of the outcome and b) the use to be made of the investigation and report
- the report should be as short as possible, and as long as necessary for its purpose(s).

The report
For all purposes, the report which emerges from the investigation must provide answers to the following questions (only the amount of detail provided should vary in response to the different needs of the recipients):
- what was the immediate cause of the accident/injury/loss?
- what were the contributory causes?
- what is the necessary corrective action?
- what system changes are either necessary or desirable to prevent a recurrence?
- what reviews are needed of policies and procedures (eg risk assessments)?

It is not the task of the investigation report to allocate individual blame, although some discussion of this is almost inevitable. Reports are usually 'discoverable'; this means they can be used by the parties in an action for damages or criminal charges. It is a sound policy to assume that accident investigation reports will

be seen by solicitors and experts acting on behalf of the injured party. They are entitled to see the factual report, and this will include anything written in it which might later prove embarrassing – so it should certainly not contain comments on the extent of blame attaching to those concerned, or advice given to management.

It is appropriate, necessary and quite proper that professional advice is given, but it should be provided in a covering letter or memorandum suitably marked 'Confidential'. The Civil Procedure Rules (see Part 4 Section 3) introduced many changes in the discovery process, such that a wider range of documents may become 'discoverable' in civil claims.

Whether the report is made on a standard form or specially written, it should contain the following:
- a summary of what happened
- an introductory summary of events prior to the accident
- information gained during investigation
- details of witnesses
- information about injury or loss sustained
- conclusions
- recommendations
- supporting material (photographs, diagrams to clarify)
- the date and the signatures of the person or persons carrying out the investigation.

Accident recording
Which injuries and incidents should be investigated and recorded? All those which can give information useful to prevent a recurrence of the incident giving rise to loss or injury. Regulations may also define requirements, although reporting requirements are usually limited to the more serious injuries, and those incidents with the most serious consequences or potential consequences. Counting only these may mask the true extent of injuries and losses by ignoring the potential consequences of incidents which by chance have led to relatively trivial injuries or damage.

Standardised report forms kept at each workplace should be used, and returned to a central point for record-keeping and analysis. It is important that supervisory staff at the workplace carry out preliminary investigations and complete a report, as they should be accountable for work conditions and need to have personal involvement in failure (accidents and damage incidents). This demonstrates their commitment and removes any temptation to leave 'safety' to others who may be seen as more qualified.

Lessons learned

Following investigation of an incident, it is always appropriate to make sure that everyone in the organisation is aware of the basic findings so that they are able to place the incident in the context of their own work areas and apply any relevant lessons to practices, procedures, training and the like. The checklist in Figure 1 may be helpful as a reminder.

Figure 1: Post-investigation checklist

Have all relevant people been advised of the nature of the incident, issues identified and solutions emerging from it?	YES	NO
Was the written procedure for investigation of the incident adequate?	YES	NO
Were the people assigned tasks able to perform them as required?	YES	NO
Were senior management informed promptly of what happened?	YES	NO
Has an accurate record of the investigation been maintained?	YES	NO

Accident analysis

The incoming reports will need to be categorised and statistics collected so that meaningful information on causes and trends can be obtained. There are several ways of doing this, including sorting by: the nature of the injury or body part involved; age group; trade; work location or work group; or type of equipment involved. Selection of categories will depend on the workplace hazards, but use of the categories

and format of the Reporting of Injuries, Diseases and Dangerous Occurrences Regulations 1995 (RIDDOR) (see Part 4 Section 20) can be used, which will assist in making comparisons between works and national or industry group figures. One classification which will be found particularly helpful is breakdown by cause. The following gives heading examples; these can be broken down further if required:

- falls
  - from a height
  - on the same level, including trips and slipping
- struck
  - by moving object
  - by vehicle
  - against fixed or stationary object
- manual handling or moving of loads
- machinery
- contact with harmful substances
- fire or explosion
- electricity
- other causes.

Each of these categories or, more usually, the aggregate numbers are converted into totals for a period, annually or at more frequent uniform intervals. Presenting the information in pictorial form will make it easier to understand; bar charts and pie charts are examples. Statistics used for comparison purposes are expressed in recognised ways using simple formulae to produce rates rather than raw numbers. Comparisons are best restricted to period-on-period within the same department or organisation, rather than between the organisation and national or international rates. An exception could be for an organisation sufficiently large to generate a statistically significant sample for comparison with national figures. Local comparative use of statistics can help with goal-setting.

Benchmarking is becoming popular in some industries, but this can produce its own problems. True comparability of data is hard to achieve, which, in turn, can lead to disappointing results and friction between the benchmarking partners. Other performance measures which do not necessarily involve injury statistics are the use of Key Performance Indicators within the health and safety management system, and the integration of health and safety management into a 'balanced scorecard' measuring system. Both of these techniques can be used advantageously by organisations with large numbers of departments, sectors or premises, and by bodies with autonomous subsidiaries.

There are significant difficulties in using and comparing accident statistics, especially between different countries. This is because of underreporting to the authorities collecting the data (which can be as high as 70 per cent), differences between countries as to which accidents in working time count as recordable (UK road traffic accidents do not count at present), and differences in the formulae used to calculate rates.

There are no standard or formal requirements in the UK for the statistical formulae used for the analysis of accident data. Those most often encountered are:

Frequency rate:

$$\frac{\text{Number of injuries x 100,000}}{\text{Total number of hours worked}}$$

Incidence rate (helpful where the number of hours worked is either low or not available):

$$\frac{\text{Number of injuries x 1,000}}{\text{Average number employed during the period}}$$

Confusingly, the incidence rate most commonly used by the UK government is:

$$\frac{\text{Number of fatal or major injuries x 100,000}}{\text{Number at risk in the particular industry sector}}$$

The **severity rate** shows the average number of days lost per time unit worked. Often, the time unit is expressed as 1,000 hours, but any suitable unit of measure can be used as long as it is stated clearly. For example:

$$\frac{\text{Total number of days lost x 1,000}}{\text{Total number of hours worked}}$$

A common experience for those new to health and safety management is to find that as the reporting and recording system is improved, the numbers of injuries and incidents reported rises. To the uninitiated this can be worrying, as it seems at first that there are more injuries following improvements in the control system than before the work started. What is usually happening here is that the system is simply capturing more information about the failures that are occurring.

Legal requirements

Employers and other 'responsible persons' with control over work premises as well as employees have a duty to report certain injuries suffered while at work, occupational diseases and defined dangerous occurrences. RIDDOR gives detailed requirements (see Part 4 Section 20). The employer is also required to keep an accident book on his/her premises, into which details of all injuries must be entered as required by the Social Security Administration Act 1992.

Revision

8 **features of the accident investigation process**:
- purpose
- procedure
- equipment
- investigation
- report
- recording of results
- analysis of results
- presentation of results in meaningful format

Selected references

Legal

Reporting of Injuries, Diseases and Dangerous Occurrences Regulations 1995

Social Security Administration Act 1992

Social Security (Industrial Injuries) (Prescribed Diseases) Regulations 1985

Guidance (HSE)

L73   *A guide to RIDDOR 95*

Self-assessment questions

1 Find out the accident frequency rate for your workplace and compare it with the national average for your industrial classification. Can you calculate frequency rates for individual causes of injury? Is this knowledge useful?

2 Is 'carelessness' an adequate sole conclusion on an accident report?

10 Techniques of inspection

Introduction

Inspection for health and safety purposes often has a negative implication associated with fault-finding. A positive approach based on fact-finding will produce better results and co-operation from all those taking part in the process.

The objectives of inspections are:
- to identify hazardous conditions and start the corrective process
- to improve operations and conditions.

There are a number of types of inspections:
- statutory – for compliance with health and safety legislation
- external – by enforcement officials, insurers, consultants
- executive – senior management tours
- scheduled – planned at appropriate intervals, by supervisors
- introductory – checks on new or reconditioned equipment
- continuous – by employees, supervisors, which can be formal and preplanned, or informal.

For any inspection, knowledge of the plant or facility is required, as well as knowledge of applicable regulations, standards and codes of practice. Some system must be followed to ensure that all relevant matters have been considered, and an adequate reporting system must be in place so that necessary remedial actions can be taken and the results of the inspection made available to management. Some experts believe that 'assurance' is a better description of the activity – there is a need to assure that the system is working properly (safely). To be effective, inspection of this type needs measurement of how good or bad things are, which can then be compared with standards set either locally, corporately or nationally. Corrective action can then be taken.

Audits look at systems and the way they function in practice; **inspections** look at physical conditions. So, while inspections of a workplace could be done weekly, an audit of the inspection system throughout an organisation would examine whether the required inspections were being carried out, the way they were being recorded, action taken as a result, and so on. More information on audits can be found in Part 1 Section 2.

Safety tours are carried out in teams, led by a senior member of management. They are not inspections in the sense that they seek to find

out and record all significant health and safety matters in the area under study, but rather they demonstrate the interest and involvement of senior management. Safety tours aim to find examples of both bad and good practice, perhaps to focus attention on current issues or topics under debate. They also offer opportunities for individual discussion between managers and all those directly involved in the work process.

Principles of inspection

Before any inspection, certain basic decisions must have been taken about aspects of it, and the quality of these decisions will be a major influence on the quality of the inspection and on whether it achieves its objectives. The decisions are reached by answering the following questions:

1 **What needs inspection?** Some form of checklist, specially developed for the inspection, will be helpful. This reminds those carrying out the inspection of important items to check; it also serves as a record. By including space for 'action by' dates, comments and signatures, the checklist can serve as a permanent record. Inspection requirements for work equipment are reviewed in Part 2 Section 2.

2 **What aspect of the items listed needs checking?** Parts likely to be hazards when unsafe – because of stress, wear, impact,

vibration, heat, corrosion, chemical reaction or misuse – are all candidates for inspection, regardless of the nature of the plant, equipment or workplace.

3   **What conditions need inspection?** These should be specified, preferably on the checklist. If there is no standard set for adequacy, then descriptive words give clues to what to look for – items which are exposed, broken, jagged, frayed, leaking, rusted, corroded, missing, loose, slipping, vibrating, etc.

4   **How often should the inspection be carried out?** In the absence of statutory requirements or guidance from standards and codes of practice, this will depend on the potential severity of the failure if the item fails in some way, and the potential for injury. It also depends on how quickly the item can become unsafe. A history of failures and their results may give assistance.

5   **Who carries out the inspection?** Every worker has a responsibility to carry out informal inspections of his or her part of the workplace. Supervisors should plan general inspections and take part in periodic inspections of aspects of the workplace considered significant under the above guideline. Workers' representatives may also have rights of inspection and their presence should be encouraged where possible. Management inspections should be made periodically; the formal compliance inspections should take place in their presence.

Techniques of inspection

The following observations have been of assistance in improving inspection skills:

- those carrying out inspections must be properly equipped to do so, having the necessary knowledge and experience, and knowledge of acceptable performance standards and statutory requirements. They must also comply fully with local site rules, including the wearing or use of PPE, as appropriate, so as to set an example

- develop and use checklists, as above. They serve to focus attention and record results, but must be relevant to the inspection

- the memory should not be relied on. Interruptions will occur, and memory will fade, so notes must be taken and entered onto the checklist, even if a formal report is to be prepared later

- it is desirable to read the previous findings before starting a new inspection. This will enable checks to be made to ensure that previous comments have been actioned as required

- questions should be asked and the inspection should not rely on visual information only. The 'what if?' question is the hardest to answer. Workers are often undervalued as a source of information about actual operating procedures and of opinions about possible corrections. Also, systems and procedures are difficult to inspect visually, and their inspection depends on those involved being asked the right questions

- items found to be missing or defective should be followed up and questioned, not merely recorded on the form. Otherwise, there is a danger of inspecting a series of symptoms of a problem without ever querying the nature of the underlying disease

- all dangerous situations encountered should be corrected immediately, without waiting for the written report, if their existence constitutes a serious risk of personal injury or significant damage to plant and equipment

- where appropriate, measurements should be taken of conditions. These will serve as baselines for subsequent inspections. What cannot be measured cannot be managed

- any unsafe behaviour seen during the inspection should be noted and corrected, such as removal of machine guards, failure to use PPE as required, or smoking in unauthorised areas

- risk assessments should be checked as part of the inspection process.

Revision

2 **objectives of inspections:**
- identification of hazards
- improvement of operations/conditions

6 **types of inspection:**
- statutory
- external
- executive
- scheduled
- introductory
- continuous

10 **techniques of value:**
- have necessary experience and knowledge
- use checklists
- write things down
- read previous reports first
- ask questions
- follow up on problems
- correct dangerous conditions at once
- measure and record where possible
- correct unsafe behaviour seen
- check risk assessments

Selected references

Legal

Construction (Health, Safety and Welfare) Regulations 1996
Lifting Operations and Lifting Equipment Regulations 1998
Pressure Systems Safety Regulations 2000
Work at Height Regulations 2005

Guidance (HSE)

HSG54   *The maintenance, examination and testing of local exhaust ventilation*

British Standard

BS 7121:1991   Safe use of cranes. Part 2: Inspection, testing and examination

Self-assessment questions

1 On pages 51 and 52 you will find headings for a safety inspection checklist, with more detailed subheadings. Tick the boxes in pencil to indicate the kinds of inspections now made in your workplace for each category and the standards which apply, putting a cross to show the interval at which each is made. Then repeat, using a pen, indicating what the ideal inspections would be for each.

2 In your workplace, what are the main factors that dictate whether and how often inspections should be made?

Safety inspection checklist and frequencies

	Inspection before each use	Weekly	Monthly	Annually	Company/group standard	National standard	International standard	No current standards
Environmental factors								
Lighting								
Dusts								
Gases								
Fumes								
Noise								
Hazardous substances								
Flammables								
Acids and bases								
Toxics								
Carcinogens								
Production equipment								
Machinery								
Pipework								
Conveyors								
Power generation equipment								
Steam equipment								
Gas equipment								
Generators								
Electrical equipment								
Cables, circuits								
Switches, sockets								
Extensions								
Personal protective equipment								
Eye protection								
Clothing								
Helmets/head protection								
RPE								
Footwear								
Hand protection								
First aid and welfare								
Washing facilities								
Showers								
Toilets								
first aid facilities								

Safety inspection checklist and frequencies (continued)

	Inspection before each use	Weekly	Monthly	Annually	Company/group standard	National standard	International standard	No current standards
Fire protection equipment								
Hoses and extinguishers								
Alarm systems								
Sprinklers								
Access routes, roadways								
Roads								
Pavements								
Crossing points								
Signs								
Lifts								
Stairways								
Emergency routes								
Access equipment								
Ladders								
Stepladders								
Trestles								
Cradles								
Material handling equipment								
Chains								
Ropes								
Fork-lifts								
Specialised equipment								
Transport								
Cars								
Internal vehicles								
Road load vehicles								
Containers								
Disposal								
Storage								
Cylinders								

11 Information sources

Introduction

There is a great deal of information available on health and safety topics. The problem is that it is mostly unco-ordinated, in many places, and often written by specialists so that it cannot be easily understood by people who have to work with it. Information technology is moving towards the production of solutions to problems by the combination of many of these sources.

Authoritative guides on particular topics are known as **primary sources**. The collected references to these guides are **secondary sources**, and include bibliographies, reading lists, abstracts and indexes. Most information sources specialise as primary or secondary sources, with a small amount of overlap – some sources have both facilities. Technical articles in journals have reading lists, data sheets refer to larger databases, and databases often do not carry complete texts. Systems for information provision designed since about 1980 recognise the need for 'one-stop' information shopping for answers to problems and avoid the temptation to cross-reference to a number of other sources.

Because of the numbers of sources, it is not possible to do more than provide a list of groups (see Table 1 overleaf) with some well-known examples and a commentary on the material which is now available to carry the information.

Material

Most of our information is still provided on **paper** and this is not likely to change despite the introduction of new technologies. Some of these have problems associated with them – **photocopies** and **facsimile** transmission paper both suffer from ultraviolet light, so the image gradually disappears over time. Other photographic record systems for holding information such as **microfiche** are more permanent, but can still be damaged. They are inconvenient to read and copy, although they carry complete primary source material in a small space.

Computer files held at the workplace are widely used as information storage, although new generations of computers are often unable to decipher material stored on older systems. **Computer networks**, set up between offices using telephone or fibre-optic lines, are able to share much information, which is often stored centrally on a mainframe. Access to such a system enables a large amount of information to be accessed at short notice at a local site. Disadvantages are the cost and the need to service the system regularly.

The expense of using computer data resources can be minimised by gaining access to someone else's information by using an electronic database through a modem connection, and there are now a number of computer resources which accept worldwide connections through the Internet. Most are secondary sources, although the ability to carry full text and graphics is spreading. NIOSHTIC and HSELINE are the best known English-language databases.

It is also possible to access versions of these and other databases at the workplace using a **CD-ROM** reader. Compact disc technology allows up to 280,000 pages of information to be stored on a single compact disc, which can be read by a personal computer. Recording onto CD-ROM and **DVD** is now achieved cheaply. The main advantage of doing so is having, close at hand, a huge volume of material which can be stored and quickly accessed. CD-ROM constructed databases are available for purchase through providers such as Technical Indexes, who also update and extend their products regularly. In recent times, the trend for these providers is to hold their information on Web servers that can be accessed by password (and subscription). This avoids the need for the user to hold very large quantities of data personally.

International data resources mostly store information on US requirements. A major supplier of chemical data sheets and other health and safety information on CD-ROM is the Canadian Centre for Occupational Health and Safety (CCINFO).

Information can also be stored in learning programmes and combined with video into an interactive system, available on the Web and on CD-ROM, and played through specially adapted visual display screens. The use of multimedia, which combines still and moving images with sound and text information, has increased rapidly. It can be produced relatively easily and accessed through simple computer configurations.

Table 1: Providers of information

Sources	
Company safety policy	Organisation and arrangements
In-company safety services	Company safety staff, library
Corporate safety services	Central group resource, database
Enforcement agencies	HSE and local authorities (EHOs) – advise and enforce
Government bodies and departments	HMSO
Manufacturers	Product literature, updates, MSDS
Trade associations	Handbooks, advice to members
Standards organisations	BSI, CEN, CENELEC, ISO
Subscription services	Magazines, journals, newsletters
Consultancies	External audits, information, advice
Voluntary safety bodies	RoSPA, British Safety Council
Professional bodies	IOSH, IChemE
International safety bodies	ILO
Educational institutions	University programmes, colleges

The Internet

Searching the Internet for health and safety information can be extremely rewarding. There are now many thousands of websites giving access to safety, health and environmental information, but it is important to remember that not all of it can be relied on. Just as trade associations can be expected to publish information and views favourable to the members of the association, it should not be forgotten that many interest groups have websites and are anxious to promote a particular view in the guise of unbiased information. A list of websites which have been reviewed for their technical content and usefulness can be found through CCINFO.

Other tools on the Internet include **newsgroups**, which enable questions to be asked and answered in front of an international audience, and facilitate the exchange of software. The user should be aware that most postings to newsgroups will be picked up by search engines, and could result in a flood of unwanted emails. Mostly these will be 'spam' messages (the electronic equivalent of junk mail), but some may be hostile and contain viruses.

It is essential to use a virus-checking program to screen incoming material from the Web. Also, unexpected email attachments (files that accompany an email) should never be opened directly from the Web until screened in this way. Failure to observe this basic principle can result in the entire computer system being scrambled, as well as the invading message being passed on automatically to anyone emailed subsequently from the infected system.

Unless a unique Web address has been given as a reference, the searcher will have to use one of almost 2,000 available search engines to track down a particular subject. These are electronic dictionaries which automatically search the Web for new material and log its presence. Examples of search engines which work well for health and safety searching include Google (www.google.com), Lycos, Hotbot and Excite. When the word (or words) being searched for is entered, the search engine will show a list of 'hits' – Web pages which contain the word.

Many of these will be useless, and some refinement of the search is often needed to avoid being swamped by irrelevant material. Information on how to refine a search can be obtained from the search engine's 'Help' link. There are also free translation systems which can be used to provide instant interpretations of pages in unfamiliar languages.

The Web changes daily, expanding rapidly. In an earlier edition of this book, published only three years ago, an example of a search was given. The text is repeated here, and in brackets are the figures returned from the same search engine in mid-2005.

For example, information is needed about school playground safety issues. A search on Google for references to 'safety' produced a claimed figure of about 24 million (223 million) Web pages containing the word, within 0.15 seconds (0.07 seconds). Refining the search by adding the word 'school' reduced the number of hits to 'about 2.35 million' (49.1 million). Adding the third word 'playground' reduced the hits to a more manageable 168,000 (796,000).

It can be seen that the speed of the search engines has also increased. But the most important factor is not the number of hits but whether the material required can easily be found within the first few pages offered. Search engines tend to place the most recent material at the top, but it can still take considerable time to trawl through the pages offered to find relevant information. A very precise set of keywords is the best tool to use.

The searcher also needs to be aware that a significant percentage of information on the Web is put there by organisations and people with an interest – commercial of course, but also political, lobbying and sometimes mischievous. Much is opinion rather than fact.

It is also necessary to remember that the majority of Web pages are US-based, and that there are language differences. In the above example, the word 'playtime' was used initially instead of 'playground'. The number of hits reduced to 13,700 (97,900) but some of the pages inspected were not suitable for viewing by anyone easily offended – and some are doubtless illegal to view. It is also possible to confine your search to UK-based websites.

Table 2 overleaf contains a selected list of websites that provide useful information on a wide variety of topics likely to be of interest to the general reader. Once the address has been typed into a computer's Web browser, it can be saved to visit again by using the 'Favorites' menu. (This allows repeat visits to be made without having to type the full address of the page each time.) But it should be remembered that Web pages often change their addresses, and even disappear altogether, so the only guarantee that can be given is that all the selections in Table 2 were open to access in mid-2005.

NB: Not all Web browsers require the address to start with 'www'. If an address does not load, re-enter the address without the 'www'. Alternatively, type the main part of the address into a Google search and use the hyperlinks (underlined words) to get to the page you need. For example, if 'www.workcover.nsw.gov.au' does not appear to work, type 'Workcover' into Google.

Table 2: Web references

Web address	Topic and comments
Search engines	
www.google.com	The best search engine
UK Government sites	
www.hse.gov.uk	HSE home page
www.hse.gov.uk/condocs	Download consultative documents from here
www.hse.gov.uk/hsestats.htm	HSE statistics
www.dti.gov.uk/er/work_time_regs/index	The Working Time Regulations explained
www.dti.gov.uk/strd/strdpubs.htm	Download PDF documents relating to product and electrical safety
European sites	
www.europe.osha.eu.int	European Agency for Health and Safety at Work
www.occuphealth.fi/e/eu/haste	European health (and safety) database
Organisations	
www.iosh.co.uk	Links to open discussion forum for help from professionals
www.bohs.org	British Occupational Hygiene Society
www.who.int	World Health Organisation
www.ccohs.ca	CCINFO home page, huge database of links to other health and safety sites
www.rospa.co.uk	RoSPA home page, good for general interest, especially for small organisations
www.eevla.ac.uk/vts/healthandsafety	Free training on Internet use for safety searches
www.blpc.bl.uk	British Library Public Catalogue entry page
www.joule.pcl.ox.ac.uk/MSDS	The best place for pure chemicals data sheets
www.stress.org.uk	Stress Management Society
www.dstress.com	Stress Education Centre
www.asthma.org.uk	Asthma and general occupational health
www.bad.org.uk	Professional society for dermatology
Good foreign (non-EU) sites	
www.safetyline.wa.gov.au	Helpful Australian site of general use
www.workcover.nsw.gov.au	Good free downloads
www.dermnet.org.nz/index	Dermatitis and other skin conditions
www.innerbody.com	Learn anatomy here

Self-assessment questions

1 Outline the steps you take at present to find health and safety information. How could these be improved?

2 Explain with examples the difference between a primary and secondary source of information. Mark the following sources as primary or secondary:
   - your company's annual report
   - full proceedings of an annual technical conference
   - your shopping list
   - an international standard on eye protectors
   - a reference library.

12 Human factors and change

Introduction

Health and safety at work are heavily influenced by change. Few things remain the same over time. In organisations, priorities and values change as do the people who put them into practice. The challenge is to maintain continuity of purpose and to seek continual improvement. Because of this, the health and safety professional is an agent of change: a facilitator, not just an expert. It follows that the skills of the professional need to include the ability to recognise the signs of change and influence it within the organisation. For successful health and safety management, input into the management process is required from a number of disciplines – engineering, human factors, industrial relations, occupational health as well as 'safety'. Just as accidents are described as the results of multiple causes, so is change the product of many influences.

Change

Achieving real, meaningful change within any organisation is very difficult. It requires recognition at the most senior level that change is both necessary and inevitable, and that the direction and extent of it can be influenced. Planned change needs a focus, and a directed and managed plan, complete with measured milestones on the way to the stated objectives, is essential.

In order to put in place the reforms that have been described elsewhere in this book, the most senior people must put their personal support behind the plan. Without this, the plan is sure to fail to achieve its objectives, as no-one will be persuaded that the proposed change has been fully sanctioned. Without getting the buy-in from all levels, it will fail.

Success can be and has been achieved where such a 'health and safety business plan' is presented to the board or 'most senior' level for approval. This should show the present costs to the organisation of failure to manage health and safety adequately, the proposed cost of implementing the plan, and the timescale for carrying it out. The cost of research and audit should be included, together with training for all levels. With acceptance of the plan, it should then be driven down to the next layer of management by insisting on their attendance at a workshop designed to show precisely what steps those attending are required to take. Change initiatives that fail usually have in common an inability to persuade the second tier of management that they can and must make a difference.

Change must be measured, so the plan should include details of parameters that will measure the extent of success. The use of Key Performance Indicators, in addition to the counting of injuries, will be useful. These can show the degree to which the plan is being followed in each unit of the business, by agreed measurement of key events required by the plan, and progress towards agreed targets.

Human factors

Human factors are those that affect performance, including obvious elements such as the social and inherited characteristics of work groups and their members, and the individual capabilities of workers. They include:
- mental, physical and perceptual capability
- interaction between people, their jobs, the environment and the employer
- system and equipment design influences
- characteristics of organisations that affect safety-related behaviour
- individual social and inherited characteristics.

The cumulative effects of all of these can result in **human failure**. There are many ways of categorising behaviours that are significant for health and safety at work. The reader is referred to the HSE publication *Reducing error and influencing behaviour* for an excellent account of human factors in safety and a full discussion of why people fail.

We can distinguish between **active failure**, when there is the potential for immediate unwanted consequences, and **latent failure**, as a result of which the stage is set for something to go wrong. For example, taking a chance in not isolating a machine before delving inside it is an active failure, and a manager ignoring persistent breaches of safe working practice shows a latent failure. Individual incorrect or inappropriate behaviour can also be grouped into **errors**, which are unintended deviations from safe practices, and deliberate **violations**, where rule-breaking becomes standard practice.

Peer influence is a powerful motivator for behaviour, especially in small work groups. The inexperienced are likely to take their leads from the behaviour of others. Many other factors can influence concentration on the task, such as age and the effects of drugs, including alcohol.

Knowledge of what constitutes safe behaviour does not guarantee that it will be forthcoming. It is a truism that 'to err is human', and we recognise this when we accept that one of the long-term objectives of health and safety management is to attempt progressively to reduce the opportunities for people to make mistakes. Recent legislative initiatives, particularly from the European Union, have incorporated the human factors approach. This can best be seen in the Manual Handling Operations Regulations 1992 and the Health and Safety (Display Screen Equipment) Regulations 1992.

Safety culture

Producing and maintaining a positive safety culture requires recognition of the significance of the above elements, and planning strategies to identify and resolve human factors issues. Key points are:
- good communications between and with employees and management
- ensuring a real and visible commitment to high standards by senior management
- maintaining good training standards to achieve competence
- achievement of good working conditions.

Stress

Stress can be defined as the reaction people have to excessive pressures, traumatic experiences or any abnormal demand placed on them because of their life circumstances. Therefore, bereavement produces stress, as does overwork. Everyone has to deal with pressures and there is some evidence that pressure is necessary for maximum effectiveness to be achieved. However, response to pressure can be physically and mentally damaging if sustained over long periods. Illness, including depression and heart disease, can result from excessive exposure to stress.

Surveys indicate that stress is a significant workplace issue. Nearly a fifth of managers surveyed in 1997 admitted to taking time off during the previous 12 months because of work-related stress. Stress poses a significant risk, but is certainly the least well controlled of all workplace risks.

Factors which contribute to stress include:
- the physical work environment, including noise and thermal effects
- 'office politics' – relationships with work colleagues and with third parties encountered while at work
- volume of work – too much or too little
- worries about job security
- lack of ability to control the pace and nature of work activities
- changes in working practices – these can lead to alterations in the sense of self-worth and ability to cope
- shift or other unusual work timetables
- poor communication and lack of input into decision-making
- non-work environment.

Stress can cause sleeplessness. The importance of sleep to mental wellbeing is significant, as it affects memory, learning and physical condition. Studies at the Finnish Institute of Occupational Health have shown that sleep deprivation affects the body's immune system, metabolism and hormonal functions.

Attempts should be made by the employer to identify and manage workplace stress, which is thought to reduce the effectiveness of individuals, increase labour turnover and sickness absence, and increase the chance of injury.

In 1999, a former Birmingham City Council housing official was forced to retire on the grounds of ill health, and his employer was ordered to pay £67,000 compensation following the admission of liability for the stress caused. In this landmark case, lack of experience and qualifications were said to be major factors, indicating the need for improvement in the selection process. This was the first time that a UK employer had admitted liability for causing stress to employees. Recent cases (2002) appear to have made it more difficult for victims to gain compensation following work-related stress. The rule is currently that the suffering employee must advise his/her employer accordingly, and give the opportunity to remodel the work and work pattern to make them less stressful. Failure to advise in this way can lower the chance of success for a claim.

Communicating safety

Various forms of propaganda selling the 'health and safety message' have been used for many years – posters, flags, stickers, beer mats and so on. They are now widely felt to be of little measurable value in changing behaviour and influencing attitudes to health and safety issues. Because of the long tradition of using safety propaganda as part of safety campaigns, however, there is a reluctance to abandon them. Possibly, this is because they are seen as constituting visible management concern while being both cheap and causing minimal disturbance to production. In contrast, much money is spent on advertising campaigns and measuring their effectiveness in selling products.

How can safety messages be effective?

Avoid negativity – Studies show that successful safety propaganda contains positive messages, not warnings of the unpleasant consequences of actions. Warnings may be ineffective because they fail to address the ways in which people make choices about their actions; these choices are often made subconsciously and are not necessarily 'rational' or logical. Studies also show that people make poor judgements about the risks involved in

activities, and are unwilling to accept a perceived loss of comfort or money as a trade against protection from a large but unquantified loss – which may or may not happen at some point in the future.

'It won't happen here' – the non-relevance of the warning or negative propaganda needs to be combated. The short-term loss associated with some (but not all) safety precautions needs to be balanced by a positive short-term gain such as peace of mind, respect and peer admiration.

Safety propaganda can be seen as management's attempt to pass off the responsibility for safety to employees. Posters, banners and other visual aids used in isolation without the agreement and sanction provided by worker participation in safety campaigns can easily pass the wrong message. The hidden message can be perceived as: "The management has done all it can or is willing to do. You know what the danger is, so it's up to you to be safe and don't blame us if you get hurt – we told you work is dangerous."

Management can have high expectations of the ability of safety posters to communicate the safety message. This will only be justified if they are used as part of a designed strategy for communicating positive messages.

Expose correctly – The safety message must be perceived by the target audience. In practice, this means the message must be addressed to the right people, be placed at or near to the point of danger, and have a captive audience.

Use attention-getter techniques carefully – Messages must seize the attention of the audience and pass their contents quickly. However, propaganda exploiting this principle can fail too readily to give the message intended – sexual innuendo and horror are effective attention-getters, but as they may be more potent images, they may only be remembered for their potency. Other members of the audience may reject the message precisely because of the use of what is perceived as a stereotype, eg 'flattering' sexual imagery can easily be rejected by parts of the audience as sexist and exploitative.

Strangely, the attention-grabbing image may be too powerful to be effective. An example to consider is the 'model girl' calendar. If the pictures are not regarded as sexist and rejected, they may be remembered. But who remembers the name of the sender? This is, after all, the point of the advertising.

Comprehension must be maximised – For the most effect, safety messages have to explain problems either pictorially or verbally in captions or slogans. To be readily understood, they must be simple and specific, as well as positive. Use of too many words or more than one message inhibits communication. Use of humour can be ineffective; the audience can reject the message given because only stupid people would act as shown and this would have no relevance to themselves or their work conditions.

Messages must be believable – The audience's ability to believe in the message itself and its relevance to them is important. Endorsement or approval of the message by peers or those admired, such as the famous, enhances acceptability. The 'belief factor' also depends on the perceived credibility of those presenting the message. If the general perception of management is that health and safety has a low priority, then safety messages are more likely to be dismissed because the management motivation behind them is questioned.

Action when motivated must be achievable – Safety propaganda has been shown to be most effective when it calls for a positive action, which can be achieved without perceived cost to the audience and which offers a tangible and realistic gain. Not all of this may be possible for

any particular piece of propaganda, and the major factor to consider is the positive action. Exhortation simply to 'be safe' is not a motivator.

Does safety propaganda work?

There is little evidence for the effectiveness of health and safety propaganda. This is mainly because of the difficulty of measuring changes in attitudes and behaviour which can be traced to the use of propaganda. For poster campaigns, experience suggests that any change in behaviour patterns will be temporary, followed by a gradual reversion to previous patterns, unless other actions such as changes in work patterns and environment are made in conjunction with the propaganda.

Limited experimental observations by the author show that the effectiveness of safety posters, judged by the ability of an audience to recall a positive safety message, is good in the short term only. One week after exposure to a poster, 90 per cent of a sample audience could recall the poster's general details and 45 per cent could recall the actual message on it. Two weeks after exposure, none of the audience remembered the message, and only 20 per cent could recall the poster design. In both cases, the attention of the audience was not drawn specifically to the poster when originally exposed to it.

Safety propaganda can be useful in accident prevention, provided its use is carefully planned in relation to the audience, the message is positive and believable, and it is used in combination with other parts of a planned safety campaign. Safety posters which are not changed regularly become part of the scenery, and may even be counter-productive by giving a perceived bad image of management attitude ('All they do is stick up a few old posters!').

Revision

Change should be:
- planned
- measured

4 **types of human failure:**
- active
- latent
- errors
- violations

6 **important elements in safety propaganda:**
- positive message
- correct exposure
- attention-getting
- comprehension
- belief
- motivation

9 **stress factors:**
- physical work environment
- relationships
- volume of work
- job security concerns
- lack of ability to control work
- changes in working practices
- work timetables
- poor communication
- non-work environment

Selected references
Guidance (HSE)

HSG48	*Reducing error and influencing behaviour*
HSG65	*Successful health and safety management*

Self-assessment questions

1 List the types of safety propaganda used in your workplace. Can you think of more effective ways of communicating the safety message?

2 If effective change must start at the top, what do you think are the barriers to successful change at lower levels in your organisation?

Part 2
Workplaces and
work equipment

1 Workplace health and safety issues

Introduction

This Section covers hazards in the workplace that are independent of the work processes and equipment used. It is plainly not possible to review all of the many different kinds of workplace that will be encountered – from shops to airport baggage handling areas, from control rooms to assembly lines, from hospitals to hotels – in principle, these are simply workplaces, where people spend most if not all of their working time.

The main legal provisions covering health and safety in workplaces are to be found in the Workplace (Health, Safety and Welfare) Regulations 1992. These are discussed in summary in Part 4 Section 7 of this book. Definitions of terms such as 'workplace' and 'work' are given in the Regulations together with details of exceptions to the application of the Regulations, but for the purpose of this Section the words take their everyday meaning. Some of the topics discussed below have more detailed or more specific requirements within detailed regulations, which should be consulted.

Ventilation

The basic requirement for every enclosed workplace is for sufficient fresh or purified air. Human body wastes are liberated into the workplace atmosphere – carbon dioxide, heat, smells and water vapour – which rapidly become unpleasant unless the ventilation is adequate. Air pollution from the processes in the workplace can include traces of ozone and smoke particles, and raised humidity levels may be found. Pollution can also enter from the outside, from processes there or exhausts from plant items.

Where opening windows does not give sufficient ventilation, mechanical systems should be considered. The fresh air supply rate should not fall below five litres per second per occupant, but this will depend on processes and equipment, the cubic capacity of the workplace, and the level of physical activity carried out there.

Temperature

Generally, where people normally work for more than a short period, the temperature should be at least 16 degrees Celsius, and where much of the work involves severe physical effort, the temperature should be at least 13 degrees Celsius. Other factors may impinge on these, such as air circulation and relative humidity. Where it is not physically practicable to maintain these temperatures (eg rooms open to the external air or where food is kept in cold storage), the actual temperatures should be as close as practicable to these. Lower maximum temperatures may be required by law.

High temperatures in the workplace can be achieved by the action of sunlight on the building or structure which exceeds the building's designed ability to protect the occupants. The law does not provide a maximum beyond which work cannot continue. Management action should be taken when discomfort becomes apparent to limit the exposure time to uncomfortable temperatures. A continuous temperature of 30 degrees Celsius is widely used as the point at which work should stop. Thermometers should be available so that temperatures can be monitored.

Lighting

The general need is for adequate and appropriate lighting, preferably by natural light. To achieve this, several topics need to be reviewed:

- lighting design, including emergency lighting
- the type of work to be done
- overall work environment
- health issues and individual requirements
- maintenance, replacement, disposal.

Lighting design for **internal areas** needs to deliver illuminance that is reasonable for the needs of the work environment and at uniform levels in each work area. It must minimise glare and the casting of shadows. **General** lighting does this for the entire work area, **localised** lighting aims to provide different levels in different parts of the same area, and **local** lighting combines background and a close local light source (luminaire) close to where the work is done to produce a high level of illuminance in a small area. The term 'luminaire' describes a lamp holder and its lamp. **External lighting** designs are also expected to deliver uniform levels of illuminance, as well as to minimise glare to users and third parties. **Emergency lighting** is either standby (so people can continue to work in safety) or escape (so people can leave in safety).

Light levels are measured directly in **lux**, using a simple meter. The average illuminance across a work area can be deceptive because the figure may conceal low readings in some critical areas. Therefore, it is necessary to know the measured illuminance at critical points. The Chartered Institute of Building Services Engineers (CIBSE) Codes give recommended values, some of which are shown in Table 1.

The **type of work** being done may require special local lighting to augment general lighting for specific tasks. Higher illuminance is required for fine detail. Where colour perception is necessary, for example, provision of lighting that does not alter the natural appearance of colours will be required. The strobe effect of some lighting can make moving parts appear to be stationary.

Overall **work environment** considerations include the level of natural light. Most people prefer to work in daylight, which often needs to be supplemented. Glare is an important issue which can often be minimised by use of reflected light from walls and ceilings, eg by employing uplighters. The presence of dust or flammable material may call for special luminaire designs.

Health aspects of lighting include the potential for symptoms of eyestrain and the use of unsuitable postures to get closer to the work. The human eye (and the rest of the body) accommodates to low light levels, so that sufferers may be unaware of the cause of their discomfort. Individual preferences and special needs must not be ignored. Flickering lights can produce seizures in some epileptics.

A lamp's light output decreases over time, so regular planned **maintenance** should include a replacement policy as well as arrangements for:
- cleaning of all lighting, including emergency lighting and control equipment
- replacing damaged lighting
- safe access
- safe disposal of lamps and luminaires (some contain substances hazardous to health such as mercury and sodium).

Table 1: Task and area light levels

Work activity	Example locations and types of work	Average illuminance (lux)	Minimum measured illuminance (lux)
People, vehicle, machine movement	Lorry park, traffic route, corridor	20	5
Rough work, detail perception not necessary	Loading bays, construction site clearance	50	20
Work requiring limited perception of detail	Kitchens, assembly lines	100	50
Work requiring perception of detail	Offices, bookbinding	200	100
Where perception of fine detail is necessary	Drawing office, proofreading, assembly of electronic components	500	200

Windows

In addition to their contribution to lighting the interior, windows make an important contribution to ventilation and a general sense of wellbeing at work. Windows, and glazed openings in doors and partitions to improve visibility, need to be protected against breakage unless they are made of 'safety materials' where necessary for safety. This is important where glass is present below shoulder level, and every such case should be assessed.

'Safety materials' are either:
- inherently robust, eg blocks of glass
- glass which breaks safely if it breaks. 'Safety glass' is flat glass or plastic sheet that breaks so as not to leave large sharp pieces
- ordinary annealed glass of adequate thickness in relation to its maximum size.

Where necessary, glass should be marked to make it apparent. This marking is known as 'manifestation', which can take any form as long as it is conspicuous.

Windows, skylights and ventilators need a safe way of opening them, and must not expose anyone to risks when in the open position. They also need to be cleaned in a safe way if this cannot be done from the ground. Ways of achieving this include:
- selecting pivoting windows that can be cleaned from inside
- provision of access equipment, limiting the height of ladders used to 9m unless landing areas are provided
- provision for the use of access equipment including ladder tie points/facilities
- adequate anchor points for safety harnesses.

Cleanliness

Standards of cleanliness that should be achieved depend on the activities in the workplace. A pharmaceutical 'clean room' will be cleaner than a factory floor. All workplaces need to be kept clean, specifically to avoid accumulation of dirt or refuse on any surface. An appropriate minimum cleaning frequency for floors and traffic routes is once a week, but some environments such as food stores will need a much tighter regime to remove debris quickly. The surfaces concerned should have been designed for easy cleaning, and must be free of holes and worn areas. Keeping to an appropriate cleaning regime is part of what is often called 'good housekeeping'; other parts include fostering safe and tidy stacking habits, clearing desks at night, and generally removing the small but important primary causes of most work injuries.

Working space

The minimum allowance should be 11m³ for each person normally working in a workplace. In rooms with high ceilings, the approximate minimum floor area should be 3.7m² per person. There should be enough height and space around objects in the workplace to allow access to work areas, and these figures should apply when the room is empty and not when full of furniture and equipment. Some rooms are necessarily small (eg ticket booths) but in these cases layout planning will be needed in order to meet the test of 'sufficient' space for those working there.

Seating

Most people at work will need to sit down at some point, if not all the time. Seating design for the workplace must relate to individual needs, what work is being done and the dimensions of the workplace. Special needs may include providing special seating for disabled employees where necessary, special backrests, and access for wheelchair users. A guide weight limit for users of gas lift chairs is 100kg. Risk assessments must include the issue of seating, usually a part of the work system.

Important considerations in seating include:
- supportive upholstery – padded edges and sharp points
- adjustability – range, seat height and forward tilt for some users such as checkout staff, back rest height and tilt, forward movement
- back rest – presence, adjustability for real support
- foot rests – should be provided where needed

- arm rests – as an option where requested
- training provision – people need to know how to adjust their seating.

Slips, trips and falls

Slips, trips and falls on the same level rival manual handling injuries as the most common cause of non-fatal major injuries to employees. Solutions are often inexpensive, easy to devise and put in place, and their benefits are considerable. They need to be part of a health and safety management system that includes regular inspection, maintenance and good housekeeping. When they occur, investigation will often expose failures in the system as well as primary causes of the incident. Table 2 overleaf gives a summary of the main hazards and control measures.

Traffic routes

The potential for slips, trips and falls has been outlined above, and the same general considerations for floors apply to 'traffic routes', a term which includes routes for vehicles as well as pedestrians.

Wherever practicable, separation of pedestrians from vehicles should be achieved by making the route wide enough, and traffic routes should be indicated by road or other marking as appropriate. On enclosed or constricted routes such as bridges, doorways and tunnels, there should be physical separation by a solid barrier or kerb. Doors, especially those that swing in both directions, should have a transparent section.

Welfare facilities

The key points for **toilet and washing facility** provision are:
- privacy
- ventilation and lighting
- cleanliness
- location
- quantity.

Toilets should be in separate closed areas, with doors that can be secured from within. Windows should be obscured unless they cannot be looked through from outside. Urinals should not be visible when entrance or exit doors are open. Rooms with toilets should not communicate directly with a food preparation or consumption area. Walls and floor surfaces should be such that wet cleaning can be carried out, and a cleaning system should be in place.

Washing facilities must include soap or other means of cleaning, and towels or other means of drying. Running hot and cold or warm water should be provided to allow effective washing of the face, hands and forearms, although workers facing dirty or strenuous work, or work which contaminates the skin significantly, should be provided with showers or baths. Hot water supplies should be fitted with means of reducing the effective temperature so that inadvertent scalding cannot occur. A warning notice is a poor substitute for this control measure.

Welfare facilities should be located so as to be reasonably available to those who may use them. The specified numbers of washing facilities, toilets and urinals are given in the ACoP. Broadly, one of each should be provided for every 10 to 15 people at work, although the exact needs should be based on the use made of them so as to allow for congestion at certain times, and also where the general public may share them.

Drinking water must be readily accessible at suitable places and marked with a sign where necessary (to warn people which is drinkable water and which is not). Except where drinking fountains or jets are supplied, cups or beakers are required.

Clothing worn at work only and not taken home (eg hats worn for food hygiene reasons and uniforms) should be stored at work in suitable **accommodation**. There should be a storage area for personal clothing worn travelling to and from work, and a means of drying such clothing. The extent of these facilities will vary, but adequate personal clothing space and security is important. Changing rooms may be needed, especially to prevent contamination of personal clothing by

Table 2: Guide to slip, trip and fall hazards and controls

	Examples	Common control measures following risk assessment	Aggravating factors
Slip hazards			• Inadequate design specifications
Inappropriate footwear	Platform-soled shoes, stiletto-heeled shoes	Footwear policy where necessary	
Wet walking surfaces	After cleaning, walking inside from wet conditions	Spill clean-up procedures, signs, barriers, door mats	• Poor or uneven lighting
Highly-polished or other unsuitable walking surfaces	Over-waxing, accumulation of cleaning product(s)	Follow suppliers' instructions for cleaning procedures, change cleaning product, consider change of surface	• Lack of premises maintenance
			• Lack of equipment maintenance
Insecure walking surfaces	Loose carpets, rugs	Securing strips	
Contaminated walking surfaces	Spillage of dry material from process	Clean-up and rubbish removal policy	• Wrong or no cleaning procedure
Sudden gradient change	Unmarked change from flat to steep slope	Insert steps, add visible tread nosings	• Effects of alcohol/drugs
			• Fatigue
Trip hazards			• Moving around too quickly
Loose floor surfaces	Loose, worn or frayed carpets	Replace floor covering	
Loose materials on floor	Objects left in walk routes	Improve housekeeping, inspections	• Disability not taken into account, especially impaired vision or lack of sight
Uneven surface	Raised paving slabs, surface cracks in car parks	Identify for maintenance, regular inspections	
Cables across routes	Trailing telephone leads	Remove leads or provide covers for cable runs	
Low fixtures	Door stops	Review design	
Low socket outlets	Floor socket boxes left open	Inspection, training on need to close boxes, consider repositioning	
Fall hazards			
Temporary access	Standing on chairs to reach high objects	Reposition commonly used objects, provide steps or kick-stools	
Carrying objects down stairs	Missing footing while carrying delicate object	Training, review handrail provision	
Missing parts of floor surfaces	Floor raised by contractors	Regular inspection, preplanning, barriers, signs	
Unguarded edges	Unprotected return slopes	Review all changes of level for fall potential, handrails	

work materials. Separate facilities for men and women may be necessary for reasons of propriety.

The key points for the provision of **rest facilities** are:
* separate rest rooms/facilities are required when contamination from the workplace is likely
* rest rooms and areas require seats with backrests and tables
* non-smokers must be protected from tobacco smoke
* pregnant women and nursing mothers require convenient facilities, including somewhere suitable to lie down
* seats in work areas are not excluded as rest and eating areas provided they are clean and there is a suitable surface
* means of heating water and food is required where hot food and drink cannot be bought in or near the workplace.

Facilities for **smoking** are to be arranged so that non-smokers do not experience discomfort from smoke. Usually, this is done by banning smoking altogether in the workplace, or in rest rooms and other rest areas, or by providing a specially ventilated room for smokers to use.

Legal requirements

Every employer must ensure, so far as is reasonably practicable, the health and safety and welfare of his/her employees while at work (Health and Safety at Work etc Act 1974, Section 2). The workplace overall is covered by the Workplace (Health, Safety and Welfare) Regulations 1992 (WHSWR), and a study of the ACoP is recommended for detailed requirements. The Building Regulations are not retrospective in action, but provide a framework to allow designers and others to prepare workplaces where conditions are optimised. They are also revised in accordance with new techniques and opinions.

Breaches of the WHSWR are doubly significant, as they can provide substance for civil claims against employers for failing to satisfy their statutory obligations as well as for enforcement action by regulatory bodies.

Revision

5 **considerations for workplaces:**
* design
* what work is to be done
* work environment
* health issues and individual requirements
* maintenance

5 **principles for welfare facilities:**
* privacy
* ventilation and lighting
* cleanliness
* location
* adequacy and quantity of provision

Selected references

Legal

Health and Safety at Work etc Act 1974
Management of Health and Safety at Work Regulations 1999
Workplace (Health, Safety and Welfare) Regulations 1992
Provision and Use of Work Equipment Regulations 1998
Personal Protective Equipment at Work Regulations 1992
Lifting Operations and Lifting Equipment Regulations 1998
Disability Discrimination Act 1995

Guidance (HSE)

HSG38	*Lighting at work*
HSG57	*Seating at work*
HSG129	*Health and safety in engineering workshops*
HSG121	*A pain in your workplace. Ergonomic problems and solutions*

HSG155	*Slips and trips*
HSG202	*General ventilation in the workplace*
INDG63(rev)	*Passive smoking at work*
L24	Workplace (Health, Safety and Welfare) Regulations 1992: ACoP and guidance

British Standards

BS 4800:1989	Schedule for paint colours for building purposes
BS 5266:1999	Part 1: Code of practice for emergency lighting of premises other than cinemas and certain other specified premises used for entertainment
BS 5489:2003	Code of practice for the design of road lighting. Lighting of roads and public amenity areas
BS EN 1838:1999	Lighting applications. Emergency lighting

CIBSE codes

Code for interior lighting, ISBN 0 900 953 64 0
Lighting for offices, LG07, ISBN 0 900 953 63 2
Emergency lighting, Technical Memorandum 12, and August 1999 Addendum

Self-assessment questions

1 Define a 'workplace'.

2 Consult the ACoP to the Workplace (Health, Safety and Welfare) Regulations and verify that the toilets and washing facilities provided in your workplace are in accordance with it.

2 Work equipment

Introduction

Hazards from work equipment can arise in two physical ways. The first is possibly the easier to recognise – **equipment and machinery hazards**, including traps, impact, contact, entanglement or ejection of parts of and by the equipment, or failure of components. The second way is by **non-equipment hazards**, which can include electrical failure, exposure to chemical sources, pressure, temperature, noise, vibration and radiation. Hazards can also arise from the 'software element' – computer control, human intervention by the person carrying out the task using the equipment, and lack of maintenance. It is important to distinguish between continuing hazards associated with the normal working of the equipment, such as not having guards fitted where necessary to protect the operator, and those arising from the failure of components or safety mechanisms, such as breakdown of the guarding mechanism.

This Section mainly covers identification and control of work equipment and machinery hazards. Obviously, not all work equipment is machinery; the principles of guarding described below apply to all work equipment, but 'machine guarding' is the term normally used to describe the topic and this convention will be followed here. The principles apply to all work equipment. The amount of detailed consideration required in respect of individual pieces of equipment will be proportionate to the risks of use.

Work equipment safety

The definition of 'work equipment' is noted in the ACoP to the Provision and Use of Work Equipment Regulations 1998 (PUWER 98) as 'extremely wide'. Almost any equipment used at work falls within the scope of the Regulations, from hand tools to machinery and apparatus, even motor vehicles not owned privately. The major exclusions are livestock, substances and structural items. The principles for the safe use of work equipment of all kinds are:

- it must be **suitable** for the purpose in the particular circumstances of the use. To establish this, a suitable and sufficient risk assessment is required. For example, electrical equipment is not suitable for use in flammable atmospheres unless specifically designed for that purpose. Risk assessments need to consider ergonomic factors associated with equipment use
- it must be **installed**, **located** and **used** so as to minimise risk to operators and others. This means the risk assessment has to take into account the physical circumstances of use. For example, there may not be adequate space between particular pieces of equipment and the environment (see below for a summary of layout considerations)
- all forms of energy and all substances used or produced by the equipment must be capable of **safe supply and/or removal**
- it must be **maintained** in an efficient condition and working order (with respect to its safety, not its production capacity), and in good repair. In order to achieve this, the equipment needs to be **designed** and installed appropriately. High-risk equipment should have a detailed maintenance log
- it must be **inspected** by competent persons at suitable intervals and after safety-critical situations. This is to assure that the equipment can be operated, adjusted and maintained safely where there is a significant risk (of major injury) as a result of incorrect installation, deterioration or exceptional circumstances that could affect safe operation of the equipment. Some work equipment, including lifting equipment such as cranes and some ventilation installations, will need specific examinations under other legislation. Examples are given in Table 1
- there must be adequate **information, instruction and training** given to the operator or user and their supervisors, covering safe use, the risks and their control measures.

Table 1: Sample inspection requirements

Type of work equipment	Inspection	Thorough examination	Recording of inspection	Legal requirement
All types, general requirement	At installation or reassembly in new location, before use (may be visual only)	As determined by specific regulation, manufacturer, insurer or local policy	Yes, where significant risk is present, keep until next inspection	Provision and Use of Work Equipment Regulations 1998, Regulation 6
Lifting equipment, general	Weekly – (from HSE guidance on adequacy, not statutory)	• Before first use, unless already supplied with a current thorough examination report or new with an EC declaration of conformity • After installation or assembly where those conditions will determine safety • Every 12 months • When exceptional circumstances are liable to affect safety	• Yes, notify employer of defects that could put people at risk • Written report needed	Lifting Operations and Lifting Equipment Regulations 1998, Regulations 9 and 10
Lifting equipment used for lifting persons	As above	Every 6 months and as above, *or* in accordance with an examination scheme prepared by a competent person	As above	As above
Work platforms and supports of all kinds, including scaffolding	• Before first use and thereafter at least every 7 days • After alteration or any event likely to affect strength or stability	No	Written report required, no statutory form	Construction (Health, Safety and Welfare) Regulations 1996, Regulations 29 and 30
Pressure systems (as defined)	As specified in a written scheme of examination	In accordance with a written scheme of examination prepared by a competent person	As specified in a written scheme of examination	Pressure Systems Safety Regulations 2000, Regulations 8 and 9

Figure 1: Work equipment safety verification checklist

Work equipment type:				
Make and model:				
Supplier/source:				

No	Health and safety elements			Comments
1	Suitability			
a				
b				
c				
2	Installation or location			
a				
b				
3	Physical circumstances of use			
a				
b				
c				
4	Safe use requirements	Yes	No	
a				
b				
c				
5	Maintenance requirements			
a				
b				
c				
d				
6	Design sign-off	Yes	No	(name)
a	The design of the equipment has been reviewed and is considered appropriate for the equipment, its likely users and work location(s)			
7	Inspection requirements			(name/agency)
a	Who will undertake statutory inspections?			
b	Are the frequencies of all inspections, tests and thorough examinations required known and recorded?	Yes	No	
8	Information, instruction and training requirements			
a				
b				
c				
9	Legislation			
	What are the relevant regulations?			
10	Health issues identified	Yes	No	
a				
b				
c				
d				
11	Risk assessment carried out	Yes	No	
Date:	Done by:		Date of review:	

The checklist has been completed in respect of the work equipment identified
Signed: Position: Date:

For a detailed review of the legal requirements of PUWER, the reader is referred to Part 4 Section 8.

Pressurised equipment

In addition to PUWER's general requirements, the main specific Regulations which apply to systems under pressure are the Pressure Equipment Regulations 1999 and the (more general) Pressure Systems Safety Regulations 2000. These Regulations contain definitions of relevant systems covered by them, which are generally those at a pressure of more than 0.5 bar above atmospheric. Common examples of such equipment with high potential for injury are boilers and air receivers.

In addition to the general matters mentioned in this Section, the Pressure Systems Safety Regulations also require a competent person to draw up a new, or certify an existing, written scheme of examination. Such a scheme is not required for components of gas welding sets. The best short summary of the position on written schemes is contained in the HSE leaflet INDG178(rev1).

Maintenance of work equipment

Maintenance can be:
- planned preventive – adjusting or replacing at predetermined intervals
- condition-based – centred on monitoring safety-critical parts and maintaining them as necessary to avoid hazards
- breakdown – fix when broken only, not appropriate if a failure presents an immediate risk without correction, and needs an effective fault-reporting system.

Risk assessment

The amount of risk depends on several factors. These include the type of equipment and what it is used for (its suitability), the need for approach to or contact with it, its age and condition, the ease of access, the quality of supervision of the operator's behaviour, and the complexity of its operation. Risk levels also depend on the knowledge, skills and attitude of the person(s) present at a particular time, and the general awareness of any danger and the skills needed to avoid it. The identification of

those at risk and when the risk occurs is important, and applies to management as well as to operators. The process of risk assessment is described in Part 1 Section 3.

Methods of preventing equipment accidents

Equipment hazards can be controlled by:
1 eliminating the cause of the danger ('intrinsic safety' – see below)
2 reducing or eliminating the need for people to approach any dangerous part(s) of the equipment
3 making access to the dangerous parts difficult (or providing guards or other safety devices so that access does not lead to injury)
4 appropriate maintenance and inspection regimes
5 the provision of protective clothing or equipment.

(These are listed in order of preference and effectiveness, and may be used in combination.)

The safety of operators and those nearby can be achieved by:
- training to improve people's ability to recognise danger
- redesigning to make dangers more obvious (or use warning signs)
- training to improve skills in avoiding injury
- improving motivation to take the necessary avoiding action.

Safety by design

Intrinsic safety

Intrinsic safety is a process by which the designer eliminates dangers at the design stage with consideration for the elimination of dangerous parts, making parts inaccessible, reducing the need to handle work pieces in the danger areas, provision of automatic feed devices, and enclosure of moving parts.

Control selection

Controls should be designed and provided which:

- are in the correct position
- are of the correct type
- remove the risk of accidental start-up
- have a directional link (control movement matched to equipment movement)
- are distinguished by direction of movement
- possess distinguishing features (eg size, colour, feel).

Failure to safety

Designers should ensure that mechanical and electrical equipment fails to safety and not to danger. Examples of this are clutches scotched to prevent operation, provision of arrestor devices to prevent unexpected strokes and movement, fitting of catches and fall-back devices, and fail-safe electrical limit switches.

Maintenance and isolation procedures

Designers should consider the safety of operatives during cleaning and maintenance operations. Routine adjustments, lubrication and so on should be carried out without the removal of safeguards or dismantling of components. Where frequent access is required, as in the case of large fixed machinery, interlock guards can be used. Access equipment, where essential, should be provided, as long as access to dangerous parts is prevented. Self-lubrication of parts should be considered if access is difficult. Positive lock-off devices should be provided to prevent unintentional restarting of the equipment. Permit-to-work systems (see Part 1 Section 5) will be required in some circumstances, notably where a third party could energise a system without the knowledge of persons exposed to risks from the equipment or system when in operation.

Safety by position

Parts of machines and other work equipment which are out of reach and kept out of reach are called 'safe by position'. However, designers must consider the likelihood of dangerous parts normally out of reach becoming accessible in some circumstances. An example of when this could happen is during painting of a factory machinery area from ladders.

Work equipment layout

The way in which equipment such as static machinery is arranged in the workplace can reduce accidents significantly. Safe layout will take account of:

- spacing – to facilitate access for operation, supervision, maintenance, adjustment and cleaning
- lighting – both general lighting to the workplace (natural or artificial but avoiding glare) and localised for specific operations
- cables and pipes – should be placed to allow safe access and to avoid tripping, with sufficient headroom
- safe access for maintenance.

Machine guarding

Many serious accidents at work involve the use of machinery. In circumstances where intrinsic safety (see above) has not been achieved, machinery guarding will be the final option to eliminate the remaining risks. Whether a particular guard is effective or not will depend on its design and construction, and the way in which it relates to the operating procedures and what the machine is used for.

People can be injured directly by work equipment in **five** different ways. Some machines can injure in more than one way. These are:

Traps – the body or limb(s) become trapped between closing or passing motions of a machine. In some cases, the trap occurs when the limb(s) are drawn into a closing motion, eg in-running nips.

Impact – injuries can result from being struck by moving parts of machinery, or the machine itself.

Contact – injuries can result from contact of the operator with sharp or abrasive surfaces. Alternatively, contact with hot or electrically live components will cause injury.

Entanglement – injuries resulting from the entanglement of hair, rings and items of clothing in moving (particularly rotating) parts of equipment.

Ejection – Injuries can result from elements of the workplace or components of the work equipment being thrown out during the operation of the equipment, eg sparks, swarf, chips, molten metal splashes and broken components.

Machinery guard material

The selection of the material from which the guards will be constructed is determined by **four** main considerations. These are:

- strength, stiffness and durability
- effects on reliability (eg a solid guard may cause the equipment to overheat)
- visibility (there may be operational and safety reasons for needing a clear view of the danger area)
- the need for control of other hazards (eg limiting the amount of noise output by choice of special materials).

Ergonomics and the design of machinery guards

Guards should be designed with people in mind. The study of the relationship between man and work equipment is called **ergonomics**, and the study of the relationship between differing body shapes and the requirements for reaching and vision in the safe operation of a machine is called **anthropometry**. For an outline of the topic, see Part 3 Section 9.

Stress and fatigue contribute significantly to the causes of accidents and designers should aim to reduce this to a minimum by considering a range of layout issues and facilities:

- the correct placing of controls
- the positioning of operating stations and height of work tables
- the provision of seating
- suitable access to the workstation.

Similar thought will be needed to ensure that guards are fully functional and are seen by equipment operators and users as aids rather than obstructions.

Types of guard

Fixed guard

A fixed guard should be fitted wherever practicable and should, by design, prevent access to dangerous parts of work equipment. It should be robust to withstand the process and environmental conditions. Its effectiveness will be determined by the method of fixing and the size of any openings, allowing for an adequate distance between the opening and the danger point. This may be determined by national standards or regulations. The guard should only be removable with the use of a special tool, such as a spanner or wrench.

Interlocked guard

The essential principles of an interlocked guard are that the machine cannot operate until the guard is closed, and the guard cannot be opened until the dangerous parts of the machine have come fully to rest. Interlocked guards can be mechanical, electrical, hydraulic, pneumatic or a combination of these. Interlocked guards and their components have to be designed so that any failure of them does not expose people to danger.

Control guard

If the motion of the machine can be stopped quickly, control guards can be used. The principle of control guarding is that the machine must not be able to operate until the guard is closed. When the guard is closed by the operator, the machine's operating sequence is started. If the guard is opened, the machine's motion is stopped or reversed, so the machine must be able to come to rest or reverse its motion quickly for this technique to be effective. Generally, the guard in a control system is not locked closed during the operation of the machine.

Automatic guard

An automatic guard operates by physically removing from the danger area any part of the body exposed to danger. It can only be used where there is adequate time to do this without causing injury, which limits its use to slow-moving work equipment.

Distance guard

A distance guard prevents any part of the body from reaching a danger area. It could take the form of a fixed barrier or fence designed to prevent access.

Adjustable guard

Where it is impracticable to prevent access to dangerous parts (eg they may be unavoidably exposed during use), adjustable guards (fixed guards with adjustable elements) can be used. The amount of protection given by these guards relies heavily on close supervision of the operative, and on correct adjustment of the guard and its adequate maintenance.

Self-adjusting guard

This guard is automatically opened by the movement of the workpiece and returns to its closed position when the operation is completed.

Trip devices

These automatically stop or reverse the machine before the operative reaches the danger point. They rely on sensitive trip mechanisms and on the machine being able to stop quickly (the stopping action may be assisted by a brake). Examples include trip wires and mats containing switches which stop the machine when they are trodden on.

Two-hand control devices

These devices force the operator to use both hands to operate the machine controls. However, they only provide protection for the operator and not for anyone else who may be near the danger point. Guards should be arranged to protect all persons. Where these devices are provided, the controls should be spaced well apart and/or shrouded, the machine should only operate when both controls are activated together, and the control system should require resetting between each cycle of the machine.

Which guard?

Fixed guards provide the highest standard of protection and should be used where practicable when access to the danger area is

not required during normal operation. The following gives guidance on the selection of safeguards (in order of merit):

- where access to the danger area is not required during normal operation:
  1. fixed guard, where practicable
  2. distance guard
  3. trip device
- where access to the danger area is required during normal operation:
  1. interlocking guard
  2. automatic guard
  3. trip device
  4. adjustable guard
  5. self-adjusting guard
  6. two-hand control.

The basic requirements for all types of guards and safety devices are that they should be compatible with the process, of adequate strength for their function, and that they should not be easily bypassed or defeated.

Supply of machinery

Manufacturers, designers, importers and suppliers have a duty to ensure that their equipment is safe so far as is reasonably practicable (Health and Safety at Work etc Act 1974, Section 6). A detailed extension of these duties is also contained in the Management of Health and Safety at Work Regulations 1999. Manufacturers and others in the supply chain have further duties to comply with the specific safety requirements set out in minimum safety standards for equipment manufactured, designed, imported and supplied in the European Union.

The main provisions of the Supply of Machinery (Safety) Regulations 1992 (as amended in 1994) took effect on 1 January 1995. The result is that most machinery supplied in the United Kingdom must satisfy a broad range of health and safety requirements covering moving parts as well as other issues, and carry 'CE' marking and other information. In some cases, the machinery must have undergone type-examination by an approved body. The maker or the importer must also have records of relevant technical information related to the machine. Machines which do not comply cannot be supplied legally within the United Kingdom.

It should be noted that these rules apply throughout the European Economic Area – the present member states of the EC plus Norway, Iceland and Liechtenstein. There are many types of work machinery which are excluded from the coverage of the Regulations. They include means of transport, seagoing vessels, steam boilers, fairground equipment and agricultural tractors. The reader interested in a complete guide to the Regulations will find the DTI publication *Machinery. Guidance notes on UK regulations* (URN 95/650) an excellent reference.

The relevant UK Standard is BS EN ISO 12100. Its purpose is to give an overall framework for designers and manufacturers to produce machines that are safe for their intended purpose. Part 1 sets out basic terminology and overall methodology for producing machine safety standards, and Part 2 gives technical principles and specifications with recommendations on how the philosophy can be applied using available techniques.

Legal requirements

Every employer must ensure, so far as is reasonably practicable, the health and safety of his/her employees and others (Health and Safety at Work etc Act 1974, Sections 2(1) and (3)). This extends to the provision and maintenance, by the employer, of equipment, plant and systems of work that are safe (Section 2 (2)(a)). This obligation covers work equipment,

guarding and layout. More specific guarding (fencing) requirements are contained in PUWER. For a discussion of the requirements of these Regulations, see Part 4 Section 8. Most of the old-style regulations relating to specific processes or industries have now been revoked by PUWER 98.

Breaches of the law dealing with work equipment are viewed seriously by the courts. Where injuries are severe, cases are increasingly referred to Crown Courts for sentence. Besides breaches of specific requirements within PUWER 98, enforcing authorities commonly link allegations of failure to carry out a suitable and sufficient risk assessment. After an accident in late 2000 involving an exposed rotating reel, a carpet manufacturing company was fined £15,000 for a failure to maintain the machinery, £15,000 for a failure to prevent access to a dangerous machine part, and £15,000 for failing to make a suitable risk assessment as required by the Management of Health and Safety at Work Regulations 1999. Also in 2000, the lack of a guard to a conveyor belt's drive roller resulted in multiple arm fractures to an agency worker. There had been a similar incident at another of the company's depots, from which appropriate lessons were not learned. A magistrates' court fined the depot owner £20,000 under Section 3 of the Health and Safety at Work etc Act 1974, £5,000 under PUWER 98 (Regulation 11) for a failure to prevent access to dangerous machinery, and £5,000 for not making risk assessments in relation to people not directly employed.

Revision

Danger can arise from work equipment hazards – such as traps and entanglement, and non-work equipment hazards – the operator, noise.

6 **ways to design safety:**
- intrinsic safety
- control selection
- failure to safety
- maintenance and isolation procedures
- safety by position
- machine layout

3 **kinds of maintenance:**
- planned preventive
- condition-based
- breakdown

5 **dangers associated with work equipment:**
- traps
- impact
- contact
- entanglement
- ejection

6 **principles for work equipment:**
- suitability for purpose
- design, installation, location and use
- inputs and outputs evaluated
- efficient condition maintained
- inspection
- information, instruction and training

9 **types of guard:**
- fixed
- interlocked
- control
- automatic
- distance
- adjustable
- self-adjusting
- trip
- two-hand control

Selected references

Legal

Health and Safety at Work etc Act 1974
Management of Health and Safety at Work Regulations 1999
Workplace (Health, Safety and Welfare) Regulations 1992
Provision and Use of Work Equipment Regulations 1998
Manual Handling Operations Regulations 1992
Personal Protective Equipment at Work Regulations 1992
Lifting Operations and Lifting Equipment Regulations 1998
Construction (Health, Safety and Welfare) Regulations 1996
Pressure Equipment Regulations 1999
Pressure Systems Safety Regulations 2000

Guidance (HSE)

L22	*Safe use of work equipment. ACoP and guidance on Regulations*
L114	*Safe use of woodworking equipment*
HSG89	*Safeguarding agricultural machinery. Advice for designers, manufacturers, suppliers and users*
HSG129	*Health and safety in engineering workshops*
INDG178(rev1)	*Written schemes of examination*

British Standards

BS EN ISO 12100:2003	Safety of machinery
BS EN 294:1992	Safety distances to prevent danger zones being reached by the upper limbs
BS EN 349:1993	Minimum gaps to avoid crushing of parts of the human body
BS EN 418:1992	Emergency stop equipment, functional aspects, principles for design
BS EN 811:1997	Safety distances to prevent danger zones being reached by the lower limbs
BS EN 953:1998	Guards. General requirements for the design and construction of fixed and movable guards
BS EN 1050:1997	Safety of machinery. Principles for risk assessment
BS EN 60204-1:1993	Safety of machinery. Electrical equipment of machines

Department of Trade and Industry

Machinery: guidance notes on UK regulations, URN 95/650 (1995)

Self-assessment questions

1 Select a piece of work equipment of your choice, and identify the ways in which its controls have been designed to minimise risk to the user or operator.

2 In what circumstances would plastic sheeting not be an adequate material to use in machine guarding?

3 Select a piece of static (fixed) work equipment of your choice and complete the verification checklist.

3 Mechanical handling

Introduction
Mechanical handling techniques have improved efficiency and safety, but have introduced other sources of potential injury into the workplace. Cranes, powered industrial trucks and fork-lifts, and conveyors are the primary means for mechanical handling. In all circumstances, the safety of the equipment can be affected by operating conditions, workplace hazards and the operator.

Cranes
Basic safety principles for all mechanical equipment apply to cranes. More details are given in the previous Section, but in summary the principles are that the equipment should be of good construction, made from sound material, of adequate strength and free from obvious faults. All equipment should be tested and regularly examined to ensure its integrity. The equipment should always be properly used.

What can go wrong
Overturning can be caused by weak support, operating outside the machine's capabilities and by striking obstructions.

Overloading by exceeding the operating capacity or operating radii, or by failure of safety devices.

Collision with other cranes, overhead cables or structures.

Failure of support – placing over cellars and drains, outriggers not extended, made-up or not solid ground – or of **structural components** of the crane itself.

Operator errors from impaired/restricted visibility, poor eyesight or inadequate training.

Loss of load from failure of lifting equipment, lifting accessories or slinging procedure.

Hazard elimination
Matters which require attention to ensure the safe operation of a crane include:

Identification and testing: Every crane should be tested and a certificate should be issued by the seller and following each test. Each should be identified for reference purposes, and the safe working load clearly marked. This should never be exceeded, except under test conditions.

Maintenance: Cranes should be inspected regularly, with any faults repaired immediately. Records of checks and inspections should be kept.

Safety measures: A number of safety measures should be incorporated for the safe operation of the crane. These include:
- load indicators – of two types:
  - load/radius indicator
  - automatic safe load indicator, providing audible and visible warning
- controls – should be clearly identified and of the 'dead-man' type
- overtravel switches – limit switches to prevent the hook or sheave block being wound up to the cable drum
- access – safe access should be provided for the operator and for use during inspection/maintenance and emergencies
- operating position – should provide clear visibility of both hook and load, with the controls easily reached. Communication through a banksman by signal or radio is a potential weak link as messages may not be received or properly understood
- passengers – should not be carried without authorisation, and never on lifting tackle
- lifting tackle – chains, slings, wire ropes, eyebolts and shackles should be tested/examined, be free from damage and knots as appropriate, and be clearly marked for identification and safe working load, and be properly used (no use at or near sharp edges, or at incorrect sling angles).

Operating area: All nearby hazards, including overhead cables and bared power supply conductors, should be identified, and removed or covered by safe working procedures such as

Figure 1: Sample basic checklist – common lifting appliances

		Yes	No	Comments
Premises address:				
Completion by:				
Date:				
No		Yes	No	Comments
	Public protection and information			
1	Are the work areas fenced off, or is there other protection for the public against being struck by moving equipment or loads?			
2	Is access to the work areas restricted to authorised visitors?			
3	When work finishes, are the lifting appliances immobilised effectively?			
	Cranes and lifting appliances			
4	Are crane erection and dismantling method statements available for all cranes and lifting appliances?			
5	Are the safe working loads and other constraints known to operators?			
6	Are all operators trained and competent?			
7	Where appropriate, do all operators possess the appropriate certification or authority to operate equipment?			
8	Are safe load indicators fitted where appropriate to all lifting appliances?			
9	Have slingers or banksmen been appointed and identified to operators as the only persons entitled to give signals?			
10	Have all slingers or banksmen been adequately trained in signalling and slinging?			
11	Have all slingers or banksmen been specifically instructed in the identification of weight and centre of gravity before lifting a load?			
12	Are all lifting appliances inspected at least weekly by a competent person, with the inspection results recorded?			
13	Do all lifting appliances have appropriate in-date test certificates issued by a competent authority?			
14	Are drivers of visiting mobile cranes and lifting appliances required to produce inspection documents and evidence of competence?			
15	Are all mobile lifting appliances required to operate outriggers only with additional timber placed beneath to spread the load?			
16	Is all plant and equipment in good repair?			
17	Is a maintenance log maintained to record any defects and the date of their repair as well as preventive maintenance?			
	Hoists and lifts			
18	Are all hoists and lifts inspected visually at least weekly by a competent person and the results recorded?			
19	Do all hoists have appropriate in-date test certificates issued by a competent person?			
20	Are effective gates and barriers in place at all landings, including ground level?			
21	Are the gates kept shut other than when the platform or cage is at the landing?			
22	Are the controls arranged so that operation is from one position only?			
23	Are all operators trained and competent?			
24	Are materials hoists and goods-only lifts marked to prevent people riding on them?			
25	Is the safe working load clearly marked on all lifting appliances?			
26	Are all hoists protected to prevent anyone being struck by any moving part, or materials falling down the hoistways?			
27	Are the arrangements for maintenance appropriate?			
28	Are arrangements for the release of trapped passengers appropriate and in place?			

locking off and permit systems. Solid support should be available and on new installations the dimensions and strength of support required should be specified. The possibility of striking other cranes or structures should be examined.

Operator training: Crane operators and slingers should be fit and strong enough for the work. Training should be provided for the safe operation of the particular equipment.

Powered industrial trucks

Trucks should be of good construction, free from defects and suitable for the purpose in terms of capacity, size and type. The type of power supply to be used should be checked, because the nature of the work area may require one kind of power source rather than another. For example, in unventilated confined spaces, internal combustion engines will not be acceptable because of the toxic gases they produce.

Trucks should be maintained so as to prevent failure of vital parts, including brakes, steering and lifting components. Special facilities may be required for some tasks, such as mast replacement. Specific risk assessments should be made, which will take into account local conditions and availability of appropriate equipment. Any damage should be reported and corrected immediately. Overhead protective guards should be fitted for the protection of the operator. Powered industrial trucks and their attachments should only be operated in accordance with the manufacturer's instructions.

What can go wrong

Overturning from manoeuvring with load elevated, driving too fast, sudden braking, striking obstructions, use of forward tilt with load elevated, driving down a ramp with the load in front of the truck, turning on or crossing ramps at an angle, shifting loads, or unsuitable road or support conditions.

Overloading by exceeding the maximum lifting capacity of the truck.

Collision with structural elements, pipes, stacks or with other vehicles.

Floor failure because of uneven or unsound floors, or by exceeding the load capacity of the floor. The capacity of floors other than the ground level should always be checked before using trucks on them.

Loss of load can occur if devices are not fitted to stop loads slipping from forks.

Explosions and fire may arise from electrical shorting, leaking fuel pipes, dust accumulation (spontaneous combustion) and from hydrogen generation during battery charging. The truck itself can be the source of ignition if operated in flammable atmospheres.

Passengers should not be carried unless seats and other facilities are provided for them.

Hazard elimination

Matters which require attention for the safe operation of powered industrial trucks include:

Operating area: The floor should be of suitable construction for the use of trucks. It should be flat and unobstructed where movement of machines is expected, with gullies and openings covered. Storage and stacking areas should be properly laid out, with removal of blind corners. Passing places need to be provided where trucks and people are likely to pass each other in restricted spaces, and traffic routes need to be clearly defined with adequate visibility. Pedestrians should be excluded from operating areas; alternatively, clearly defined gangways should be provided – with trucks given priority. Suitable warning signs will be required to indicate priorities.

Lighting should be adequate to facilitate access and stacking operations. Loading bays should be appropriately designed and stable, with chocks provided to place behind or beneath wheels. Ramps and slopes should not exceed 1:10 unless the manufacturer specifies that use on steeper gradients under load is acceptable. Battery charging areas should be separate, well-ventilated and lit, with no smoking or naked lights permitted. Battery lifting facilities should be provided. Where possible, reversing lights and/or sound warnings should be fitted, especially where pedestrians may share floor space with trucks.

Training should be provided for operators in the safe operation of their equipment, followed by certification.

Conveyors
The most common types of conveyor are belt, roller or screw conveyors.

What can go wrong
Trapping – limbs can be drawn into in-running nips.

Contact with moving parts such as drive elements or screw conveyors.

Entanglement with rollers or drive mechanisms.

Striking – materials falling from heights or incorrectly handled.

Hazard elimination
Belt conveyors require guards or enclosures at the drums, which are the main hazard because of the presence of trapping points between belt and drum. These are also required where additional trapping points or nips occur as the belt changes direction or at guide plates or feed points. Guards may be required along the length of a conveyor in the form of enclosures or trip wires to cut off supply. Safe access at appropriate intervals should be provided over long conveyor runs.

Roller conveyors: Where rollers may be either power-driven or free-running, guards at power drives are required. Other hazards can be present, which also require guarding – these include areas where in-running nips are created, where intermediate rollers are power-driven or when a belt is fitted. Walkways should be provided if access is required over the conveyor mechanism.

Screw conveyors should be guarded to prevent access at all times. Repairs and maintenance should only be undertaken when the drive is locked off.

Legal requirements
Mechanical handling is covered by the general duties provisions of the Health and Safety at Work etc Act 1974, and the more detailed requirements of the Management of Health and Safety at Work Regulations 1999. The Provision and Use of Work Equipment Regulations 1998 (PUWER 98) apply to all work equipment. Many kinds of mechanical handling equipment fall within the scope of the Lifting Operations and Lifting Equipment Regulations 1998 (LOLER), but others do not.

In most cases, LOLER applies to work equipment which has as its principal function a use for lifting or lowering of the type associated with traditional lifting equipment, such as cranes, or accessories, such as chains. So LOLER applies to fork-lift trucks and cranes, but not to conveyor belts moving articles on a horizontal level. LOLER does not define the term 'lifting equipment'; however, similar standards of safety are always required by PUWER 98. These Regulations are discussed in Part 4 Sections 8 and 9. Inspection and examination requirements were covered in the previous Section.

Revision

8 causes of crane failure:
- overturning
- overloading
- collision
- foundation failure
- structural failure
- loss of load
- operator error
- lack of maintenance

7 causes of powered truck failure:
- overturning
- overloading
- collision
- floor failure
- loss of load
- operator error
- presence of unauthorised passengers

5 conveyor hazards:
- trapping
- contact
- entanglement
- non-machinery hazards, including noise, vibration
- striking by objects

Selected references

Legal
Lifting Operations and Lifting Equipment Regulations 1998

Guidance (HSC/HSE)
HSG6	*Safety in work with lift trucks*
HSG136	*Workplace transport safety: guidance for employers*

British Standards
BS 7121:1989	Code of practice for the safe use of cranes
BS 4531:1986	Specification for portable and mobile troughed-belt conveyors
BS 4430:1969	Recommendations for industrial trucks. Part 2: Operation and maintenance

Self-assessment questions

1 What checks should be made before a crane is used in a workplace?

2 A powered lift truck overturns and the operator is injured. List the potential causes of the overturning.

4 Manual handling

Introduction

It has long been recognised that the manual handling of loads at work contributes significantly to the number of workplace injuries, with approximately a quarter of all reported accidents attributed to these activities. The majority of these injuries result in more than three days' absence from work, almost half of the total being sprains or strains, often of the lower back, with other types of injury including cuts, bruises, fractures and amputations. Many of the injuries are of a cumulative nature rather than being attributable to any single handling incident.

Until recently, the training of employees to lift concentrated on methods which would allow them to minimise the risks from moving and lifting the heavy loads expected of them. Films and videos explained how these techniques would help them to do the work without injury. Now, however, the aim is to reduce the opportunities for injury by reducing the amount of lifting and moving that the human body is required to do in the work environment. We should be asking not 'How do I lift this safely?' but 'Do I really need to lift this at all?' In the same way, people who ask 'What is the maximum weight people are allowed to lift?' – the concept in outdated and revoked regulations – are missing the point because weight is only one of many factors contributing to the risks inherent in manual handling.

The casualties of the old approach are all around us – about 80 per cent of the working population will suffer some form of back injury requiring them to take time off work at some point in their lives. Those who are injured are three times as likely to be injured again in the same way as those without a back injury. This situation needs to be addressed by informed employers and employees, seeking to identify and remove hazardous lifting and to find new ways of doing work that has traditionally involved human effort to move loads.

Injuries resulting from manual handling

Some of the more common types of injury resulting from manual handling are considered here. It is important to remember that the process of ageing will affect the integrity of the spine. For example, loss of 'spinal architecture' (ie the natural curvature of the spine) produces excessive pressure on the edges of the discs, wearing them out faster. These effects are not considered here.

Disc injuries – 90 per cent of back troubles are attributable to **disc lesions**. The discs lie between the vertebrae, acting as shock absorbers and facilitating movement. They do not 'slip' in the conventional sense, as they are permanently fixed to the bone above and below. Roughly circular, the discs are made up of an outer rim of elastic fibres with an inner core containing a jelly-like fluid. When someone stands upright, forces are exerted directly through the whole length of the spine and in this position it can withstand considerable stress. However, when the spine is bent, most of the stress is exerted on only one part (usually at the part where the bending occurs). Also, due to the bending of the spine in one place, all the stress is exerted on one side of the intervertebral disc, thus 'pinching' it between the vertebrae. This 'pinching effect' may scar and wear the outer surface of the disc so that at some time it becomes weak and eventually, under pressure, it ruptures. Many authorities believe that all disc lesions are progressive, rather than sudden. It is also important to remember that the disc cores dry out with age, making them less flexible and functional, and more prone to injury. The disc contents are highly irritating to the surrounding parts of the body, causing an inflammatory response when they leak out.

Ligament/tendon injuries – Ligaments and tendons are connective tissues, and hold the back together. Ligaments are the gristly straps that bind the bones together while tendons attach muscles to other body parts, usually bones. Repetitive motion of the tendons may cause inflammation. Both can be pulled and torn, resulting in sprains. Any factor, such as the effects of age effects and cold weather, that produces tightness in ligaments and tendons predisposes the back to sprains. Two main ligaments run all the way down the spine to support the vertebrae.

Muscular/nerve injuries – The muscles in the back form long, thick bands that run down each side of the spine. They are very strong and active, but tiring of these muscles can result in aches and pains, and can create stress on the discs. Postural deformities can result from damage to the muscles. Fibrositis (rheumatic pain) can also result. Nerves can be become trapped between the elements of the spine causing severe pain and injury.

Hernias – A hernia is a protrusion of an internal organ through a gap in a wall of the cavity in which it is contained. For example, any compression of the abdominal contents towards the naturally weak areas may result in a loop of intestine being forced into one of the gaps or weak areas formed during the development of the body. When the body is bent forward, possibly during a lift, the abdominal cavity decreases in size causing a compression of internal components and increasing the risk of hernias.

Fractures, abrasions and cuts – These can result from dropping the objects that are being handled (possibly because of muscular fatigue), falling while carrying objects (perhaps as a result of poor housekeeping), from other inadequacies in the working environment such as poor lighting, or from the contents of the load.

Injury during manual handling
The injuries highlighted above can result from lifting, pushing, pulling or carrying an object during manual handling.

Lifting – Compressive forces on the spine, its ligaments and tendons can result in some of the injuries identified above. High compressive forces in the spine can result from lifting too much, poor posture and incorrect lifting

technique. Prolonged compressive stress causes what is known as 'creep-effect' on the spine, squeezing and stiffening it. If the spine is twisted or bent sideways when lifting, the added tension in the ligaments and muscles when the spine is rotated considerably increases the total stress on the spine to a dangerous extent. The elastic disc fibres are also put under tension during repetitive twisting and individual fibres are damaged.

Pushing and pulling – Stresses are generally higher for pushing than pulling. Because the abdominal muscles are active as well as the back muscles, the reactive compressive force on the spine can be even higher than when lifting. Pushing also loads the shoulders and the ribcage is stiffened, making breathing more difficult.

Carrying – Carrying involves some static muscular work which can be tiring for the muscles, the back, shoulders, arms and hands depending on how the load is supported. A weight held in front of the body induces more spinal stress than one carried on the back. Likewise, a given weight held in one hand is more likely to cause fatigue than if it was divided into equal amounts in each hand. As with pushing, carrying objects in front of the body or on the shoulders may restrict the ribcage. Thus, the way in which a load is carried makes a great difference to the fatiguing effects. This is why getting a good grip is important, as well as keeping the load close to the body – which places it closer to the body's centre of gravity. The 'power grip', where the load is taken deep into the hand by holding the fingers straight and essentially at right angles to the palm, involves less strain on tendons and joints than where the fingers alone are used to support the object's full weight. The potential for back injury is higher for obese people and pregnant women, because the extent to which the centre of gravity of the load can approach the centre of gravity of the body is necessarily limited.

Manual handling assessments
The Manual Handling Operations Regulations 1992 (see Part 4 Section 10) establish a clear

hierarchy of measures to reduce the risk of injury when performing manual handling tasks. To summarise, manual handling operations which present a risk must be avoided so far as is reasonably practicable; if these tasks cannot be avoided, then each such task where there is a risk of physical injury must be assessed. As a result of that assessment, the risk of injury must be reduced for each particular task identified so far as is reasonably practicable.

It is important to remember that what is required initially is not a full assessment of each of the tasks, but an appraisal of those manual handling operations which involve a risk that cannot be dismissed as trivial in order to determine if they can be avoided. Consideration of a series of questions will be useful in completing this stage of the exercise. These include:

Is there a risk of injury? An understanding of the types of potential injury will be supplemented with the past experiences of the employer, including accident/ill health information relating to manual handling and the general numerical guidelines contained in official guidance.

Is it reasonably practicable to avoid moving the load? This question will be useful in establishing 'authorised' manual handling tasks within a certain workplace or department. Some work is dependent on the manual handling of loads and cannot be avoided (as, for example, in refuse collection). If it is reasonably practicable to avoid moving the load, then the initial exercise is complete and further review will only be required if conditions change.

Is it reasonably practicable to automate or mechanise the operation? Introduction of these measures can create different risks (eg introduction of fork-lifts creates a series of new risks) which require consideration.

The aim of the detailed **full assessment** is to evaluate the risk associated with a particular task and identify control measures which can be implemented to remove or reduce the risk (possibly mechanisation and/or training). A sample format for the assessment, plus a worked example, can be found in the guidance to the Regulations, which is essential reading.

For varied work (such as done in the course of maintenance, construction or agriculture), it will not be possible to assess every single instance of manual handling. In these circumstances, each type or category of manual handling operation should be identified and the associated risk assessed. Assessment should also extend to cover those employees who carry out manual handling operations away from the employer's premises (such as delivery drivers).

The assessment must be kept up to date. It needs reviewing whenever it may become invalid, such as when the working conditions or the personnel carrying out those operations have changed. Review will also be required if there is a significant change in the manual handling operation, which may affect the nature of the task or the load, eg changing to different bulk sizes.

Before beginning an assessment, the views of staff and employees can be of particular use in identifying manual handling problems. Involvement in the assessment process should be encouraged, particularly in reporting problems associated with particular tasks.

Records of accidents and ill health are also valuable indicators of risk, together with absentee records, poor productivity and morale, and excessive product damage. Individual industries and sectors have also produced information identifying risks associated with manual handling operations, which will be a useful source of information.

Schedule 1 to the Manual Handling Operations Regulations specifies four interrelated factors which the assessment must take account of. The answers to the questions specified about the factors form the basis of an appropriate assessment. These are:

The tasks

Do they involve:

- holding or manipulating loads at a distance from the trunk?
- unsatisfactory bodily movement or posture, especially:
  - twisting the trunk?
  - stooping?
  - reaching upwards?
- excessive movement of loads, especially:
  - excessive lifting or lowering distances?
  - excessive carrying distances?
  - excessive pushing or pulling of loads?
- risk of sudden movement of loads?
- frequent or prolonged physical effort?
- insufficient rest or recovery periods?
- a rate of work imposed by a process?

The loads

Are they:

- heavy?
- bulky or unwieldy?
- difficult to grasp?
- unstable, or with contents likely to shift?
- sharp, hot, or otherwise potentially damaging?

The working environment

Are there:

- space constraints preventing good posture?
- uneven, slippery or unstable floors?
- variations in levels of floors or work surfaces?
- extremes of temperature or humidity?
- conditions causing ventilation problems or gusts of wind?
- poor lighting conditions?

Individual capability

Does the job:

- require unusual strength, height, etc?
- create a hazard to those who might reasonably be considered to be pregnant or to have a health problem?
- require special information or training for its safe performance?

Other factors

Is movement or posture hindered by PPE or clothing?

Reducing the risk of injury

In considering the most appropriate controls, an ergonomic approach to designing the manual handling operation will optimise health and safety as well as productivity levels associated with the task. The task, the load, the working environment, individual capability and the interrelationship between these factors are all important elements in deciding optimum controls designed to fit the operation to the individual rather than the other way around. Techniques of risk reduction include:

Mechanical assistance. This involves the use of handling aids, of which there are many examples. One could be the use of a lever, which would reduce the force required to move a load. A hoist can support the weight of a load while a trolley can reduce the effort needed to move a load horizontally. Chutes are a convenient means of using gravity to move loads from one place to another.

Improvements in the task. Changes in the layout of the task can reduce the risk of injury by, for example, improving the flow of materials or products. Improvements which will permit the body to be used more efficiently, especially if they permit the load to be held closer to the body, will also reduce the risk of injury. Improving the work routine by reducing the frequency or duration of handling tasks will also have a beneficial effect. Using teams of people and, where appropriate, PPE, such as gloves, can also contribute to a reduced risk of injury. All equipment supplied for use during handling operations (eg handling aids and PPE) should be maintained and there should be a defect reporting and correction system.

Reducing the risk of injury from the load. The load may be made lighter by using smaller packages/containers or specifying lower packaging weights. Additionally, the load may be made smaller, easier to manage, easier to grasp (eg by the provision of handles), more stable and less damaging to hold (clean, free from sharp edges, etc). On the other hand, the introduction of smaller and lighter loads may carry a penalty in the form of further bending and other repetitive movements to handle the load than were necessary before. This could result in no change in the overall risk.

Improvements in the working environment. This can be done by removing space constraints, improving the condition and nature of floors, reducing work to a single level, avoiding extremes in temperature and excessive humidity, and ensuring that adequate lighting is provided.

Individual selection. Clearly, the health, fitness and strength of an individual can affect his/her ability to perform manual handling tasks. Health screening is an important selection tool. Knowledge and training have important roles to play in reducing the number of injuries resulting from manual handling operations. There is little point in enquiring about any previous back injuries on a job application form if no attempt is made to ensure that those who do admit to previous problems are not given work which is foreseeably likely to produce them.

Back belts. These are claimed to offer protection against injury. As a sole solution, they are not likely to satisfy legal requirements. Their actual effectiveness is controversial; some authorities believe that for certain individuals a back belt may render them more likely to suffer an injury. Designs and functions vary widely, some preventing or limiting movement. Abdominal and back support belts provide protection, if at all, to individuals and not the group involved. They are, therefore, less desirable than controls which reduce the risks to all, using techniques described above.

Manual handling training

A training programme should include mention of:
- dangers of careless and unskilled handling methods
- principles of levers and the laws of motion
- functions of the spine and muscular system

- effects of lifting, pushing, pulling and carrying, with emphasis on harmful posture
- use of mechanical handling aids
- selection of suitable clothing and PPE for lifting
- techniques of:
  - identifying slip/trip hazards
  - assessing the weight of loads and how much can be handled by the individual without assistance
  - bending the knees, keeping the load close to the body when lifting (but avoiding tension and knee-bending at too sharp an angle)
  - breathing, and avoiding twisting and sideways bending during exertion
  - using the legs to get close to the load, making best use of body and load weight
  - using the 'power grip'.

There are several guiding principles for the safe lifting of loads. These are:
- secure grip
- proper foot position
- bent knees and comfortably straight back
- keep arms close to the body
- keep the chin tucked in
- body weight used to advantage.

All attending the training should have an opportunity to practise under supervision.

Other factors which should be discussed with trainees include:
- personal limitations (age, strength, fitness, girth)
- nature of loads likely to be lifted (weight, size, rigidity)
- position of loads
- working conditions to minimise physical strain
- the requirements of the Manual Handling Operations Regulations 1992 and risk assessments.

Revision

5 types of injury:
- disc injuries
- ligament/tendon injuries
- muscular/nerve injuries
- hernias
- fractures, abrasion, cuts

7 items to include in training programmes:
- the dangers of bad lifting technique
- the principles of leverage
- the functions of the body in lifting
- demonstration of good technique
- use of mechanical aids
- opportunity to practise
- requirements of the Regulations and risk assessments

Hierarchy of measures:
- avoid hazardous manual handling work altogether as far as reasonably practicable by redesigning or automating
- suitable and sufficient recorded risk assessments of remaining hazardous work
- reduce the risks as far as reasonably practicable

8 points for safe lifting:
- check load characteristics – weight, size, position, destination
- be aware of personal limitations, ask for assistance if necessary
- take secure grip
- keep back straight and knees bent
- keep arms close to body
- keep the chin tucked in
- be aware of body weight and how to use it to advantage
- co-ordinate two or more persons handling an object

4 main factors in assessment:
- task
- load
- working environment
- individual capability

Selected references

Legal

Manual Handling Operations Regulations 1992

Guidance (HSE)

L23	*Manual handling – guidance on Regulations*
HSG48	*Reducing error and influencing behaviour*
HSG60(rev)	*Upper limb disorders in the workplace*
HSG115	*Manual handling – solutions you can handle*
HSG121	*A pain in your workplace: ergonomic problems and solutions*
HSG155	*Slips and trips: guidance for employers on identifying hazards and controlling risks*
INDG171L	*Upper limb disorders*

Self-assessment questions

1 What factors should be considered by management before manual lifting of loads is authorised?

2 Identify areas in your workplace where mechanised handling techniques could be used instead of manual handling techniques.

5 Working at height

Introduction

Many home and work injuries involve falls from heights. The human body is not designed to resist impacts well, and the resulting injuries are unpredictable in their extent. This is where luck has a place in safety – the outcome of fall injuries is mostly dictated by chance. People have failed to survive a fall from as little as a metre; others have survived unbroken falls from 10m. Because the outcome in any individual case cannot be known in advance, the only course to follow must be one of prevention.

Readers are referred to the Work at Height Regulations 2005 (see Part 4 Section 23), which have incorporated the principles and much of what follows in this Section into law. Table 1 will assist in interpreting the Regulations.

Fall prevention and protection

An understanding of the difference between these two concepts is essential. Fall **prevention** aims to remove the need for people to work exposed to falls. This is done by design and planning work. Fall **protection** is the use of techniques to protect those who are necessarily exposed to fall hazards so as to minimise the risks.

Changing luminaires in very high ceilings can be done from access equipment, but a safer solution is to design a fixed way of access above the ceiling space. Programming the early erection of fixed and final stairways during a building's construction removes the need for temporary access up ladders and also the need to protect an open stairwell. These solutions need to be considered before other access choices. The safety precedence sequence, mentioned in Part 1, shows why this is so (Table 1). Several of these controls may be used in combination to increase effectiveness.

For temporary access to heights, as in construction work, the principle is to provide protected access for every person likely to be at risk in preference to provision of personal protection. Thus, the use of a working platform with edge protection is always preferred to methods which do not prevent falls but provide protection when falls occur.

Access equipment

Each task should be assessed and a suitable means of access chosen based on an evaluation of the work to be done, the duration of the task, the working environment (and its constraints), and the capability of the person or people carrying out the task.

There are many different types of access equipment. This Section covers general principles, and the following:
- ladders, stepladders and trestles
- general access scaffolds
- scaffold towers
- suspended cradles
- mast-elevated work platforms
- power-operated work platforms
- personal suspension equipment (abseiling equipment and boatswain's chairs).

Other, highly specialised, equipment is available and the general principles will apply to their use. Usually, they have been specially designed for particular tasks and manufacturers' information should be used in operator training.

General principles

Accidents using access equipment occur because one or more of the following common problems have not been controlled in advance, or was thought to be an acceptable risk under the circumstances:
- faulty design of the access structure itself
- inappropriate selection where safer alternatives could have been used
- subsidence or failure of base support
- structural failure of suspension system
- structural failure of components
- structural failure through overloading
- structural failure through poor erection, inspection or maintenance

Table 1: Safety precedence sequence applied to working at height

Safety precedence ranking (the relative effectiveness of the control method)	Control principle	Application of the principle to working at height
1	Hazard elimination	Take steps to remove the need to work at height at all Design improvements, eg design the erection sequence to remove need to bolt up at height
2	Substitution options	Use of access equipment, eg replace ladder access to steelwork with mobile elevated work platforms Use a fixed work platform properly guarded
3	Use of barriers	Supply suitable guardrails to open edges where people approach
4	Isolation	Supply barriers to walkways well back from exposed edges
5	Segregation	Lock access doors to areas where falls could occur, with limited keyholders, eg access to roof and liftshafts
6	Use of procedures	Introduce safe systems of work, procedures such as equipment maintenance and inspection
7	Use of warning systems	Fix signs, give instructions not to approach edges
8	Use of personal protective equipment	Must be used as a sole measure only when all other options have been exhausted, eg safety harnesses

- structural failure through overbalancing
- instability through misuse or misunderstanding
- overreaching and overbalancing
- climbing while carrying loads
- slippery footing – wrong footwear, failure to clean
- falls from working platforms and in transit
- unauthorised alterations and use
- contact with obstructions and structural elements
- electrical and hydraulic equipment failures
- trapping by moving parts.

Ladders, stepladders and trestles

The key points to be observed when selecting and using this equipment are summarised below.

Ladders

1 See whether an alternative means of access is more suitable. Take into account the nature of the work and its duration, the height to be worked at, what reaching movements may be required, what equipment and materials may be required at height, the angle of placement and the foot room behind rungs, and the construction and type of ladder.
2 Check visually whether the ladder is in good condition and free from slippery substances.
3 Check facilities available for securing against slipping – tied at top, secured at bottom, or footed by a second person if no more than 3m height access is required.
4 Ensure the rung at the step-off point is level with the working platform or other access

point, and that the ladder rises a sufficient height above this point (at least 1.05m or five rungs is recommended), unless there is a separate handhold.

5   A landing point for rest purposes is required every 9m.

6   The correct angle of rest is approximately 75 degrees (corresponds to a ratio of one unit horizontally at the foot for every four units vertically).

7   Stiles (upright sections) should be evenly and adequately supported.

8   Ladders should be maintained free from defects and should be inspected regularly.

9   Ladders not capable of repair should be destroyed.

10 Metal ladders (and wooden ladders when wet) are conductors of electricity and should not be placed near or carried beneath low power lines.

11 It is important to ensure that ladders are positioned the correct way up. Timber pole ladders often have stiles thicker at the base than at the top and should have metal tie rods underneath the rungs. Metal ladders often have rungs with both flat and curved surfaces – the flat surface is the one on which the user's feet should rest.

Stepladders

1   Stepladders are not designed to accept side loading.

2   Chains or ropes to prevent overspreading are required, or other fittings designed to achieve the same result. Parts should be fully extended.

3   Stepladders should be levelled for stability on a firm base.

4   Work should not be carried out from the top step.

5   Overreaching should be avoided by moving the stepladder – if this is not possible, another method of access should be considered.

6   Equipment should be maintained free from defects. Regular inspection is required.

7   No more than one person should use a stepladder at one time.

Working platforms and trestles

1   Trestles are suitable only as board supports.

2   They should be free from defects and inspected regularly.

3   Trestles should be levelled for stability on a firm base.

4   Platforms based on trestles should be fully boarded, adequately supported and provided with edge protection where appropriate.

5   Safe means of access should be provided to trestle platforms, usually by a stepladder.

6   Working platforms in construction work must by law be no less than 600mm in width, so many older trestles may no longer be suitable to support such platforms as they will be too narrow.

General access scaffolds

There are three main types of access scaffold, commonly constructed from steel tubing or available in commercial patented sections. These are:

* independent tied scaffolds, which are temporary structures independent of the structure to which access is required but tied to it for stability
* putlog scaffolds, which rely on the building (usually under construction) to provide structural support to the temporary scaffold structure through an arrangement of putlog tubes (with special flattened ends) placed into the wall
* Birdcage scaffolds, which are independent structures normally erected for interior work that have a large area and normally only a single working platform.

The key points to be observed when specifying, erecting and using scaffolds are:

* select the correct design with adequate load-bearing capacity
* ensure adequate foundations are available for the loads to be imposed
* the structural elements of the scaffold should be provided and maintained in good condition
* structures should be erected by competent persons or under the close supervision of a competent person, in accordance with any design provided and applicable regulations and codes

- all working platforms should be fully boarded, with adequate edge protection, including handrails or other means of fall protection, nets, brickguards and/or toeboards to prevent materials or people falling from the platforms
- all materials resting on platforms should be safely stacked, with no overloading
- adequate and safe means of access should be provided to working platforms
- unauthorised alterations of the completed structure should be prohibited
- inspections of the structure are required, prior to first use and then at appropriate intervals afterwards, which will include following substantial alteration or repair, after any event likely to have affected stability, and at regular intervals not exceeding seven days. Details of the results should be recorded on an inspection form.

Scaffold towers

Scaffold towers are available commercially in forms comparatively easy to construct. They may also be erected from traditional steel tubing and couplers. In either form, competent and trained personnel are required to ensure that all necessary components are present and in the right place. Many accidents have occurred because of poor erection standards; a further common cause is overturning.

The key points to be observed in the safe use of scaffold towers are:
- erection should be in accordance with the manufacturer's or supplier's recommendations
- erection, alteration and dismantling should be carried out by experienced, competent persons
- towers should be stood on a firm, level base, with wheel castors locked if present
- scaffold equipment should be in good condition, free from patent defects including bent or twisted sections, and properly maintained
- the structure should be braced in all planes to distribute loads correctly and prevent twisting and collapse
- the ratio of the minimum base dimension to

the height of the working platform should not exceed 1:3 in external use, and 1:3.5 in internal use, unless the tower is secured to another permanent structure at all times. Base ratios can be increased by the use of outriggers, but these should be fully extended and capable of taking loads imposed at all times
- free-standing towers should not be used above 9.75m unless tied. The maximum height to the upper working platform when tied should not exceed 12m
- a safe means of access should be provided on the narrowest side of the tower. This can be by vertical ladder attached internally, by internal stairways, or by ladder sections designed to form part of the frame members. It is not acceptable to climb frame members not designed for the purpose
- trapdoors should be provided in working platforms where internal access is provided
- platforms should be properly supported and fully boarded
- guardrails, toeboards and other appropriate means should be provided to prevent falls of workers and/or materials
- mobile scaffold towers should never be moved while people are still on the platform. This is a significant cause of accidents
- ladders or stepladders should not be placed on the tower platform to gain extra height for working.

Suspended access (cradles)

A **suspended access system** includes a working platform or cradle, equipped with the means of raising or lowering when suspended from a roof rig. The key points to be observed in the safe installation and use of this equipment are:
- it should be capable of taking the loads likely to be imposed on it
- experienced erectors only should be used for the installation
- supervisors and operators should be trained in the safe use of the equipment, and in emergency procedures
- inspections and maintenance are to be carried out regularly

- suspension arrangements should be installed as designed and calculated
- all safety equipment, including brakes and stops, should be operational
- the marked safe working load must not be exceeded and the effects of wind should also be considered
- platforms should be free from obstruction and fitted with edge protection
- the electrical supply is not to be capable of inadvertent isolation and should be properly maintained
- adverse weather conditions should be defined so that supervisors and operators know what is not considered acceptable
- all defects noted are to be reported and rectified before further use of the equipment
- safe access is required for the operators and unauthorised access is to be prevented
- necessary protective measures for those working below, as well as the public, should be in place before work begins.

Personal suspension equipment

A **boatswain's** (or **bosun's**) **chair** is a seating arrangement provided with a means of raising or lowering with a suspension system. This should only be used for very short duration work, or in positions where access by other means is impossible.

Abseiling equipment is used by specialists to gain access where the duration of work is likely to be very short indeed and the nature of the work lends itself to this approach.

The key points to be observed in using personal suspension equipment are:
- the equipment must be suitable and of sufficient strength for the loads which are anticipated and the purpose it is to be used for. A specific risk assessment should be made in every case
- the equipment must be securely attached to plant or a structure strong and stable enough for the circumstances
- suitable and sufficient steps must be taken to prevent falls or slips from the equipment
- the equipment must be installed or attached so as to prevent uncontrolled movement.

Mast-elevated work platforms

Generally, this equipment consists of three elements:
- mast(s) or tower(s) which support(s) a platform or cage
- a platform capable of supporting persons and/or equipment
- a chassis supporting the tower or mast.

The key points to be observed in the erection and use of this equipment are:
- only trained personnel should erect, operate or dismantle the equipment
- the manufacturer's instructions on inspection, maintenance and servicing should be followed
- firm, level surfaces should be provided, and outriggers are to be extended before use or testing, if provided
- repairs and adjustments should only be carried out by qualified people
- the safe working load of the equipment should be clearly marked on it, be readily visible to the operator, and never be exceeded
- raising and lowering sequences should only be initiated if adequate clearance is available
- the platform should be protected with edge guardrails and toeboards, and provided with adequate means of access
- emergency systems should be used only for that purpose and not for operational reasons
- unauthorised access into the work area should be prevented using ground barriers
- contact with overhead power cables should be prevented by preliminary site inspection and by not approaching cables closer than a given distance. Where necessary, this distance can be obtained from the power supply company concerned.

Power-operated mobile work platforms

A wide variety of equipment falls into this category, ranging from small, mobile tower structures with self-elevating facilities to large vehicle-mounted, hydraulically operated platforms.

The key points to be observed in their use are:

Self-assessment question 1

	Ladders, stepladders, working platforms	General access scaffolds	Scaffold towers	Suspended cradles	Personal suspension equipment	Mast-elevated work platforms	Power-operated work platforms
Faulty design of the access structure itself							
Inappropriate selection (safer alternatives available)							
Subsidence or failure of foundations/footings							
Structural failure of suspension system							
Structural failure of component strength							
Structural failure through overloading							
Structural failure (poor erection/maintenance)							
Structural failure through overbalancing							
Climbing carrying loads							
Slippery footing – wrong footwear, failure to clean							
Fall from working platforms and in transit							
Unauthorised alterations and use							
Contact with obstructions and structural elements							
Electrical and hydraulic equipment failures							
Trapping by moving parts							

6 Transport safety

Introduction

Any piece of mobile equipment that moves within the workplace is classed as a 'vehicle', including cars, trucks and self-propelled machinery. The safe operation of vehicles results from planning and activity, not chance. In the majority of vehicle accidents, the principal factors are driver failure and vehicle failure, both of which can be controlled. A high proportion of accidents involves those who have no direct control of vehicles, chiefly pedestrians.

Causes of transport accidents

Transport accidents occur because of:

- contact – with structures or services
- overturning – through incorrect loading, speeding or poor surface conditions
- collision – with other vehicles or pedestrians
- impact – materials falling or the vehicle overturning onto the operator
- entanglement – in dangerous parts of machinery or controls
- explosion – when charging batteries or inflating tyres
- operator/supervisor error – through inadequate training or experience.

Preventing transport accidents

These accidents are preventable by good management, involving the use of a planned approach, including the following considerations:

- traffic control in the workplace – includes making decisions on needs, priorities, rights of way, and the separation of pedestrian and motor traffic
- maintenance procedures – safety, economy and efficiency all benefit from periodic vehicle checks and inspections, in addition to local or national requirements of regulations or codes
- driver selection, training, certification and supervision – the use of vehicles by unauthorised persons must be prevented. Authorisation should depend on the individual's progress through a training programme on the specific type(s) of vehicle to be driven and selection of the individual for the task
- control of visiting drivers – any local rules must be communicated to visiting drivers, either in writing, as a contract condition, or by the provision of suitable traffic safety signs or markers
- accident investigation – including reporting system, subsequent analysis of reports and corrective action follow-up.

Traffic control in the workplace

The key to traffic control lies in planning safe routes, wide enough for necessary movements. Adequate clearance should be provided for safe movement of vehicles, identifying overhead or floor-level obstructions. Traffic should be kept away from vulnerable structures such as tanks and pipes. Vehicles and pedestrians should be separated where practicable, with warning lights and/or signs displayed. Signs will also be needed where there are unavoidable height limitations.

The basic principle is to keep pedestrians away from vehicles by providing separate routes where practicable. Barriers at building entrances and exits, and corners of buildings, should be considered. Safe pedestrian crossing places should be clearly marked and fitted with mirrors to improve visibility into blind areas. Indoor routes should be clearly marked on the floor to alert pedestrians.

Vehicle speeds should be controlled with speed limits, backed where necessary with speed ramps. Movements should be properly supervised when reversing or where access is difficult or blind, using recognised signals. Access routes and parking areas should be lit where practicable, especially where pedestrians and vehicles share routes.

Loading and unloading of vehicles should take place in designated areas without obstruction to other traffic. Vehicles not in use or broken down should be left where obstruction is minimised. Vehicle loads must be both stable and secure. Drivers should be adequately protected against falling objects and rollover. Keys should be removed from vehicles when not in use, to immobilise them and guard against theft or misuse. Parking areas should be located in safe places so that drivers leaving vehicles are not put at risk of crossing dangerous work areas.

The vehicles should be checked to ensure that dangerous parts of machinery are guarded, and that there is safe means of access and egress to driving positions. Reminders of the correct means of access, and the need to wear appropriate clothing, should be part of the training of drivers and maintenance staff.

Maintenance procedures

Maintenance of the workplace itself is an important aspect of transport safety. Spilled or dropped loads and other loose items should be removed from traffic routes as soon as possible. The road surface must be kept in good condition, with worn markings renewed and potholes filled. Signs and lighting also need maintenance.

Planned maintenance procedures for vehicles prevent accidents and delays due to mechanical failures, minimise repair downtime, and prevent excessive wear and breakdown. Drivers and operators are usually the first to notice when defects develop, and should check their vehicles against a basic checklist before work starts. Vehicle checklists should cover the following items:

* brakes
* headlights
* stoplights and indicators
* tyres
* screens and wipers
* steering wheel
* glass
* horn
* mirrors
* instruments
* exhaust system
* emergency equipment
* ignition
* connecting cables.

In addition to other health and safety considerations, **vehicle repair** requires attention to the following:

* brakes must be applied and wheels chocked, especially before entry under vehicle bodies
* raised bodies must be propped unless fitted with proprietary devices for stability in the raised position
* axle stands must be used in conjunction with jacks; hydraulic jacks should not be relied on alone
* when charging batteries, the risk of explosion or burns should be eliminated
* precautions are required to prevent explosion risks when draining, repairing or carrying out hot work near fuel tanks
* tyre cages should be provided for inflating vehicle tyres
* steps must be taken to eliminate exposure to dangerous dusts and fumes, eg from operating internal combustion engines in confined areas
* only trained personnel should carry out maintenance work.

Driver selection, training and supervision

Selection will require an evaluation of age, experience, driving record, maturity and attitude. People who drive safely also have other qualities such as a courteous manner and the ability to get on well with others. These can be used to pick potential safe drivers. Local or national driver qualification requirements must also be observed; the usual requirement for heavy goods vehicles is that drivers must be aged 21 or over.

Training of drivers may be for remedial, refresher or special reasons in addition to basic instruction for drivers unfamiliar with the vehicle to be driven. Generally, driver training courses should cover applicable local and national driving rules, company driving rules, what to do in the event of an accident to comply with local, national and company requirements, and defensive driving techniques. The risk assessment should give guidance on the amount and level of training that will be needed. Drivers' experience and training should be checked and verified so that training can supplement and fill any gaps. Refresher training should be planned.

A standard training package for new employees should include:

- job information, including route layout and how to report faults or incidents
- information specific to the vehicle, speed limits and loading areas, and dangers of unsafe working practices
- information on the management structure and supervision given, including penalties likely to be imposed for failure to comply with safe working practices.

Supervisory staff must ensure that all vehicle drivers engage in safe practices. They should be aware of their drivers' safety performance, and how this compares with other company areas and with equivalent industry figures. They will also be responsible for the investigation and recording of accidents.

Control of visiting drivers

For any workplace, visitors are unlikely to be familiar with work practices, layout and local rules, and will require to have this information presented to them in an appropriate way. This can be done by publication of written material, or by positioning appropriate signs. Control of visitors must be exercised, since if the breaking of local rules is condoned, this will have a negative effect on attitudes of those who observe them.

Transport accident investigation

In addition to the requirements of local or national laws and codes concerning reporting and recording of accidents, feedback of information gained from investigations is very useful. Information of this kind will have consequences for the driver training programme, trigger discussions with the persons involved, add to company experience and records, and provide a means of assessing each driver.

Each driver should be required to complete a standard report form for each accident involving a vehicle. Personal investigation by supervisory staff should also be made, where possible, to verify the driver's data and obtain any necessary extra information.

A record should be maintained for each driver, as well as for each vehicle. Driver records can form the basis for awards, and reviews of them will identify repetitions of inappropriate behaviour which may be rectified by further training.

Carrying hazardous loads

Manufacturers of hazardous substances should have final responsibility for safety, in the sense that they are likely to have most knowledge of the specific properties of the substances they make, and to be best able to recommend safe handling and emergency procedures. This information should be provided in data sheets (at least) and be made available to all concerned, including drivers and dispatch points.

There are many codes and regulations covering this subject. They contain specific, detailed requirements about containers and packaging, and their marking. They also have requirements on identifying substances by their properties, including flammability, and ways in which information has to be given to those handling and carrying the loads, and to the public. This is done in a number of ways, including by the display of special signs on the packaging and on the vehicle. Requirements differ slightly between countries, and there are also international standards; reference to these and to national codes and regulations is essential if loads are carried between countries.

The instruction and training of drivers is also covered extensively by codes and regulations. The provisions usually include the need to alert emergency services in the event of accidents, use of any emergency equipment required to be carried on the vehicle, and the steps which can be taken safely following an accident to minimise the consequences before the arrival of emergency services. The driver should also be able to give these services the information they need to know about the load, including methods of treating spillages and/or fire. Special attention is required to the selection of mature drivers for hazardous loads, with appropriate skills and abilities.

Legal requirements

Generally, the Health and Safety at Work etc Act 1974 and the more detailed requirements of the Management of Health and Safety at Work Regulations 1999 apply to all aspects of transport safety. Provisions relating to risk assessment, training, capability and two or more employers sharing the same workplace are especially relevant.

A number of specific requirements of the Workplace (Health, Safety and Welfare) Regulations 1992 apply to all workplaces (except construction sites) regardless of their age and purpose. No review of workplace conditions can be considered complete without making a verification check for compliance with them. Organisation of traffic routes is covered by Regulation 17 and the condition of roads by Regulations 5 and 12. For the construction industry, the Construction (Health, Safety and Welfare) Regulations 1996 contain similar provisions for construction sites, which include the need to plan loading bays and traffic routes, and stipulations on vehicle use.

The Provision and Use of Work Equipment Regulations 1998 apply to transport vehicles (as work equipment).

The carrying of hazardous loads is covered in detail in legal provisions. In a work of this length, it is not possible to do more than indicate the enormous number of regulations, codes and standards which can apply nationally and internationally to the movement of transport and goods of various kinds. The selected references at the end of this Section provide access to information on most of the more common requirements.

For employers whose main or secondary activity is the transport of dangerous goods, the Transport of Dangerous Goods (Safety Advisers) Regulations 1999 came into effect on 1 March 1999. These require the appointment by an employer of a safety adviser qualified in both the substance and the mode of transport where the Regulations apply; appointed safety advisers must have a vocational training certificate. The Regulations do not apply to other employers for whom such transport is only an occasional activity not creating a significant risk.

'Dangerous goods' are those which fall into the UN hazard categories 1 to 9:
1 explosive substances and articles
2 gases – flammable, non-flammable compressed, and toxic
3 flammable liquids
4 flammable solids – flammable on contact with water, or liable to spontaneous combustion
5 oxidising agents and organic peroxides
6 toxic or infectious substances
7 radioactive materials
8 corrosive materials
9 miscellaneous (includes asbestos).

Revision

8 causes of transport accidents:
- contact
- overturning
- collision
- impact
- entanglement
- explosion
- operator/supervisor error
- poor maintenance

5 main elements of transport safety:
- driver selection, instruction and supervision
- control of visiting drivers
- workplace traffic control and its general condition
- accident investigation
- preventive maintenance and vehicle repair

Figure 1: Sample basic summary checklist – transport safety

Premises address:				
Completion by:				
Date:				
No		Yes	No	Comments
	Workplace			
1	Are vehicle traffic routes suitable for the traffic that uses them? (Surfaces well maintained? Wide enough? One-way system?)			
2	Are the traffic routes appropriate to keep pedestrians and vehicles apart? (Crossing points, barriers in place? Are people using them?)			
3	Are all necessary safety features in place? (Mirrors, road markings, signs?)			
	Vehicles			
4	Are the vehicles safe and suitable for their intended use? (Lights, horns, guards, rollover protection, good cab access/egress?)			
5	Is a preventive maintenance programme in place for each vehicle? (Vehicle log, driver's pre-use checks?)			
	Drivers			
6	Is there a procedure for selection of new drivers? (Experience check, test?)			
7	Is there an active training programme for drivers, containing specific information on job hazards and the workplace?			
8	Is there a refresher training programme for drivers?			
9	Is there adequate supervision of drivers during their work? (Speeding, stress, safe practices observed?)			
	High-risk transport tasks			
10	Has attention been given to reversing as an issue in the premises? (Scope for improvement? All vehicles left braked and secured?)			
11	Are parking facilities appropriate for work and private vehicles? (Location, condition, lighting, and are they actually used?)			
12	Are vehicles ever loaded beyond their capacity? (System in place for checks on this?)			
13	Are loading and unloading operations carried out safely, with loads secured as necessary?			
14	Are tipping operations carried out safely? (Checks on this? Power lines, pipework, people put at risk?)			
15	Are sheeting and unsheeting operations carried out safely? (Risk assessment, checks on this?)			
16	Is there a system in place which avoids the need for anyone to climb on a vehicle or load?			

Selected references

Legal
Health and Safety at Work etc Act 1974
Management of Health and Safety at Work Regulations 1999
Workplace (Health, Safety and Welfare) Regulations 1992
Provision and Use of Work Equipment Regulations 1998

Guidance (HSC/HSE)
HSG67	*Health and safety in motor vehicle repair*
HSG136	*Workplace transport safety – guidance for employers*

Self-assessment questions

1 A vehicle maintenance fitter insists he is not exposed to danger when carrying out his work. Identify the hazards to which he is exposed.

2 Almost a quarter of all deaths involving vehicles at work happen while vehicles are reversing, mostly at low speeds. Apply basic principles of hazard control to prepare a list of ways in which reversing accidents can be prevented.

7 Classification of dangerous substances

Introduction

Dangerous substances can be categorised in many ways. When discussing or describing them in health and safety terms, they are categorised according to the type of harm they can cause. Many can fall into more than one category. Some can cause harm after a single exposure or incident, others may have long-term effects on the body following repeated exposure.

Classification

Generally, substances can be placed into one or more of the following broad categories:

Explosive or flammable – dangerous because of their potential to release energy rapidly, or because the product(s) of the explosion or combustion are harmful in other ways. Examples of these are organic solvents.

Harmful – substances which, if inhaled, ingested or enter the body through the skin, present a limited risk to health.

Irritant – these adversely affect the skin or respiratory tract, eg acrylates. Some people overreact to irritants like isocyanates, which are sensitisers and can cause allergic reactions.

Corrosive – substances which will attack chemically either materials or people, eg strong acids and bases.

Toxic – substances which prevent or interfere with bodily functions in a variety of ways. They may overload organs such as the liver or kidneys. Examples are chlorinated solvents and heavy metals (eg lead).

Carcinogens, mutagens and teratogens – these prevent the correct development and growth of body cells. Carcinogens cause or promote the development of unwanted cells as cancer. Teratogens cause abnormal development of the human embryo, leading to stillbirth or birth defects. Mutagens alter cell development and cause changes in future generations.

Agents of anoxia – vapours or gases which reduce the oxygen available in the air, or prevent the body using it effectively. Carbon dioxide, carbon monoxide and hydrogen cyanide are examples.

Narcotics – produce dependency, and act as depressors of brain functions. Organic solvents are common narcotics.

Oxidising agents – substances which give rise to exothermic reactions (give off heat) when in contact with other substances, particularly flammable ones.

Labelling

In most jurisdictions, substances with potential for harm are required to be safely packaged and labelled according to various codes of practice and regulations. These involve the use of descriptive labels showing the hazard in the form of pictograms, accompanied by required descriptive phrases which convey necessary information in a shortened form. The labels also carry the details of the manufacturer. In all cases, management should not be content to rely on the packaging to provide sufficient information to make decisions about safe storage, use, handling, transportation and disposal of the substance. The data sheet that must be supplied with the substance will provide the necessary information for these purposes.

Legal requirements

A number of 'carriage' regulations first introduced in 1996 impose 'carriage requirements' on those who consign dangerous goods for transport by various methods, requiring them to identify chemical hazards by classification, to give label information about the hazards, and to package the chemicals safely. These can be found in the selected references, together with the excellent guidance material to which the interested reader is referred. The Chemicals (Hazard Information and Packaging) Regulations 1999 (CHIP 99) cover the 'supply' requirements for dangerous chemicals, implementing directives on packaging aimed at the supplier. This short Section is intended only to give a summary of the categories usually found in regulations. The classification, packaging and labelling system established by CHIP 99 is very comprehensive, and beyond the scope of this book, but a further mention of it occurs in Part 4 Section 27.

See the previous Section for a short summary of the requirement to appoint a safety adviser for the routine transport of dangerous goods, and for some further references.

Revision

Substances are categorised according to the type of harm they can cause.

9 classifications:
- explosive/flammable
- harmful
- irritant
- corrosive
- toxic
- carcinogens/mutagens/teratogens
- agents of anoxia
- narcotic
- oxidising

Selected references

Legal

Chemicals (Hazard Information and Packaging for Supply) Regulations 1999
Classification and Labelling of Explosives Regulations 1983
Notification of New Substances Regulations 1993 (as amended)

Guidance (HSC/HSE)

L93	*Approved tank requirements*
L131	*Approved classification and labelling guide*
HSG117	*Making sense of NONS*

Self-assessment questions

1 In your own workplace, find examples of each of the nine categories of classification for substances. If none are present for a category, consult a reference book and find a new example apart from those mentioned in the text.

2 Can you think of other, simple, ways in which substances could be classified in your workplace so as to give an indication of their potential for harm? Are there any drawbacks to your classification?

8 Chemical and biological safety

Introduction

Chemical and biological incidents at work can affect many people including operators, nearby workers, others on site and members of the public. Systems for the control and safe use of chemicals, substances and biological agents are crucial to the safe operation of any workplace which handles, uses, stores, transports or disposes of them, regardless of the quantity involved. Planning for safety requires the setting out of broad principles which can then be translated into specific detail at the workplace.

Planning for chemical safety

The way in which we plan for chemical safety is based on the principles for the control of all types of risk, which will be familiar to you by the time you complete working through this book. Part 1 Section 3 discusses the principles of risk assessment, which are applicable to chemical safety. The stages of control include risk assessment as the basis for control measures. The six stages of control are summarised here:

- identification of the hazard to be controlled
- assessment of the risk
- control of the risk
- training of staff
- monitoring the effectiveness of the strategy
- necessary record-keeping.

Identification of the hazard

In the present case, we already know the hazard, as 'chemical safety' is the objective. **Hazard** means the inherent property or ability of something to cause harm – it is not the same as **risk**, which is the likelihood of the harm occurring in given circumstances. The extent of the hazard posed by a chemical or substance can be established by the use of information – from the supplier, the manufacturer, from records, historical knowledge and other sources of information. Some substances are labelled on their containers; others, such as reaction products, are not brought into the workplace from outside, have no containers, and must be identified and checked against reference material.

Assessment of the risks

The risks presented by a chemical or substance will not be dependent only on its physical and chemical properties, but will also be a function of the way it is used, where it is used, and how it may be misused or mishandled. Assessment of these factors is essential in determining local risk at a particular location.

Controlling the risks

This can be achieved by elimination of the hazardous substance entirely, by its substitution with a less hazardous alternative, mechanical/remote handling, total enclosure, exhaust ventilation, special techniques such as wet methods, or use of PPE. These control techniques may be used in combination, but they are mentioned in order of effectiveness and desirability. There is also the possibility of reducing exposure time and numbers of people at risk.

Training of operators

Operators need to be instructed in the nature of the hazards they face, in the degree of risk associated with the procedures intended to eliminate hazards, and in the effective control of any remaining risks. In particular, this must ensure that operators do not use their own solutions to problems. In-house procedures involving chemicals are often written down for all to see and understand; where this is done, it is important that the procedures are publicised, available and readable. Language difficulties with minority groups should be resolved and translations provided where necessary.

Monitoring effectiveness

Regular monitoring of the effectiveness of control measures is needed to ensure that change – in work conditions, quality, information and systems – is taken into account and anticipated in procedures. This can include regular monitoring of work practices, sampling of air/process quality, and purchasing procedures. Except for workplaces where little change is expected, monitoring and appropriate measuring should take place at regular intervals specified in the procedures. This is especially important where PPE is relied on as a major method of controlling risks.

Record-keeping

Keeping records helps to ensure the regular maintenance of plant at prescribed intervals and enhances awareness of change. Medical surveillance of operators may be required by local or national codes and regulations.

Risk control issues

The following topics should be considered and adopted as appropriate to reduce exposure.

Plant design

Chemical processes can be controlled by introducing containment and process control – the safe design of plant processes and the use of automated and/or computer-controlled systems. Plant should not be altered without specific authority and review.

Safe systems of work

These include the use of detailed operating and emergency instructions, permit-to-work systems (especially for maintenance operations), and the installation, use and maintenance of protective equipment.

Transport

Controlled movement of chemicals in containers or vehicles intended for that purpose reduces the likelihood of exposure to operators and third parties. Adequate emergency procedures will be required.

Storage

Chemicals should not be affected by adjacent processes or storage. External storage of chemicals should be the method of choice, as this gives quick dispersion, minimises sources of ignition, and provides secure storage which reduces the risk of damage to containers. Features of storage facilities should include a hard standing, bund retention, adequate separation, adequate access for mechanical handling, contents gauges for fixed tanks, relief valves (if required), security of stacking, firefighting facilities and warning signs in accordance with applicable codes and regulations.

Waste disposal

Adequate steps must be taken to ensure the correct disposal of waste, preventing harm to members of the public and damage to the environment, in accordance with legislative requirements.

Emergency procedures

These must be drawn up for production and storage areas, and for the site as a whole. Emergency services and operatives alike should be briefed on the procedures to be followed, and adequate time must be given for training and rehearsal at regular intervals.

The relative importance of each of these topics will depend on the assessed degree of risk at the workplace.

Biological safety

Pathogenic micro-organisms that could affect health are known as **biological agents**. They can affect people who work with them directly, as in a laboratory environment, or indirectly, as a result of exposure to them during the course of normal work. An example of the latter means of exposure is infection of people who work with animals by micro-organisms that cross the 'species gap'. The diseases that these produce are known as zoonoses. There are about 40 known zoonoses in the UK. They include Weil's disease (leptospirosis), which is caused by bacteria passed from rats via their urine, and psittacosis, which is acquired by contact with affected birds. Most infections are relatively mild, but evidence is growing that some long-term health effects may result from them.

Some biological agents produced, liberated or encouraged by work processes can affect the public as well as employees. Examples include legionellosis and hospital-acquired infections (HAIs), such as MRSA. Legionella bacteria are found in natural water sources, but can become trapped and will multiply rapidly within water systems such as wet cooling towers and hot water systems in buildings. They can cause a number of diseases which resemble pneumonia in their symptoms, including legionnaires' disease. HAIs are a serious and growing problem: at least 10 per cent of all British hospital patients are estimated to acquire infections from medical interventions or just from being in hospital. About 1 per cent of all deaths annually occur as a direct result of HAIs, and they may also be a contributory factor in a further 3 per cent of deaths, according to the HSE's Health Directorate. Procedures to control infection are well developed, and subject to the Management of Health and Safety at Work Regulations 1999 and the Control of Substances Hazardous to Health Regulations 2002 (COSHH).

Work in laboratories where exposure to biological agents may occur is subject to the standard requirements for the control of hazards and risk assessment. To further the assessment, COSHH places biological agents in one of four hazard groups, ranging from 1 (lowest risk) to 4 (highest). Categorising is done by the Advisory Committee on Dangerous Pathogens, which is responsible for maintaining the HSE Approved List of Dangerous Pathogens. Table 1 gives some examples of the classification.

Table 1: Hazard group examples

Biological agent	Hazard group
Ebola, Lassa fever	4
Hepatitis B and C, rabies, CJD	3
Legionella, hepatitis A	2

It is of interest to note that some of the pathogenic organisms placed in Group 3 (which requires a full containment regime) are thought to have an infective dose of less than 100 organisms. Discussion of the principles of laboratory safety with micro-organisms is beyond the scope of this book, but the basic hierarchy of controls required by COSHH apply:
- risk assessment
- substitution with a safer biological agent where possible

- prevention of exposure
- use of engineering or physical controls to minimise release into the workplace
- selection of other control measures, including personal protective equipment
- maintenance, inspection and testing of control measures, such as safety cabinets
- information, instruction and training for employees
- immunisation and health surveillance, keeping of records
- required notification of new premises using biological agents in Category 2 and above.

Legal requirements

The general duties sections of the Health and Safety at Work etc Act 1974 apply to chemical and biological safety in the same way that they apply to all work activities, and provide protection for those affected by work activities as well as those at work. More detailed requirements are contained in the Management of Health and Safety at Work Regulations 1999.

In addition, specific regulations controlling substances hazardous to health (the Control of Substances Hazardous to Health Regulations 2002 (as amended) (COSHH), covered in Part 4) apply to all work activities. Work sites handling, storing or using certain chemicals in quantity are covered by the Control of Major Accident Hazards Regulations 1999 (COMAH), which are not discussed in this book in detail. Broadly, COMAH came into force on 1 April 1999 to implement the Seveso II Directive, aimed at preventing and mitigating the effects of those major accidents involving dangerous substances such as chlorine, LPG, explosives and the like. The chemical industry is most likely to be affected, but some other storage activities and sites may involve quantities of dangerous substances specified in the Regulations and, thus, come within their scope.

The scope of the COSHH Regulations, both as to substances and the quantities of them which cause the duties to be applied, can be found in Regulation 3 and Schedule 1. COSHH also applies to biological agents. Schedule 3 contains information on categorisation. Further information can be obtained from the guidance material, which is summarised below. This contains information on COSHH control measures.

Revision

6 **stages of control:**
- hazard identification
- risk assessment
- risk control
- staff training
- monitoring
- record-keeping

6 **topics for control of chemical risk:**
- plant design
- systems of work
- transport
- storage
- disposal
- emergencies

Hazard – the inherent ability to cause harm
Risk – likelihood and severity of harm in particular circumstances

6 **examples of recognised zoonoses:**
- anthrax
- bovine tuberculosis
- leptospirosis
- psittacosis
- Lyme disease
- rabies

Selected references

Legal
Management of Health and Safety at Work Regulations 1999
Control of Substances Hazardous to Health Regulations 2002 (as amended)
Control of Major Accident Hazards Regulations 1999
Dangerous Substances (Notification and Marking of Sites) Regulations 1990
Notification of Installations Handling Hazardous Substances Regulations 1982 (as amended)

Guidance (HSC/HSE)
L5   *COSHH and control of carcinogenic substances* ACoP
L111   *A guide to the Control of Major Accident Hazard Regulations*
HSG97   *A step-by-step guide to COSHH assessment*
HSG110   *Seven steps to successful substitution of hazardous substances*

The occupational zoonoses, ISBN 0 11 886397 5

Self-assessment questions

1 Is operator training sufficient by itself to prevent chemical accidents?

2 Give an example of a substance with different levels of risk when used and when stored.

3 What zoonoses might be relevant to work in industrial premises?

9 Electricity and electrical equipment

Introduction

The ratio of fatalities to injuries is higher for electrical accidents than for most other categories of injury – if an electrical accident occurs, the chances of a fatality are about one in 30 to 40. Despite the beliefs of some, including some electricians, the human body does not develop tolerance to electric shock. The consequences of contact with electricity are: electric shock, where the injury results from the flow of electricity through the body's nerves, muscles and organs and causes abnormal function to occur (eg the heart stops); electrical burns resulting from the heating effect of the current which burns body tissue; and electrical fires caused by overheating or arcing apparatus in contact with a fuel (see the next Section).

Causes of electrical failures

Failures and interruption of electrical supply are most commonly caused by:

- damaged insulation
- inadequate systems of work
- inadequate overcurrent protection (fuses, circuit breakers)
- inadequate earthing
- carelessness and complacency
- overheated apparatus
- earth leakage current
- loose contacts and connectors
- inadequate ratings of circuit components
- unprotected connectors
- poor maintenance and testing.

Preventing electrical failures

These failures can be prevented by regular attention to the following points:

Earthing – providing a suitable electrode connection to earth through metal enclosure, conduit, frame and so on. Regular inspections and tests of systems should be carried out by a competent person.

System of work – when working with electrical circuits and apparatus, switching and locking off the supply. After this, the supply and any apparatus should be checked personally by the worker to verify that it is 'dead'; permit-to-work systems should be used in previously identified high-risk situations. Working on live circuits and apparatus should only be permitted under circumstances which are strictly controlled and justified in each case. Rubber or other non-conducting protective equipment may be required. Barriers and warning notices should be used to ensure that persons not directly involved in the work do not expose themselves to risk. The use of insulated tools and equipment will always be necessary.

Insulation – where work is required near uninsulated parts of circuits. In all circumstances, making the apparatus 'dead' must be considered as a primary aim, and rejected only if the demands of the work make this impracticable. A variety of permanent or temporary insulators may be used, such as cable sheathing and rubber mats.

Fuses – these are strips of metal placed in circuits which melt if the circuit overheats, effectively cutting off the supply. Different fuses melt at different predetermined current flows. A factor in the selection of fuses is that there is a variable and appreciable delay in their action, which may expose those at risk to uninterrupted current for unacceptable periods.

Circuit breakers – detect electromagnetically and automatically any excess current flow and cut off the supply to the circuit.

Residual current devices – detect earth faults and cut off the supply to the circuit.

Competency – only properly trained and suitably experienced people should be employed to install, maintain, test and examine electrical circuits and apparatus.

Static electricity

Electrical charging can occur during the movement of powders and liquids, and this can cause sparking which may ignite a dust cloud or a flammable vapour. Less critically, static electricity in other working environments may be a cause of annoyance to workers, and can be a factor in other types of accident where attention has been distracted by static discharge. Protection against static electricity can be achieved by earthing and not using or installing equipment which can become statically charged (see the next Section). Operators should also wear anti-static shoes.

Electrical equipment

Electrical tools used should be selected and operated bearing in mind the following considerations.

Substitution – electrical tools and equipment may be replaced with pneumatic alternatives – which have their own dangers.

Switching off circuits and apparatus – this must be readily and safely achievable.

Reducing the voltage – the lowest practicable voltage should be used in every circuit.

Cable and socket protection – electrical connections should be protected against physical and environmental effects, such as rain, which could have adverse consequences on the integrity of circuits and apparatus.

Plugs and sockets – these must be of the correct type and specification, and meet local and national regulations and codes of practice.

Explosive atmospheres – equipment for service in dusty or flammable environments must be considered and selected with care. The type of equipment to be used is commonly specified by law.

Maintenance and testing – checks should be carried out at regular and prescribed intervals by competent and experienced personnel. Fixed equipment and installations should be maintained at regular intervals following the advice of the supplier, manufacturer or designer. For portable and transportable equipment (operating at more than 50 volts AC or 120 volts DC), a planned maintenance regime

appropriate to the risk should be followed. This is because fixed equipment is usually better able to withstand damage, and portable equipment is more likely to receive damage and be operated in unfavourable conditions.

There are three levels of preventive maintenance for this equipment.

Visual checks by the user – for visual evidence of damage or faulty repair to connectors, leads and equipment.

Visual checks by an authorised third party – more detailed, including removal of connector covers and checks on fuses. When done at regular set intervals, these should detect misuse, abuse and incorrect selection of equipment.

Inspection and testing by a third party – periodic thorough visual examination by someone other than the normal user of the equipment, as well as use of portable appliance tester (PAT) equipment. Appropriate competency is required, and this is wider than for visual-only checks.

Recording of the results and values measured will provide a baseline figure for assessing any subsequent deterioration in performance or quality, and for monitoring the effectiveness of the maintenance programme. Labelling equipment passed as safe with a date for the next test is good practice.

Frequency of maintenance checks depends on a range of factors, including the likelihood and type of potential damage. Frequency can be altered following experience of the maintenance programme. Different pieces of equipment may require different schemes. Table 1 overleaf offers some initial intervals based on average use conditions.

The treatment of electric shock
NB: *Advice given in this Section is not intended as a substitute for, or an alternative to, attending a recognised first aid course.*

Table 1: Sample initial inspection intervals

Environment/trade	User visual check	Formal visual check by third party other than user	Inspection and electrical testing
Office premises: rarely or occasionally moved items such as computers, copiers, fax machines	No	2 to 4 years	Up to 5 years unless double-insulated, in which case this is not required
Double-insulated items: not hand-held, moved occasionally, eg fans, projection units	No	2 to 4 years	No
Double-insulated equipment: hand-held, eg floor cleaners	Yes	6 months to 1 year	No
Earthed equipment, eg kettles and some floor cleaners	Yes	6 months to 1 year	Yes, 1 to 2 years
Cables and plugs connected to the above, mains extension leads	Yes	6 months to 4 years depending on the type of equipment it is connected to	Yes, 1 to 5 years, depending on the type of equipment it is connected to
Hotels and other public access premises	As for office premises, depending on type of equipment and conditions of use		
General industrial	Daily/before first use	Before first use, thereafter every 3 months	6 to 12 months
Construction	Daily/before first use	Before first use, thereafter every month	3 months
Equipment hire	Daily/before first use	Before each issue and after hire return	Before each issue

Recognising a casualty

Apart from proximity to a source of electrical energy, casualties may show symptoms of asphyxia and have no discernible pulse. Violent muscular contraction caused by contact with high voltage supplies can throw casualties some way from the original point of contact. In cases of electric shock, breathing and heartbeat can stop altogether, accounting for the pallid blue tinge to the skin sometimes seen. There may also be burns visible, which can indicate contact with electricity.

Breaking the contact

Low voltage supplies, commonly at 240 volts, can be switched off at the mains or meter if readily accessible. If not, or if it cannot be found quickly, the supply can be turned off at the plug point or by pulling out the plug. The important point to remember when removing the casualty from the source is not to become a casualty in turn. Dry insulating material such as newspaper or wood can be used to push or pull the casualty clear by wrapping around or pushing at arms and legs near the source. Do not touch the casualty directly.

Higher voltage supplies such as those found in overhead or underground power lines produce deep burns and are frequently fatal on contact. It is important not to approach a casualty touching or lying close to one of these lines until the line has been isolated by the supply authority. Close approach can lead on occasions to electrical arcing. At higher voltage levels, insulating material requires special properties not possessed by anything likely to be close to hand.

Treatment
The order of priority for treatment is:
1 resuscitation if required
2 place in recovery position if casualty is still unconscious but breathing normally
3 treat any burns
4 treat for shock
5 remove to hospital in all cases where resuscitation was required, or where casualty was unconscious, received burns or developed symptoms of shock; pass to hospital any information on duration of electrical contact.

Resuscitation
Mouth-to-mouth ventilation should be used where a casualty is not breathing and ventilation of the lungs is required to restart breathing reflexes. **Cardiopulmonary resuscitation (CPR)** will be needed if the heart has stopped as well as breathing. Both these techniques require practice; descriptions of them are beyond the scope of this Section but they are taught on first aid courses and details can also be found in first aid manuals.

Recovery position
A casualty in this position maintains an open airway and stability of the body is improved. Placing the casualty in the recovery position is not difficult, but a description of the method is beyond the scope of this Section. The position to be achieved is one where the casualty lies on the front and side, with the head supported for comfort and slightly flexed backwards to maintain a clear airway. The uppermost hand is near the face with the arm bent at the elbow to give support, and the uppermost leg is bent at the hip and knee forming a right angle at the knee, with the thigh well forward.

Burns treatment
For severe burns, likely to be associated with electric shock, it is important to immobilise the burned part, and to cover the burned area with a dry, sterile, unmedicated dressing or anything similar, and secure with a bandage if available.

The important things **not to do** are:
* do not allow clothing to restrict any swelling of the burned part
* do not apply anything to the burn
* do not break blisters or attempt to treat or touch the injured area
* do not remove anything sticking to the burned part.

Sips of cold water will help the casualty to replace lost fluid, but give no medication.

Shock treatment
Traumatic shock can prove fatal even when other injuries have been treated. It is a consequence of a reduction in volume of body fluids. The casualty feels faint, sick and possibly thirsty, and has shallow breathing or yawning. The pulse rate may increase but may also become weak and irregular. The skin is pale and cold.

Treatment involves keeping the casualty warm, loosening tight clothing and elevating the legs if possible. Do not give anything by mouth, or apply external heat, or allow the casualty to smoke.

Do not move the casualty unless this cannot be avoided.

Legal requirements
These are covered completely by the Electricity at Work Regulations 1989, which are discussed in Part 4 Section 15 and amplify the general duties placed on employers, employees and the self-employed by the Health and Safety at Work etc Act 1974. Further requirements are contained in the Management of Health and Safety at Work Regulations 1999. Arrangements and provisions for the treatment of electric shock are contained in the Health and Safety (First Aid) Regulations 1981 (see Part 4 Section 19).

Revision

3   **results of electrical failure:**
- electric shock
- electrical burns
- electrical fires

10   **preventive measures:**
- earthing
- safe system of work
- insulation
- fuses
- circuit breakers
- residual current devices
- competency
- reduced voltage
- maintenance
- isolation

Selected references

Legal
Electricity at Work Regulations 1989
Electrical Equipment (Safety) Regulations 1994
Equipment and Protective Systems etc Regulations 1996

Guidance (HSC/HSE)

HS(R)25	*Memorandum of guidance on the Electricity at Work Regulations*
HSG47	*Avoiding danger from underground services*
HSG85	*Electricity at work: safe working practices*
HSG107	*Maintaining portable and transportable electrical equipment*
HSG118	*Electrical safety in arc welding*
HSG141	*Electrical safety on construction sites*
GS6(rev)	*Avoidance of danger from overhead electric lines*
GS50	*Electrical safety at places of entertainment*
PM38(rev)	*Selection and use of electric hand lamps*

British Standards

BS 2754:1976	Memorandum: Construction of electrical equipment for protection against electric shock
BS 7671:1992	Requirements for electrical installations – IEE Wiring Regulations, 16th edition
BS EN 60309	Plugs, socket outlets and couplers for industrial purposes
BS 6867:1993	Code of practice for maintenance of electrical switchgear over 36Kv (see also BS 6423 for switchgear up to 1Kv, and BS 6626 for switchgear between 1Kv and 36Kv)

Self-assessment questions

1   What preventive measures can be taken against electrical failure?

2   An electric shock occurs as a result of using a drill outside. What factors might have contributed to this accident?

10 Fire

Introduction

The control of the start and spread of fire is an important feature of the prevention of accidents and damage. Fire needs fuel, oxygen and a source of energy to ignite it; these are known as the 'fire elements'. Examples of these are wood, air and heat in the right combination. The ratio of fuel to oxygen is crucial; too much or too little of either will not permit a fire to start. Also, the source of ignition must be above a certain energy level. The fire elements are often illustrated as the 'fire triangle'; this offers the simple reminder that if one side of the triangle, or element, is removed or not present, the other two cannot support combustion.

There are five main hazards produced by fire: oxygen depletion, flame/heat, smoke, gaseous combustion products and structural failure of buildings.

Heat is transmitted, and fire is spread, by convection, conduction, radiation and direct burning and usually by a combination of several of the methods.

Classification of fire

There are four main categories of fire, which are based on the fuel and the means of extinction.

Class A

Fires which involve solid materials, predominantly of an organic kind, forming glowing embers. Examples are wood, paper and coal. The extinguishing mode is by cooling and is achieved by the use of water.

Class B

Fires which involve liquids or liquefiable solids; they are further subdivided into:
- **Class B1**, which involve liquids soluble in water, eg methanol. They can be extinguished by carbon dioxide, dry powder, water spray, light water and vaporising liquids
- **Class B2**, which involve liquids not soluble in water, such as petrol and oil. They can be extinguished by foam, carbon dioxide, dry powder, light water and vaporising liquids.

Class C

Fires which involve gases or liquefied gases (eg methane or butane) resulting from leaks or spillage. Extinguishment can be achieved by using foam or dry powder in conjunction with water to cool any leaking container involved.

Class D

Fires which involve metals such as aluminium or magnesium. Special dry powder extinguishers, which may contain powdered graphite or talc, are required to fight these. No other extinguisher type should be used.

Electrical fires, which involve the electricity supply to live equipment, can be dealt with by extinguishing mediums such as carbon dioxide, dry powder or vaporising liquids, but not water. Electricity is a cause of fire, not a category of fire. Electrical fires have been removed from the traditional 'categories' of fire for this reason.

Knowledge of the correct type of extinguisher to use, or install, in areas at particular risk is essential (see above).

Fire protection

There are three strategies for protecting against and dealing with fire:

Structural design precautions – providing protection through insulation, integrity and stability permitting people to escape, eg compartmentalisation, smoke control, unobstructed means of escape.

Fire detectors and alarms – activated by sensing heat, flame, smoke or flammable gas, eg heat detectors, radiation detectors, smoke detectors.

Firefighting – with portable extinguishers or fixed firefighting equipment (manually or automatically operated).

Structural and design precautions

Controlling the likelihood of fire occurring can be achieved by the removal of one of the fire elements by:

- **removing sources of ignition.** Heat is the most common source of ignition, and can result from:
  - friction – parts of a machine, which require proper lubrication
  - hot surfaces – machine panels, which can be cooled by design, insulation or maintenance
  - electricity – arcing or heat transfer, removed by selection or design of appropriate electrical protection
  - static electricity – through relative motion and/or separation of two different materials, prevented by providing an earth path, humid atmosphere, high voltage device or ion exchange
  - smoking – spent smoking materials, prevented by restricting the activity or confining it to defined areas

  Heat transfer can be by radiation, conduction or convection

- **controlling the fuel.** Supply of materials should be kept to a minimum during work or in storage areas. The main fuel sources are:
  - waste, debris and spillage
  - gases and vaporising liquids
  - flammable liquids
  - flammable compressed gases and liquefied gases
  - liquid sprays
  - dusts.

Means of fire detection and alarms

Fire, or flammable atmospheres, can be detected in the following ways:

Heat detection – sensors operate by the melting of a metal (fusion detectors) or expansion of a solid, liquid or gas (thermal expansion detectors).

Radiation detection – photoelectric cells detect the emission of infrared/ultraviolet radiation from the fire.

Smoke detection – using ionising radiations, light scatter (smoke scatters a beam of light), obscuration (smoke entering a detector prevents light from reaching a photoelectric cell).

Flammable gas detection – measures the amount of flammable gas in the atmosphere and compares it with a reference value.

Fire alarms must make a distinctive sound, audible in all parts of the workplace. The meaning of the alarm sound must be understood by all. They may be manually or automatically operated.

Firefighting

Fires can be extinguished by suppression or removal of one of the three fire elements. This can be done by fuel starvation, by preventing fuel flow or removing its source, and by containment. For example, covering a fire will prevent oxygen from reaching fuel, and cooling with water removes a primary source of ignition. Interference with the combustion process can also be done as a chemical process. Selection of the appropriate type of extinguishing agent depends on knowledge of the likely source and composition of the fire.

A fire plan should be available in even the smallest premises. Such a plan will be of value to professional firefighters, who need to know the location of flammable and hazardous material in premises. The associated evacuation procedure should also be practised at intervals.

Fire extinguishers

The selection of appropriate extinguishing media has been discussed above. All portable fire extinguishers and hose reels should be inspected regularly for signs of damage or obstruction which might affect their operation, and also to make sure that extinguishers have not been discharged or operated. Their useful life is about 20 years. They should receive an annual service, and extended service every five years.

Signs and training

Signs should be provided to identify courses of action to be taken, means of escape and location of firefighting equipment. Training should be given in the use of systems and equipment provided, and to minimise undesirable or apparently illogical behaviour.

Means of escape in case of fire

The purpose of a means of escape is to enable people confronted by fire to proceed in the opposite direction to an exit away from the fire, through a storey exit, a protected staircase and/or a final exit to ground level in open air away from the building. Any means of escape must not rely on rescue facilities from the fire service.

Means of escape in buildings other than those with only a ground floor generally consist of three distinct areas. These are:
- any point on a floor to a staircase
- the route down a staircase
- the route from the foot of the staircase to the open air, clear of the building.

Essentially, there are two physical areas associated with means of escape:
- the area in which people escaping from a fire are at some risk (the unprotected zone)
- areas where risk is reduced to an acceptable minimum (protected zones). Protected zones at an exit must be fully protected, using walls or partitions which have a fire-resisting capability.

Emergency escape lighting may be needed in areas without daylight or which are used at night. It should:
- indicate the escape route clearly
- provide illumination along escape routes
- ensure that fire alarm call points and firefighting equipment can be located.

Determining the means of escape

Types of means of escape will be determined by:
- occupancy characteristics – numbers, physical capability of users
- building uses – residential, commercial, manufacturing, entertainment, etc

- construction characteristics of the building
- evacuation times
- occupant movements.

Travel distance

This is a technical term used in fire safety, and is the greatest allowable distance which has to be travelled from the fire to reach the beginning of a protected zone of a means of escape or final exit. This distance must be limited, although it will vary according to the fire risk.

Protected lobbies and staircases

As parts of the protected zone, these must be of fire-resisting capability. Doors must be self-closing. The objective is to ensure that people entering lobbies and staircases on any floor are to be able to remain within their safe confines until reaching the ground. Such protection serves the purposes of preventing smoke and heat from obstructing the staircase so making it unsuitable for escape purposes, and of preventing a fire from spreading into a staircase from one storey and passing to another.

Staircases should be protected, usually by a ventilated lobby. Staircases should:
- be continuous
- have risers and treads of dimensions specified in standards
- be ventilated
- have any internal glazing and spaces underneath able to resist fire
- be fitted with handrails on both sides.

Lifts and escalators

Lifts and escalators must be disregarded as potential means of escape. Lifts have limited capacity, may be delayed in arriving and have potential for mechanical failure. Escalators have restricted width, and the stair pitch when not in operation may lead to possible injury, congestion and undesirable behaviour. Both normally require an electrical supply which may terminate at any time.

Width and capacity of escape routes

There should be no risk of overcrowding, delay or formation of 'bottlenecks' in the event of mass evacuation. Escape routes should be at least 1m wide, and 1.2m if wheelchair access is required. Protected zones must not contain stores, electrical equipment, portable heaters or other items that could compromise protection.

Fire doors

These have the main functions of preserving the safe means of escape by retarding the passage of fire (so they must be fire-resistant) and retarding the spread of smoke and hot gases in the initial stages of a fire.

Final exit doors

General principles of design and function apply to all means of escape final exit doors, which may be modified in particular cases. The operating principles are that these doors must:
* open in the direction of escape
* preferably not open onto steps
* not be revolving doors
* be capable of being opened at any time while the building is occupied, preferably without using keys
* be protected from fires in the basement or in adjacent buildings
* be of adequate width
* be provided with adequate dispersal space.

Other means of exit

Means of **escape via roof** – this is generally unacceptable, but if it must be used, then it must be protected from smoke, fire and falls. A safe means of access must be provided to the ground.

High level access – across a roof or other areas at height to protected means of escape. If required to be a means of escape, it must be protected against fire, be clearly defined and protected against falls.

Ramps – ramps may be used if they have a non-slip surface, a uniform pitch, do not exceed a slope of 1:10 and are provided with handrails.

External fire escapes – these are not desirable, as they are subject to weather conditions such as ice and snow, which make them potentially dangerous. However, they are sometimes necessary. They must be protected against fire, and windows adjacent to them must be fire-

resistant and not open outwards so as to obstruct passage down the fire escape. It is obviously important that they can be reached easily, and not sited behind or beyond significant obstructions to passage.

Action to be taken in the event of fire

In the presence of fire, the urge to get away is a natural reaction. Information about the action to take and, where possible, practice in that action, is essential to ensure the optimum response in the event of fire. This should be contained in an emergency plan for the premises.

Notices to employees and others

Copies of notices giving simple guidance on what to do in the event of fire should be displayed in all workplaces and premises where persons could be at risk from fire. The details of each notice, which may be written as a safety sign, will vary and depend on the layout, fire risk and circumstances likely to be encountered. For this reason, they should be specially written. Diagrams of layouts showing the means of escape in each case will be helpful, especially where people are likely to be unfamiliar with the layout of the structure.

In drafting notices, attention should be paid to the needs of those who are infrequent visitors to the premises and who may have had no need or opportunity to practise any fire drill there. Instructions in the notice should, therefore, be straightforward and simple to understand. They may need to be given in translated form if speakers of a number of languages are likely to be present in the building. This will be the case in hotels, for example.

Notices should explain simply what to do if a fire is discovered, or if the fire alarm sounds. In the latter case, the actual sound made by the fire alarm should be explained if there is any possibility of confusion.

Notice contents

The displayed notices should cover the following points:

- identification of fire alarm sound
- what to do when the alarm is sounded (eg close windows, switch off appropriate equipment, leave the room, close doors behind, follow signs, report to assembly point at specified location, do not re-enter building until advised)
- what to do if a fire is discovered (eg how to raise the alarm, what fires may be tackled by staff, evacuation procedure previously described).

Fire risk assessment

A fire risk assessment must be carried out for every workplace, regardless of the existence of any approval or certification which may have been given. It is a practical exercise, requiring a tour to verify its completeness and accuracy. The purposes of the assessment are:

- to identify the extent of the fire risk
- to assess the likelihood of a fire occurring
- to identify any additional precautions that may be needed, or control systems that do not function adequately.

The HSE's approach in *Five steps to risk assessment* can be used as an assessment format:

1. identify potential fire hazards
2. decide who might be harmed or at significant risk in the event of a fire, and identify their location
3. evaluate the levels of risk from each hazard, and decide whether existing precautions are adequate – or whether the hazard can be removed or controlled more effectively
4. record the findings and advise employees of the results, and prepare or revise the emergency plan
5. review the assessment as necessary, as conditions change and at regular intervals in any case.

There are also checklists and formats available commercially. The reader's attention is drawn particularly to the Home Office's *Fire safety: an employer's guide*, referenced below, as a mine of information. This publication was used to derive the assessment format and checklists which follow this Section.

The use of a simple code can be convenient to describe the extent of compliance, the condition or the degree of control. The following gives an easily understood picture:

A **fully satisfactory** – meets all requirements, does not need improvement
B **adequate** – some improvement possible
C **less than adequate** – significant improvement required, action must be taken
D **poor** – action is required urgently to improve the condition
NA **not applicable** – condition or circumstances not present or applicable.

Legal requirements

Major provisions are currently contained in the Fire Precautions Act 1971, as amended by the Fire Safety and Safety of Places of Sport Act 1987, and in detailed provisions concerning fire certification. The local fire authority enforces both the Fire Regulations and the parts of the Management of Health and Safety at Work Regulations 1999 which deal with precautions for the safety of people where fires may occur. The latter Regulations apply to the taking of fire precautions in general terms, requiring employers to identify circumstances where situations presenting serious and imminent danger to employees could occur. These must be identified in risk assessments, and procedures written down to control risks. The Fire Precautions (Workplace) Regulations 1997 (as amended in 1999) apply to the taking of fire precautions in premises not required to have a fire certificate. See Part 4 for a detailed review of these Regulations.

Revision

3 **fire elements:**
- fuel
- oxygen
- ignition

4 **classes of fire:**
- class A – wood, paper
- class B – liquid
- class C – gases
- class D – metals

(but not electrical fires, extinguishable by all media except water)

3 **protection strategies:**
- structural design precautions
- fire detectors and alarms
- firefighting

5 **fire hazards:**
- oxygen depletion
- flame/heat
- smoke
- gaseous combustion products
- structural failure

Selected references

Legal

Management of Health and Safety at Work Regulations 1999

Fire Certificates (Special Premises) Regulations 1976

Fire Precautions (Application for a Certificate) Regulations 1989

Fire Precautions (Non-certificated Factory, Office, Shop and Railway Premises) (Revocations) Regulations 1989

Fire Safety and Safety of Places of Sport Act 1987

Fire Precautions (Workplace) Regulations 1997 (as amended in 1999)

Guidance (Home Office)

Code of practice: Fire precautions in factories, offices, shops and railway premises not required to have a fire certificate

Fire Precautions Act 1971: guide to fire precautions in existing places of work that require a fire certificate

Guide to fire precautions in existing places of entertainment and like places

Fire safety: an employer's guide, ISBN 0 11 341229 0

Self-assessment questions

1 How would you handle a fire involving a flammable gas cylinder?

2 List the design features handling fire protection in your work area.

Table 1: Reducing sources of ignition

Potential problems	Possible corrective actions
Unnecessary sources of heat in the workplace	Remove Replace with safer alternative
Build-up of static electricity	Earthing, electrician to survey and advise Increase numbers of broad-leafed plants, spray for carpets
Fuses and circuit breakers not suitable or of incorrect rating	Electrician to advise on appropriate protection for the system
Faulty or overloaded electrical or mechanical equipment	PAT testing system required Maintenance system required
Machinery or equipment with risk of fire or explosion	Remove Replace with better design to minimise risk
Naked flame or radiant heaters in use	Remove Replace with fixed convector heaters or central heating system
Ducts and flues not cleaned, especially kitchen areas	Maintenance system needs review
Metal/metal impact	Consider spark-free tools containing beryllium compounds
Arson, vandalism	Seek advice from police, fire authority Protection of stored materials Prompt and efficient waste disposal Ban on burning of rubbish
Equipment not left in a safe condition after use	Identify relevant equipment, provide training to users and inspection at end of work
Maintenance workers and others carrying out 'hot' work – eg welding and flame-cutting	Permit-to-work system required
Smoking on the premises	Safe smoking policy in designated areas, prohibiting it elsewhere
Matches and lighters allowed in high-risk areas	Enforce prohibition
Smouldering material	Check work areas before leaving

Table 2: Reducing potential fuel sources

Potential problems	Possible corrective actions
Accumulated waste materials	Store flammable wastes and rubbish safely Improve housekeeping standards
Presence of flammable materials and substances	Remove, or reduce quantities to the minimum requirement for the business Replace with less flammable alternatives Review arrangements for use, handling, storage, transport
Storage of flammables	Ensure adequate separation distances Store in fire-resisting stores Minimum work-room quantities
Flammable wall and ceiling linings	Remove, cover or treat to reduce rate of flame spread
Furniture with damaged upholstery	Replace or repair where the foam lining is exposed
Workplace construction and materials	Improve fire resistance of the structure
Arson and vandalism	Make storage areas secure

Table 3: Reducing sources of oxygen

Potential problems	Possible corrective actions
Oxidising materials	Store away from heat sources and flammable materials Mark clearly
Oxygen cylinders	Control their use and storage, arranging for regular leakage checks Check that adequate ventilation is provided, measure and benchmark
Doors, openings and windows	Keep closed all those not required for ventilation, especially outside working hours
Ventilation systems	Shut down non-essential systems

Table 4: Detecting and warning about fire

Potential problems	Possible corrective actions
Means of giving warning cannot be heard throughout premises when initiated from a single point	Review means of giving warning Regular maintenance
Means of detection not quick enough to allow all occupants to escape	Review type of detection and equipment
Emergency plan	Fire detection and warning arrangements should be detailed in the emergency plan for the premises
Lack of knowledge of how the fire warning system works	Instruct employees – how to operate and respond to it Put instructions in writing
Power supply for detection and warning system	Back-up supply should be provided for electrical systems

Table 5: Safe means of escape

Potential problems	Possible corrective actions
Arrangements not included in the emergency plan	Review the emergency plan
Recently approved – Building Regulations, licence or fire certificate	Everyone should reach a place of relative safety within the time available for an escape, including their reaction time Carry out practice drill to confirm times
Length of time taken for all occupants to escape to a place of safety following detection of a fire	Everyone should reach a place of relative safety within the time available for an escape, including their reaction time Carry out practice drill to confirm times
Exits – not enough, not in the right place Exits – type and size not suitable, eg wheelchair width Fire could affect all available exits	750mm doorway is suitable for up to 40 people per minute; 1000mm doorway for up to 80 people per minute Assume largest exit doorway may not be available for use; remaining should be enough for expected evacuation use
Escape routes not easily identifiable, not free from obstruction, inadequately illuminated	Review safety signs and markings Review escape routes for presence of prohibited items Doors must open in the direction of travel where near foot of a stairway or where more than 50 people may have to use them Any securing device needs explanatory notice Review emergency lighting
Employees not trained in using the means of escape	Review safety training arrangements
No instructions for employees about the means of escape	Review need for training and drill Consider preparing a diagram to show the routes of escape

Table 6: Means of putting out a fire

Potential problems	Possible corrective actions
Extinguishers (including hose reels) not big enough or wrong type for the foreseen fire risk	Review provision of extinguishers
Not enough extinguishers	Increase number of extinguishers
Extinguishers not accessible, not close to the fire	Relocate or increase numbers of extinguishers
Locations of extinguishers are not obvious	Indicate their position using correct signs
Those likely to use extinguishers have not been trained to do so	Provide adequate instruction and training
Use of firefighting equipment not covered in emergency plan	Revise emergency plan

Table 7: Maintenance and testing of fire equipment

Potential problems	Possible corrective actions
All fire doors not regularly checked All firefighting equipment not regularly checked Fire detection/alarm equipment not regularly checked Other means of escape not regularly checked	Give instructions to employees to make checks and make or arrange for tests as required
Employees testing or maintaining equipment not trained to do so	Arrange for necessary training, or outsource tasks

Table 8: Maintenance requirements for fire precautions

Equipment	Period	Action required
Fire detection and warning systems, including smoke alarms and manual devices	Weekly	Check for operation and state of repair Test operation Repair or replace items as necessary
	Annually	Full check and test by competent service engineer Change batteries in alarms and detectors, where battery-powered
All types of emergency lighting equipment	Weekly	Operate torches, check batteries Repair or replace items as necessary
	Monthly	Check all systems for state of repair and apparent working order
	Annually	Full check and test by competent service engineer
All types of firefighting equipment including hose reels	Weekly	Check all items for correct installation and apparent working order
	Annually	Full check and test by competent service engineer

Table 9: Daily fire safety checks

1	Control panel – system is operating normally	YES	NO	N/A
2	Emergency lighting systems that include signs are lit	YES	NO	N/A
3	All defects in the above have been recorded and dealt with	YES	NO	N/A
4	All escape routes are clear, free from obstruction, free from slipping and tripping hazards, and available for use when the premises are occupied	YES	NO	N/A
5	All door fastenings along escape routes operate freely	YES	NO	N/A
6	All self-closing devices and automatic door closers work correctly	YES	NO	N/A
7	Each door on the escape route closes correctly, including any fitted flexible edge seals	YES	NO	N/A
8	All exit and directional signs are correctly positioned, unobstructed and can be seen clearly at all times	YES	NO	N/A
9	All fire extinguishers are in position and ready for service	YES	NO	N/A

Table 10: Weekly fire safety checks

1	Manually operated fire alarms are tested to ensure they can be heard throughout the workplace	YES	NO	N/A
2	Electrical detection and warning systems are tested weekly for function and audibility	YES	NO	N/A
3	Domestic-type smoke alarms are tested weekly	YES	NO	N/A
4	Portable fire extinguishers are inspected weekly: (checking the safety clip and indicator devices and checking for external corrosion, dents or other damage)	YES	NO	N/A
5	Hose reels are checked for any damage or obstruction	YES	NO	N/A

Table 11: Checklists for emergency plan and training

1	There is an emergency plan for the premises	YES	NO	N/A
2	The emergency plan takes account of all reasonably foreseeable circumstances	YES	NO	N/A
3	Employees are familiar with the emergency plan, and have been trained in its use	YES	NO	N/A
4	Drills are held regularly to demonstrate the use of the emergency plan	YES	NO	N/A
5	The emergency plan is made available to all who need to be aware of it	YES	NO	N/A
6	The procedures to be followed are clearly shown throughout the premises	YES	NO	N/A
7	The emergency plan takes account of all those likely to be present in the premises and those who may be affected by a fire emergency in the premises	YES	NO	N/A

Checklist – the emergency plan contents

1	What employees must do if they discover a fire	YES	NO	N/A
2	How people will be warned if there is a fire	YES	NO	N/A
3	How the evacuation should be carried out	YES	NO	N/A
4	Where people should assemble after leaving the premises	YES	NO	N/A
5	How the premises will be checked to ensure they have been vacated	YES	NO	N/A
6	Identification of key escape routes and how to access them	YES	NO	N/A
7	Details of the firefighting equipment supplied	YES	NO	N/A
8	Names and duties of employees given specific responsibilities in case of fire	YES	NO	N/A
9	Arrangements for the safe evacuation of people identified as especially at risk, including those with disabilities and all visitors	YES	NO	N/A
10	Any processes, machines or power supplies that must be shut down	YES	NO	N/A
11	Specific arrangements for any high-risk areas	YES	NO	N/A
12	The means of calling the emergency services and who is responsible for doing so	YES	NO	N/A
13	Procedures for liaison with the emergency services on arrival and notifying them of any special risks such as flammable material stores	YES	NO	N/A
14	What training employees need and the arrangements for providing it	YES	NO	N/A

Specimen fire risk assessment

Organisation:	
Office address:	
Person in charge:	Assessment date:

Step 1: Information gathering			
Identification of fire hazards (hazards that could ignite fuel)			
No.	Item	Code	Comments
1	**Sources of ignition**		
	Workplace sources (specify):		
	Work equipment sources:		
	Electrical equipment		
	Static electricity		
	Friction		
	Fixed or portable space heaters		
	Boilers, kitchen equipment, cooking		
	Metal/metal impacts		
	Arson, vandalism		
	Hot work processes		
	Poor, obstructed ventilation		
	People sources:		
	Smokers (cigarettes, pipes, matches)		
	Other (specify):		
2	**Sources of fuel (anything that burns enough to fuel a fire)**		
	Flammable substances:		
	Gases (eg LPG, acetylene)		
	Liquids (eg all solvents)		
	Solids (eg wood, paper, rubber, plastics, foam)		
	Flammable parts of structure, fittings:		
	Shelving		
	Upholstered items, textiles		
	Stored materials:		
	Files, archives		
	Paper stocks and packaging		
	Supplies		
	Process stores inward		
	Finished goods		
	Waste, especially finely divided material including shredder waste		
3	**Sources of oxygen**		
	Ventilation system		
	Oxidising substances		
	Oxygen sources (eg cylinders and piped supplies)		

Specimen fire risk assessment (continued)

Step 2: Categorise the fire risk level for the premises	
Based on the foregoing hazards identified and the guidance in *Fire safety – an employer's guide* (HSE Books, ISBN 00 11 341229 0), page 44, these premises are classed as:	Comments to aid definitions
High risk	Where: Flammables are stored in quantity, there are unsatisfactory structural features Permanent or temporary work has the potential to cause fires There is significant risk to life in case of fire
Normal risk	Where: Fire is likely to remain confined or only spread slowly so as to allow people to escape The number present is small and layout is simple Premises have effective automatic warning and suppression systems
Low risk	Where: There is minimal risk to people's lives *and* The risk of fires starting is low *or* The potential for fire, heat and smoke spreading is negligible

Step 3: Identify those at risk		
Category	Numbers	Location
Employees (NOTE: A diagram or floor plan may be of assistance)		
Visitors to the premises		
Contractors		
Others		

Specimen fire risk assessment (continued)

Step 4: Evaluate the risks *(these questions can be answered using the checklists in this section)*				
1	What is the chance of a fire occurring?	HIGH MEDIUM LOW INSIGNIFICANT		Comments
2	Can the sources of ignition be reduced?	YES	NO	
3	Can the potential fuel for a fire be reduced?	YES	NO	
4	Can the sources of oxygen be reduced?	YES	NO	
5	Are the means of detecting a fire adequate?	YES	NO	
6	Can everyone be warned in case of a fire?	YES	NO	
7	Are the means of escape safe?	YES	NO	
8	Are the means available to put out a fire adequate (if it is safe for people to do so)?	YES	NO	
9	Are maintenance and testing arrangements adequate?	YES	NO	
10	Are regular fire safety checks held and adequately detailed?	YES	NO	
11	Are fire procedures and training adequate for the needs of the premises?	YES	NO	
12	Are existing precautions adequate for the remaining risks?	YES	NO	

Step 5: Recording of findings and actions required					
Significant hazards	Those who are at risk from the hazards	List all existing controls	What more needs to be done	Action required by *(Name and agreed date)*	Action completed *(Signed and dated)*

Fire risk assessment carried out by:	
Date:	
Date of next review:	

11 Construction safety

Introduction

Over the last 150 years, the construction industry has had an unenviable safety record. The Victorians instituted a massive programme of building and civil engineering with little thought for the safety of the huge workforce – as a result thousands were killed. We live now in more enlightened times, where the taking of risks at work and the exposure of non-employees to risk are seen as less acceptable than before. The European Union construction sector employs about 7 per cent of the total workforce, yet accounts for 15 per cent of all occupational accidents and 30 per cent of all fatalities in the industrial sector. Significant organisational changes in the industry itself have had consequences for accident prevention, the effects of which have probably balanced out improved attitudes.

Traditionally, a contractor would be engaged directly by a client in consultation with specialist advisers such as architects. The contractor would employ the majority of workers on a site full-time, with specialist subcontractors being hired in by him/her or directly by the client. Those who were self-employed would be taken on to make up numbers or to provide specialist skills.

Changes in taxation rules, among many other social and economic factors, have increased the numbers of self-employed on sites to the extent that the majority of workers in the industry now regard themselves as such, although whether the HSE, Inland Revenue or personal injury lawyers would agree is another matter. (However, policy changes during 1996 and 1997 produced a reversal in the trend towards self-employment.) When the contractor employs fewer and fewer directly, there are more likely to be difficulties of control and, in turn, in maintaining quality, training and competence, as well as health and safety standards.

A variety of management systems – including project management by intermediaries, a lack of acceptance of responsibility for control, and the increasing complexity of design, materials and equipment – have all contributed to the industry's poor accident record.

For the most part, the hazards and technical solutions for hazard prevention and risk reduction are well known. A possible exception is the general lack of awareness (and data) on the occupational health record of the industry's workers, who are potentially exposed to the elements as well as a wide variety of hazardous substances. It seems that industry workers are aware of the dangers they face. In a 1991 survey, about half thought that their work posed a risk to their personal health and safety.

The Construction (Health, Safety and Welfare) Regulations were introduced in September 1996, replacing the majority of a venerable set of regulations (over 30 years old) which were not designed to anticipate the technological developments over the period or the considerable changes in management systems. The 1996 Regulations are the construction industry's version of the 1992 Workplace Regulations. Together with the Construction (Design and Management) Regulations 1994 (see below) and the Work at Height Regulations 2005, they are the UK transposition of the EU Temporary or Mobile Worksites Directive.

It could be expected that regulations designed for the needs of the modern industry will be more successful in controlling injuries than their predecessors. Statistics show that the injuries which happen today are largely the same as those of yesterday, and even of 50 years ago, and are happening for much the same reasons.

The need for further change in management and employee attitudes is well recognised and long overdue. Work by the HSE to regulate the 'software' aspects of the industry resulted in the Construction (Design and Management) Regulations 1994 (CDM). CDM is aimed at the organisation of and preparation for construction work as well as its execution, and places duties on those only marginally affected by previous safety legislation – such as the client or owner, and designers – as well as the traditional controllers of the process at site level.

The degree of trade union involvement in the industry is less than it once was. Consultation processes made available by CDM and later regulations should result in closer co-operation from the workforce.

The reader is referred to Part 4 for summaries of the legal controls on many aspects of site work, including CDM and consultation with employees, and to Part 1 for information on control techniques, which are as applicable to the construction industry as elsewhere. This Section will examine other aspects of the control of construction health and safety, after reviewing the available statistical evidence on causation. Readers from all industries are strongly recommended to become acquainted with CDM. Under it, almost every employer having construction, alteration or installation work done is potentially a 'client' as defined, and thus a duty holder. Also, the concepts it contains concerning competence have lessons for the wider selection of contractors of every sort.

The accident record

Studies by the HSE using the Labour Force Surveys (which send questionnaires to households) indicate that about half of all non-fatal reportable injuries in construction are actually reported to the HSE and local authorities. An improved picture can be drawn from a study of fatalities, which are less easy to ignore, although even these may be underreported because of poor diagnosis of ill health exposures, for example, and also where a significant time may elapse between the injury and the consequent death.

Commentary on the disparity between reality and what is reported is presented annually by the Health and Safety Commission (HSC). *Statistics 2000/01* states on page 9, paragraph 1.27: "These figures [for all-industry non-fatal

injuries] suggest that employers reported around 44% of the injuries that should have been reported under RIDDOR in 1999/2000." Reporting levels for the self-employed, though, suggest that fewer than 4 per cent of these are reported as required. The document contains a wealth of information, including extensive data on occupational health returns, covering all sectors of the UK economy.

Because of underreporting and the inevitable delays in data capture and presentation, the statistical information presented in this Section must be regarded only as a broad indication of the position.

Broadly, two workers are killed in the industry every week, and one member of the public is killed every six weeks by construction activities. Until the 'blip' in 2000/01, the annual fatality rate has remained roughly the same since the early 1990s.

Taking the long-term view, the accident rates overall are a quarter of those reported in the early 1960s, and less than a half of those in the early 1970s. This is thought to be only partly due to changes in employment patterns. Table 1 below illustrates the changes in the numbers of fatal injuries to employees (therefore, excluding the self-employed); again, these are

Table 1: Fatalities to employees in the construction industry, sample years between 1961 and 2002/03 inclusive

Year	Total
1961	272
1966	292
1971	156
1981	105
1990/91	96
1995/96	62
1998/99	47
1999/2000	61
2000/01	73
2001/02	60
2002/03	57

only partly due to changing numbers employed and patterns of employment.

Types of injury

The most common cause of fatalities in recent years has been the head injury, accounting for almost one third of the total. A marked reduction – about 25 per cent – in the overall head injury rate has been claimed since the introduction of the Construction (Head Protection) Regulations on 1 April 1990.

In most years, at least 40 per cent of all construction fatalities have been falls from a height. The figures for falls as a percentage of total fatalities and major injuries has been remarkably constant over past years (see Tables 2 and 3 opposite), justifying the attention now given to fall protection in its own right.

The information presented shows that the most serious construction injuries cluster into relatively few causal groups. This information is the basis for campaigns by the HSE and major employers. It is of interest that the recent rise in fatalities in the industry has not been paralleled by a corresponding rise in major injuries or over-three-day reportable injuries.

Accident causes

Canadian studies have shown that active involvement in safety management by the most senior levels in a construction company is directly correlated with reductions in numbers of accidents and injuries.

Knowledge of causation patterns provides a starting point for focusing preventive measures. Case studies and descriptions of accidents can be used to give information about prevention techniques – the HSE's publication *Blackspot – construction* is still recommended reading, although now out of print. It commented that in a sample studied, 90 per cent of fatalities were found to be preventable, and positive management action could have saved lives in 70 per cent of cases. The three worst task areas found by the study (75 per cent of all deaths) were maintenance (42 per cent), transport and mobile plant (20 per cent), and demolition/

dismantling (13 per cent) – each receives detailed treatment in this book.

Table 2 shows the distribution of causes of fatalities over the last seven years, considering all workers in the industry. Some activities, of course, are frequent sources of injury, but rarely result in a fatality – manual handling, for example. Others occur relatively infrequently, but when they do, there is a higher than usual chance of not surviving them – becoming trapped by collapse or overturning and electrocution are examples of this.

Occupations most at risk

Sample information on the occupations of those killed is shown in Table 4 (on page 136). While no data are available on the percentage distribution of occupations across the workforce, it can be seen that (a) some groups of workers in particular occupations are more at risk than others, and (b) some are at high risk relative to their assumed numbers (and may well be unaware of it). Most construction managers are surprised at the numbers of their colleagues killed and seriously injured each year. It is noteworthy that some occupations show a high figure for fatalities but a relatively low figure for reportable injuries for the same period. In 1997/98, for example, nine roofers were killed, while there were 120 major injuries and 164 over-three-day reportable injuries (nine fatalities in 193 injury incidents, about 1:21). In the same period, plumbers and glaziers suffered one fatality, 148 major injuries and 622 over-three-day reportable injuries (1 fatality in 771 injury incidents, 1:771). So for some trades it can (and should) be said that injury is comparatively rare but more likely to be fatal. The crucial point to be made is that we have no way of knowing any particular outcome in advance.

The total numbers for those killed or injured in some occupations can be expected to fluctuate with time and the amount of work available in the construction industry. We can assume, for example, that there are more bricklayers employed than steel erectors, but the two groups show about the same numbers killed.

Occupational health and hygiene

Traditionally, the construction industry's high level of injury-causing accidents has received the attention of enforcement, media publicity and management action. Arguably, the size of that problem has led to a neglect of the less tangible consequences of occupational hygiene and health problems, apart from well-publicised topics such as asbestos. Reports suggest that construction workers age prematurely due to hypothermia caused by working in the cold and wet. Respiratory diseases such as bronchitis and asthma are also thought to occur at above average levels in construction workers.

Experience with the implementation of COSHH in construction shows that there is little industry awareness of the principles of assessment, or significant appreciation of the risks to workers from substances brought onto the site – and especially from those created there. Also, there is said to be a disappointing response from the industry to the noise controls (mostly managerial action and measurement requirements) imposed by the Noise at Work Regulations 1989.

Controlling construction accidents

The successful control of hazards and risks in the construction industry depends on the same principles as in other industries. Specific organisational problems do exist, and need to be addressed, chiefly by recognising the need to appreciate health and safety matters at the very earliest stages – design, specification of materials, their delivery quantities and batch sizes, work planning and interaction between contractors, and between contractors and the public.

These matters are all covered by CDM, which requires specific control and planning to be done before and during the construction phase. Projects within the scope of CDM require health and safety plans before work starts, on which tenders will have been based.

Table 2: Percentage of construction fatal injuries by kind of accident, all workers

	1996/97	1997/98	1998/99	1999/00	2000/01	2001/02	2002/03
All falls from a height	50	50	47	48	40	43	40
Struck by moving vehicle	15	9	17	8	21	17	7
Struck by moving or falling object	14	19	15	28	12	17	18
Trapped by something collapsing or overturning	6	5	6	3	15	7	9
Electrical	9	9	4	10	4	5	11
Other	6	8	11	3	8	11	15
Total fatalities	66	58	47	61	73	60	57

Table 3: Percentage of construction major injuries by causation, employees only

	1996/97	1997/98	1998/99	1999/00	2000/01	2001/02	2002/03
All falls from a height	35	37	37	36	37	30	30
Slip, trip or fall on same level	19	19	20	21	21	26	26
Struck by moving vehicle	3	2	3	2	2	2	2
Struck by moving or falling object	21	20	18	18	18	18	17
Injured while lifting, handling or carrying	8	9	8	10	8	10	11
Contact with moving machinery	3	3	3	3	3	4	3
Other	11	10	11	10	11	10	11
Total major injuries	3,227	3,860	4,289	4,386	4,303	4,055	4,098

Table 4: Numbers of employees reported as killed in construction work over a 10-year period, in selected occupations

Occupation	Year												Total
	90/91	91/92	92/93	93/94	94/95	95/96	96/97	97/98	98/99	99/00	00/01	01/02	
Bricklayer	4	4	1	5	2	-	1	2	2	1	-	4	26
Carpenter/joiner	2	1	1	1	2	3	3	-	1	2	3	3	22
Electrician fitter	1	1	3	-	-	1	1	-	1	6	1	5	20
Painter	5	4	2	4	-	2	2	1	-	4	4	1	29
Driver	7	9	5	5	4	2	4	6	1	2	1	3	49
Management/ professional	9	6	7	8	2	5	-	1	1	4	4	7	54
Plumber/glazier	3	1	1	2	1	-	2	1	2	-	-	1	14
Steel erector	3	5	3	2	-	-	-	4	3	3	3	2	28
Scaffolder	5	4	6	2	2	4	5	3	1	1	6	3	42
Roofer	4	8	4	10	7	4	8	9	7	7	1	5	74

The plan, drawn up by a nominated planning supervisor, will be passed to the selected competent principal contractor, who has well-defined responsibilities for control of the safety of work on site, including the activities of other contractors. The plan will be revised and augmented by the principal contractor, in particular to incorporate the risk assessments provided by all contractors. Thus, it becomes a working document throughout the duration of the project. The health and safety file contains the detail of how the construction was done, for use of the client and future modifiers or demolishers of the structure concerned.

The Management of Health and Safety at Work Regulations 1999 (MHSWR) include requirements which have an impact on construction work. The CDM Regulations add further definition to the details for this industry. The reader will also find it instructive to review the Part 4 summary of the MHSWR from the aspect of construction activity.

Method statements covering high-risk operations should be obtained in advance from those doing the work. Each should contain the acknowledgement that, if operational conditions force any deviation from the method statement, this must be agreed by site supervision and accepted by them in writing. Tasks where method statements should always be used include demolition, the use of explosives, erection of steel and structural frames, deep excavations and tunnelling, lifting with more than one appliance, use of suspended access equipment, and falsework. They should also be considered for roofwork, especially if fragile roofs are identified. Roofwork method statements should detail access methods, work procedure and fall protection systems.

All method statements are plans for work which are based on risk assessments (See Part 1 Section 3), and will be incorporated into the site's health and safety plan where CDM applies.

Training in techniques and skills for workers should include a strong health and safety element. Management training is especially important; it fosters that positive commitment to health and safety and the positive safety culture within the organisation which is necessary before the construction process can be successfully managed.

Safety in excavations

Death in excavations is not uncommon. A cubic metre of earth weighs a tonne, and even those only partly buried can die because of pressure from below and the sides which results in suffocation. Next at risk are the rescuers.

A short review of excavation safety, just one of the many topics in construction work, gives the opportunity to show how organisation and planning is a key factor in ensuring health and safety of construction workers. The control of every hazard discussed below will be inadequate if there is little or no preplanning.

Lack of personal awareness and supervisory knowledge can only be combated with the training programmes and certification that the industry is beginning to push out beyond the major contractors.

The main hazards are:
- falls into the excavation
- falls of materials from the top of the excavation
- falls of the sides into the excavation
- getting in and out
- unanticipated undermining of adjacent structures
- unanticipated discovery of underground services
- lack of oxygen, or presence of fumes
- occupational health exposures.

Falls into excavations should be prevented by solid guardrails where a fall can occur. Points of public access should be protected in this way, regardless of the potential depth of fall. Preplanning must anticipate the demand for materials to achieve this.

Falls of materials can be prevented by insisting that spoil and materials are kept back from the edges and a gap 1m wide is left clear. Traffic routes must be planned to keep vehicles away from the edges, and anchored blocks are required to prevent delivery vehicles falling in as well as their materials.

Falls of the sides will occur sooner or later unless either the soil is verified by a geological survey as sound and without need of support, or the sides are sloped back at an appropriate angle (usually shallower than 60 degrees), or supporting material is provided under the supervision of a competent person. The work also needs inspection at regular intervals. It will be seen that in all cases preplanning is needed for each control – to provide the survey, the space or the equipment. Nobody exposed to the potential of collapse should be working unprotected at any time.

Safe access and egress is needed at least every 25m by means of a fixed ladder with a good handhold at the top.

Undermining is easily done, and can affect nearby foundations and traffic routes. Scaffolding that has been undermined will need special bracing support to retain load-bearing ability.

Underground services must be looked for on plans before excavation work begins. But plans are relatively unreliable and cable detectors should always be used to supplement their information. Where the presence of a cable or other service is known, digging by hand in the immediate vicinity should be insisted on rather than by machine.

Lack of oxygen is always a possibility in an excavation – it is a potential 'confined space'. Fumes and unwanted gases can also accumulate, so tests with a meter may be required. This should be noted in the risk assessment and method statement.

Occupational health exposures for workers in excavations include Weil's disease and tetanus, as well as contaminated soil as a result of previous industrial activities. Good protective clothing and washing facilities are essential, plus an awareness of the need for a high standard of personal hygiene.

Selected references

Legal

Construction (Design and Management) Regulations 1994
Construction (Head Protection) Regulations 1989
Construction (Health, Safety and Welfare) Regulations 1996 (as amended)
Work at Height Regulations 2005

Guidance (HSE)

HSG224	*Managing health and safety in construction: Construction (Design and Management) Regulations 1994* ACoP, ISBN 0 7176 2139 1
HSG150(rev)	*Health and safety in construction*, ISBN 0 7176 2106 5
HSG33	*Health and safety in roofwork*
HSG47	*Avoiding danger from underground services*
HSG66	*Protection of workers and the general public during development of contaminated land*
HSG144	*Safe use of vehicles on construction sites*
HSG151	*Protecting the public – your next move*
HSG168	*Fire safety in construction work*
HSG185	*Health and safety in excavations*

Designing for health and safety in construction, ISBN 0 7176 0807 7

British Standards

BS 6187:2000	Code of practice – demolition
BS 6913:2000	Operation and maintenance of earthmoving machinery. Parts 1 and 3–15

Other references

Construction safety manual (produced by the Construction Confederation)
Principles of construction safety, Allan St John Holt, Blackwell Science (2001), ISBN 0 632 05682 7

12 Demolition

Introduction

Research studies show that accidents during demolition work are more likely to be fatal than those in many other areas of construction work. Causes of accidents which have high potential for serious injury are premature collapse of buildings and structures, and falls from working places and access routes. Investigation of demolition accidents shows that there is usually a failure to plan the work sufficiently at an appropriate stage, which leaves operatives on site to devise their own methods of doing the work without knowledge and information about the dangers that confront them. Failure to plan is, of course, not confined to demolition or even to the construction industry, but it is difficult to think of many other situations where the consequences are visited so rapidly on the employees. For this reason, all construction work which includes an element of demolition falls within the scope of the Construction (Design and Management) Regulations 1994 (CDM).

Although there is no officially sanctioned definition of what constitutes 'demolition' for the purposes of CDM, or of what triggers the need for method statements, an unofficial rule of thumb which can be followed with some confidence is that demolition proper involves the taking down of load-bearing structures and/or the production of a substantial quantity of demolished material – about 5 tonnes as a minimum. This rule attempts a practical definition based on the level of risk attached to the work. A stricter interpretation would mean that the removal of any part of a structure could be classed as demolition work, and a result of its being followed would mean there would be very few projects which did not fall within the scope of CDM – clearly not what was intended.

Planning for safety

As much information should be obtained about the work to be done, at the earliest possible time. The extent to which a client is willing or able to provide structural information will partly depend on whether it is actually available to him or her. CDM requires the client to provide factual information about the state of the structure to be demolished, which he or she either already has or could find out by making reasonable enquiries. It should not be up to the contractor to discover a particular hazard of the structure or building to be demolished, although it is entirely possible that hazards which could not reasonably have been anticipated may reveal themselves during work. All parties to demolition work must remain alert to this possibility.

Demolition is normally carried out by specialist contractors with experience. Control of the actual work must take place under the supervision of a competent person – by people experienced in demolition – and this may not be achieved by a general contractor.

Information on storage and use of chemicals on a site due for demolition can usually be obtained from the owner, but the services of a competent analyst may be needed in cases where:

- the site has been vacant for some time
- previous ownership is unclear
- environmental contamination by a previous occupier could have occurred.

Architectural or archaeological items for retention must be defined before the preferred method of demolition is determined.

Demolition surveys

Using information supplied, prospective contractors should carry out a survey in sufficient detail to identify structural problems, and risks associated with flammable substances or substances hazardous to health. The precautions required to protect employees and members of the public from these risks, together with the preferred demolition procedure, should be set out in a **method statement** (see below).

The survey should:

- take account of the whole site; access should be permitted for the completion of surveys and information made available in order to plan the intended method of demolition
- identify adjoining properties which may be affected by the work – structurally, physically or chemically; premises which may be sensitive to the work, other than domestic premises, include hospitals, telephone exchanges and industrial premises with machines vulnerable to dust, noise or vibration
- identify the need for any shoring work to adjacent properties or elements within the property to be demolished; weatherproofing requirements for the work must also be noted
- identify the site's structural condition, as deterioration may impose restrictions on the demolition method.

Preferably, the survey should be divided into structural and chemical aspects, the latter noting any residual contamination. Structural aspects of the survey should note variations in the type of construction within individual buildings and among buildings forming a complex due for demolition. The person carrying out the survey should be competent. An assessment of the original construction method, including any temporary works required, can be very helpful. In time, all the necessary information should be available within the owner's health and safety file provided through the construction, design and management process at the completion of original construction or alteration.

Preferred method of work

The basic ideal principle is that structures should be demolished in the reverse order to their erection. The method chosen should gradually reduce the height of the structure or building, or arrange its deliberate controlled collapse so that work can be completed at ground level.

Structural information obtained by the survey should be used to ensure that the intended method of work retains the stability of the parts of the structure or building which have not yet been demolished. The aim should be to adopt methods which make it unnecessary for work to be done at height. If this cannot be achieved, then systems which limit the danger of such exposure should be employed. The use of balling machines, heavy duty grabs or pusher arms may avoid the need to work at heights. If these methods are possible, the contractor must be satisfied that sufficient space is available for the safe use of the equipment, and that the equipment is adequate for the job.

When work cannot be carried out safely from part of a permanent structure or building, working platforms can be used. These include scaffolds, towers and power-operated mobile work platforms. Where these measures are not practicable, safety nets or harnesses (properly anchored) may be used. The hierarchy of fall protection measures is set out clearly within the Construction (Health, Safety and Welfare) Regulations 1996, which is summarised in Part 4.

Causing the structure or building to collapse by the use of wire ropes or explosives may reduce the need for working at heights, but suitable access and working platforms may still be needed during the initial stages.

A knowledge of structural engineering principles is necessary to avoid premature collapse, especially an understanding of the effect of pre-weakening by the removal, cutting or partial cutting of structural members. Assessment of risks posed by hazards such as these is the responsibility of the designer(s) and the planning supervisor for the project.

Method statements

In order to comply with Section 2(2) of the Health and Safety at Work etc Act 1974 in relation to provision of a safe system of work, the production of a method statement is recognised as necessary for all demolition work. Because of the special demolition needs of each structure, an individual risk assessment must be made by the employer undertaking the work, in writing, to comply with Regulation 3 of the Management of

Health and Safety at Work Regulations 1999. The detailed method statement will be derived from the risk assessment carried out by the contractor. This must be drawn up before work starts, and communicated to all involved as part of the health and safety plan. It should identify the work procedure, associated problems and their solutions, and should form a reference for site supervision.

Method statements should be easy to understand, agreed by and known to all levels of management and supervision, including those of subcontracting specialists.

The method statement should include:

- the sequence of events and method of demolition or dismantling (including drawings/diagrams) of the structure or building
- details of personnel access, working platforms and machinery requirements
- specific details of any pre-weakening of structures which are to be pulled down, or demolished using explosives
- arrangements for the protection of personnel and public, and the exclusion of unauthorised people from the work area; details of areas outside the site boundaries which may need control during critical aspects of the work must be included
- details of the removal or isolation of electrical, gas and other services, including drains
- details of temporary services required
- arrangements for the disposal of waste
- necessary action required for environmental considerations (noise, dust, pollution of water, disposal of contaminated ground)
- details of controls covering substances hazardous to health and flammable substances
- arrangements for the control of site transport
- training requirements
- welfare arrangements appropriate to the work and conditions expected
- identification of people with special responsibilities for the co-ordination and control of safety arrangements.

Demolition techniques

Piecemeal demolition is done by hand, using hand-held tools, sometimes as a preliminary to other methods. Considerations include provision of a safe place of work and safe access/egress, and debris disposal. It can be completed or begun by machines such as balling machines, impact hammers or hydraulic pusher arms. Considerations for these include safe operation of the machines, clearances, capability of the equipment and protection of the operator.

Deliberate controlled collapse involves pre-weakening the structure or building as a preliminary, and completion by use of explosives or overturning with wire rope pulling. Considerations for the use of explosives include competence, storage, blast protection, firing programmes and misfire drill. Wire rope pulling requires a similar level of expertise, as well as selection of materials and clear areas for rope runs.

Demolition training

The importance of adequate training in demolition has been recognised by the introduction of the Construction Industry Training Board's 'Scheme for the Certification of Competence of Demolition Operatives'. Training requirements are imposed by the Construction (Health, Safety and Welfare) Regulations 1996, the Health and Safety at Work etc Act 1974, and specifically by the Management of Health and Safety at Work Regulations 1999. The selection of competent contractors to carry out the work, as required by CDM, will lead to the choice of those with demonstrable qualification(s), knowledge and/or experience. Possession of certificated formal training together with experience will satisfy the requirement.

Legal requirements

Several Acts and regulations control demolition work. The Health and Safety at Work etc Act 1974 applies whenever demolition work is done, as do the Management of Health and Safety at Work Regulations 1999 and the Manual Handling Operations Regulations 1992. The importance of CDM has already been mentioned in this Section. The Construction (Health, Safety and Welfare) Regulations 1996 apply to all demolition work. The Sections in Part 4 of this book dealing with those Regulations should be consulted for a summary. Other Acts and regulations which may be applicable include those covering COSHH, electricity, first aid, asbestos and lead. Laws covering the control of pollution and environmental protection may also apply.

The selected references given in the previous Section should be consulted, as the majority also apply to demolition.

Revision

6 **elements in planning for safety in demolition:**
- information
- demolition surveys
- preferred method of work
- method statements
- consultation
- training

2 **main demolition techniques:**
- piecemeal – by hand or machine
- deliberate controlled collapse, including pre-weakening, using explosives or overturning

Part 3
Occupational health
and hygiene

1 Introduction to occupational health and hygiene

Introduction

Occupational health services anticipate and prevent health problems which are caused by the work which people do. In some circumstances, the work may aggravate a pre-existing medical condition, and stopping this is also the role of occupational health. Health hazards often reveal their effects on the body only after the passage of time; many have cumulative effects, and in some cases, the way this happens is still not fully understood. Because the effects are often not immediately apparent, it can be difficult to understand and persuade others that there is a need for caution and control. Good occupational hygiene practice encompasses the following ideas:

- **recognition** of the hazards or potential hazards
- **quantification** of the extent of the hazard – usually by measuring physical/chemical factors and their duration, and relating them to known or required standards
- **assessment** of risk in the actual conditions of use, storage, transport and disposal
- **control** of exposure to the hazard, through design, engineering, working systems, the use of personal protective equipment and biological monitoring
- **monitoring** change in the hazard by means of audits or other measurement techniques, including periodic re-evaluation of work conditions and systems.

Historical development

There is evidence that the Greeks and Romans were aware of the hazards and risks to health, not to mention safety, in work activity, especially in the mining and extraction processes. Major milestones in the history of occupational health up to the beginning of the 20th century are as follows:

1526 Georg Bauer (in Latin texts Georgius Agricola) was appointed as physician to the miners of Joachimsthal. He recommended mine ventilation and the use of veils over faces to protect the miners from harmful dust. His treatise on mining of metals, including health aspects, *De re metallica*, was published in 1556.

1567 Von Hohenheim (in Latin texts Paracelsus) had a monograph published after his death on the lung diseases of miners and smelters.

1700 The Italian physician Bernardino Ramazzini wrote the first comprehensive document on occupational health, *De morbis artificium diatriba*, a history of occupational diseases. Working in Padua, he was the first to suggest that physicians should ask their patients about their work when diagnosing illness.

1802 The Health and Morals of Apprentices Act was passed in Great Britain, the world's first occupational health and safety legislation.

1831 Dr Charles Thackrah, an early UK occupational health pioneer, published *The effects of principal arts, trades and professions... on health and longevity*.

1898 Thomas Legge (knighted for his work in 1925) was appointed as first Medical Inspector of Factories.

Health hazards

Health hazards can be divided into four broad categories: physical, chemical, biological and ergonomic. Examples of the categories are:

- **physical** – air pressure, heat, dampness, noise, radiant energy, electric shock
- **chemical** – exposure to toxic materials such as dusts, fumes and gases
- **biological** – infection, eg tetanus, hepatitis and legionnaires' disease
- **ergonomic** – work conditions, stress, man–machine interaction.

Toxicity of substances

Toxicity is the ability of a substance to produce injury once it reaches a site in or on the body. The degree of harmful effect which a substance can have depends not only on its inherent harmful properties but also on the **route** and the **speed** of entry into the body. Substances may cause health hazards from a single exposure, even for a short time (**acute effect**), or after prolonged or repeated exposure (**chronic effect**). The substance may affect the body at the point of contact, when it is known as a **local agent**, or at some other point, when it is described as a **systemic agent**. **Absorption** is said to occur only when a material has gained access to the bloodstream and may consequently be carried to all parts of the body.

What makes substances toxic?
The effect a substance will have on the body cannot always be predicted with accuracy, or be explained solely on the basis of physical and chemical laws. The influence of the following factors combine to produce the **effective dose** (see Section 2):
- quantity and concentration of the substance
- the duration of exposure
- the physical state of the material, eg particle size
- its affinity for human tissue
- its solubility in human tissue fluids
- the sensitivity to attack of human tissue or organs.

Long- and short-term exposure
Some substances which are toxic can have a toxic effect on the body after only one single, short exposure. In other circumstances, repeated exposure to small concentrations may give rise to an effect. A toxic effect related to an immediate response after a single exposure is called an **acute effect**. Effects which result after prolonged exposure (hours or days or much longer) are known as **chronic effects**. 'Chronic' implies repeated doses or exposures at low levels; they generally have delayed effects and are often due to unrecognised conditions which are, therefore, permitted to persist.

Body response
The body's response against the invasion of substances likely to cause damage can be divided into external or **superficial** defences and internal or **cellular** defences. These

defence mechanisms interrelate, in the sense that the defence is conducted on a number of levels at once, and not in a stage-by-stage pattern.

Superficial mechanisms of defence
The superficial mechanisms work by the action of cell structures, such as organs and functioning systems.

The body's largest organ, the **skin**, provides a useful barrier against many foreign organisms and chemicals (but not against all of them). Its effect is, of course, limited by its physical characteristics. Openings in the skin, including sweat pores, hair follicles and cuts, can allow entry, and the skin itself may be permeable to some chemicals, such as toluene. The skin can withstand limited physical damage because of its elasticity and toughness, but its adaptation to cope with modern substances is usually viewed by its owner as unhelpful – dermatitis, with thickening and inflammation, is painful and prominent.

Defences against **inhalation** of substances harmful to the body begin in the **respiratory tract**, where a series of reflexes activate the coughing and sneezing mechanisms to expel forcibly the triggering substance. Many substances and micro-organisms are successfully trapped by nasal hairs and the mucus lining the passages of the respiratory system. The passages are also well supplied with fine hairs which sweep rhythmically towards the outside and pass along larger particles. These hairs form the **ciliary escalator**. The respiratory system narrows as it enters the lungs, where the ciliary escalator assumes more and more importance as the effective defence. In the deep lung areas, only small particles are able to enter the alveoli (where gas exchange with the red blood cells takes place), and cellular defence predominates there.

For **ingestion** of substances entering the mouth and **gastrointestinal tract**, saliva in the mouth and acid in the stomach provide useful defences to substances which are not excessively acid or alkaline, or present in great quantity. The wall of the gut presents an effective barrier against many insoluble materials. Vomiting and diarrhoea are additional reflex mechanisms which act to remove substances or quantities which the body is not equipped to deal with without damage to itself. Thus, there are a number of primitive defences, useful at an earlier evolutionary stage to prevent man unwittingly damaging himself, which are now available to protect against a newer range of problems as well as the old.

Eyes and **ears** are potential entry routes for substances and micro-organisms. The eyes prevent entry of harmful material by way of the eyelids, eyelashes, conjunctiva (the thin specialised outer skin coating of the eyeball), and by bacteria-destroying tears. The ears are protected by the outer shell or pinna, and the ear drum is a physical barrier at the entrance to the sensitive mechanical parts and the organ of hearing. Waxy secretions protect the ear drum and trap larger particles.

Other orifices may be invaded by micro-organisms. Generally acid environments, such as in the urethra, do not promote their growth. Sexual contact is the main source of exposure.

Cellular mechanisms of defence
The cells of the body possess their own defence systems.

Prevention of excessive blood loss from the circulation through blood clotting and coagulation prevents excessive bleeding and slows or prevents the entry of germs into the blood system.

Phagocytosis is the scavenging action of a defensive body cell (white blood cell) against an invading particle. A variety of actions can be used, including chemical, ingestion, enzyme attack and absorption.

Secretion of defensive substances is done by some specialised cells. Histamine release and heparin, which promotes the availability of blood sugar, are examples.

Inflammatory response can isolate infected areas, remove harmful substances by an increased blood flow to the area, and promote the repair of damaged tissue.

Repair of damaged tissue is a necessary defence mechanism, which includes the removal of dead cells, increased availability of defender cells, and replacement of tissue strength and soundness by means of temporary and permanent repairs, eg scar tissue.

Immune response is the ability to resist almost all organisms or toxins that tend to damage tissues. Some immunity is **innate**, such as the phagocytosis of organisms and their destruction by acid in the gut. In addition, the human body has the ability to develop extremely powerful specific immunity against invading agents. This is **acquired** immunity, also known as **adapting** immunity. Acquired immunity is highly specific, the resistance developing days or weeks after exposure to the invading agent.

Routes of entry
Substances harmful to the body may enter it by three main routes. These are:

Absorption – through the skin, including entry through cuts and abrasions, and the conjunctiva of the eye. Organic solvents are able to penetrate the skin, as a result of accidental exposure to them or by washing. Tetraethyl lead and toluene are examples.

Ingestion – through the mouth, which is generally considered to be a rare method of contracting industrial disease. However, the action of the main defence mechanisms protecting the lungs rejects particles and pushes them towards the mouth, and an estimated 50 per cent of the particles deposited in the upper respiratory tract and 12.5 per cent from the lower passages are eventually swallowed.

Inhalation – the most important route of entry, which can allow direct attacks against lung tissue which bypass other defences such as those of the liver. The lungs are very efficient in transferring substances into the body from the outside environment, and this is the way in for 90 per cent of industrial poisons.

Results of entry

Having gained entry into the body, substances can:

- cause diseases of the skin such as:
  - **non-infective dermatitis** – an inflammation of the skin especially on hands, wrists and forearms. This can be prevented by health screening, good personal hygiene, use of barrier creams and/or protective clothing
  - **scrotal cancer** – produced by workers' clothing, impregnated with a carcinogen such as mineral oil, rubbing against the scrotum. This can be prevented by substitution of the original substance, by use of splash guards, and by the provision of clean clothing and washing facilities for soiled work clothing
- cause diseases of the respiratory system such as:
  - **pneumoconioses** – resulting from exposure to dust which is deposited on the lung, such as metal dust and man-made mineral fibre. Other examples of these fibroses of the lungs are silicosis due to the inhalation of free silica, and asbestosis from exposure to asbestos fibres
  - **humidifier fever** – giving influenza-like symptoms and resulting from contaminated humidifying systems

- **legionnaires' disease** – from exposure to legionella bacteria
- cause **cancer** and **birth defects** – by encouraging cells to undergo fundamental changes by altering the genetic material within the cell. Substances which can do this are carcinogens, which cause or promote the development of unwanted cells as cancer. Examples are asbestos, mineral oil, hardwood dusts and arsenic. Teratogens cause birth defects by altering genetic material in cells in the reproductive organs, and cause abnormal development of the embryo. Examples are organic mercury and lead compounds. Mutagens trigger changes affecting future generations
- cause **asphyxiation** – by excluding oxygen or by direct toxic action. Carbon monoxide does this by competing successfully with oxygen for transport in the red cells in the blood
- cause **central nervous system disorders** – by acting on brain tissue or other organs, as in the case of alcohol eventually causing blindness
- cause **damage to specific organs** – such as kidneys and liver. An example is vinyl chloride monomer
- cause **blood poisoning** – and producing abnormalities in the blood, as in benzene poisoning, where anaemia or leukaemia is the result.

Revision

4 **main health hazards:**
- physical
- chemical
- biological
- ergonomic

6 **factors determine toxicity:**
- concentration
- duration
- physical state
- affinity
- solubility
- sensitivity

Acute effects are immediate responses to single short-term exposures

Chronic effects are long-term responses to prolonged exposures

5 **superficial responses to harmful substances:**
- respiratory tract
- mouth and gut
- skin
- eyes and ears
- other orifices

6 **cellular responses to harmful substances:**
- prevention of blood loss
- phagocytosis
- secretion of defensive substances
- inflammatory response
- repair of damaged tissue
- immune response

3 **main routes of entry:**
- absorption
- ingestion
- inhalation

Self-assessment questions

1 List the possible sources of health hazards in your workplace.

2 Operatives in your workplace report general discomfort when working with a material. Discuss how you would assess the problem.

3 Describe how the body can defend and repair itself when the skin is cut.

4 Give examples of reflexes which take part in the body's response to the presence of foreign substances.

5 'Ingestion of toxic chemicals is a rare method of contracting industrial disease.' Discuss this statement, giving examples from your own experience.

6 Outline the measures which can be taken to prevent outbreaks of dermatitis.

2 Occupational exposure limits

Introduction

An important part of an occupational hygiene programme is the measurement of the extent of the hazard. This is generally done by measuring physical and/or chemical factors, including exposure duration, and relating them to occupational hygiene standards. Authorities in several countries publish recommended standards for airborne gases, vapours, dusts, fibres and fumes. The two primary (English language) sources are:

The HSE in the United Kingdom, which publishes occupational exposure limits (OELs) annually and as necessary. The new UK standard specifies workplace exposure limits (WELs).

The American Conference of Governmental and Industrial Hygienists (ACGIH), which publishes a list of threshold limit values (TLVs) annually. The Occupational Safety and Health Administration (OSHA) publishes national standards based on recommendations from the National Institute of Occupational Safety and Health (NIOSH).

Guidance Note EH40

HSE Guidance Note EH40 is published annually, reproducing the current list of WELs. The previous system of exposure classification and regulation made use of OELs which were set as maximum exposure limits (MELs) and occupational exposure standards (OESs). These were introduced in 1989 at the commencement of the original COSHH Regulations. The Control of Substances Hazardous to Health (Amendment) Regulations 2004 replaced these two types of OEL with the WEL, with effect from 6 April 2005. This was done because of a perceived lack of understanding of the previous system. Some OESs have not been converted into WELs because of a belief that the original OES limits were not appropriate for the protection of health. On the other hand, the absence of a WEL for a substance does not mean that it is inherently safe. Exposure should always be controlled to a level to which nearly all the working population can be exposed regularly without risks to health. To achieve this, it may be necessary to establish an in-house working standard.

WELs

These are OELs set under COSHH, which are concentrations of hazardous substances in the air, averaged over a specific time period – a time-weighted average or TWA. Two time periods are used: long-term (8 hours) and short-term (15 minutes). Short term limits or STELs are set for substances when even a short exposure to them can produce harmful effects on the human body.

Layout of EH40

The document contains two important sets of tables:

Table 1: Approved WELs
Table 2: Biological monitoring guidance values.

These lists are legally binding; COSHH imposes requirements by reference to them, which are the only official record of substances assigned WELs. EH40 also contains notes on the application of OELs, a commentary on monitoring exposure, calculation methods and useful material on mixed exposures to more than one substance at a time, and has an up-to-date list of references.

Threshold limit values (US)

The TLV is a concentration of an airborne substance and represents conditions under which it is believed that nearly all workers may be repeatedly exposed day after day without adverse health effects. There are different types of TLV:

Time weighted average TLV (TLV-TWA) – limits for indefinitely continued exposure eight hours a day, five days a week.

Short-term exposure level TLV (TLV-STEL) – maximum concentrations of contaminant in air, beyond which the worker should not be exposed for more than a continuous exposure time period of 15 minutes.

Ceiling TLV (TLV-C) – this converts the TLV-TWA into a value not to be exceeded at any time.

All values quoted are for inhalation and the units are milligrams per cubic metre (mgm^{-3}) or parts per million (ppm). Skin absorption is also denoted by 'SK'. TLVs were discontinued in Great Britain because of disagreement with ACGIH over some values, and the requirement of ACGIH that the table of TLVs must be published as a whole or not at all. For most practical purposes, there is no difference between TLVs and OESs. It should be remembered that all occupational exposure limits refer to healthy adults working at normal rates over normal shift durations and patterns. In practice, it is advisable to work well below the standards set, and to bear in mind the desirable goal of progressive risk reduction over time.

Revision

2 **UK standards:**
   - WELs (since April 2005)
   - 8-hour TWA and 15-minute STEL

3 **US standards:**
   - TLV-TWA
   - TLV-STEL
   - TLV-C

Selected references

Legal
Control of Substances Hazardous to Health Regulations 2002 (as amended)
Ionising Radiations Regulations 1999
Control of Asbestos at Work Regulations 2002
Control of Lead at Work Regulations 2002

Guidance (HSE)
EH40/year   *Occupational exposure limits*
EH64   *Occupational exposure limits – criteria document summaries*

Other
American Conference of Governmental and Industrial Hygienists Incorporated: *Threshold limit values (for chemical substances and physical agents) and biological exposure indices*

Self-assessment questions

1 What are the sources of occupational exposure limits?

2 Choose two substances found in your workplace air as contaminants and look up their WELs.

3 Environmental monitoring

Introduction
The key to preventing exposure to substances which could be hazardous to health depends on the first two steps mentioned in Section 1 of this Part – recognition of the hazard or potential hazard, and evaluation of the extent of the hazard. People in the workplace may encounter hazards from several sources. An important means of evaluation is measurement to determine the extent of the threat.

Some useful definitions
Dusts are solid particles suspended in air, which will settle under gravity. They are generated usually by mechanical processes including crushing and grinding. They can be of organic or inorganic origin. Particle size lies between 0.5 and 10 microns.

Fumes are solid particles formed by condensation from the gaseous state, eg metal oxides from volatilised metals. They can flocculate and coalesce. Their particle size is between 0.1 and 1 micron.

Mists are suspended liquid droplets formed by condensation from the gaseous state or by break-up of liquids in air. They can be formed by splashing, foaming or atomising – their particle size lies between 5 and 100 microns. Fogs are fine mists comprised of suspended liquid droplets at the lower end of the particle size range.

Vapours are the gaseous forms of substances which are normally in the solid or liquid state, eg sodium vapour in luminaires. They are generated by decrease from normal pressure or temperature increase.

Gases are any substances in the physical condition of having no definite volume or shape, but tending to expand to fill any container into which they are introduced.

Aerosol is a term used to describe airborne particles which are small enough to float in air. These can be liquids or solids.

Smoke contains incomplete combustion products of organic origin, the particle sizes of which range between 0.01 and 0.3 microns.

NB: A **micron** is a unit of length corresponding to one millionth of a metre. The equivalent SI unit is micrometre (µm).

Measurement – which technique?
As discussed in Part 3 Section 1, the health effects of exposure to toxic substances can be acute or chronic. It will, therefore, be necessary to distinguish appropriate types of measurement:

- long-term measurements which assess the average exposure of a person over a given time period
- continuous measurements capable of detecting short-term exposure to high concentrations of contaminants which can cause an acute effect
- spot readings can be used to measure acute hazards if the exact point of time of exposure is known and the measurement is taken at that time; chronic hazards may be assessed if a statistically significant number of measurements are made.

Measurement techniques
The more common air quality measuring techniques are:

Grab sampling – Stain detectors are used for measuring airborne concentrations of gases and vapours. Well known as a means of assessing alcohol consumption in roadside police checks, stain tubes are sealed glass tubes packed with chemicals which react specifically with the air contaminant being measured. In use, the tube is opened at the ends, a hand pump is attached and a standard volume of contaminated air is drawn through the tube. The chemical in the tube then changes colour in the direction of the airflow. The tube is calibrated so that the colour change corresponds to the concentration of the contamination.

The **main drawbacks** of grab sampling include inability to measure personal exposure by this method, except in the most general sense, and tube errors. These may arise because of the small volumes used, conditions such as temperature which affect some reactions in the tube chemical, and the possibility that the extent of the reaction may be influenced by the presence of other substances in the sampled air which may also cause a colour change in the tube. The accuracy of this method is not high, and it is best to use it to give a rough indication of the presence of a contaminant, with an estimate of the extent of contamination. It does not provide a time-weighted average result, and a single reading may not indicate a longer-term concentration.

Long-term sampling – This involves the taking of air samples for several hours, thus giving the average concentration at which the contaminant is present throughout the sampling period. Sampling may be done by attaching equipment to the operator so as to sample air entering the breathing zone (personal sampling), or by measuring at different points in the workplace (static or area sampling). Long-term sampling can be done by the use of long-term stain detector tubes which are connected to a pump. This draws air through the tube at a predetermined constant rate. At the end of the sampling period, the tube is examined and produces a value for the average level of concentration during the period.

The **main drawbacks** are as before, except that the accuracy of measurement of the sampled volume of air is improved by the use of a pump taking small samples over a long period. For a few substances, direct measuring diffusion tubes are available, which do not require the use of a pump, as diffusion is the means by which the sample of contaminant is collected.

More accurate results are obtained from operator personal sampling by means of **charcoal tube sampling**. This involves drawing air through a tube containing activated charcoal which absorbs the contaminant. The tube is then analysed in a laboratory to find the airborne concentration of the contaminant. Diffusion badges or monitors containing absorbents are becoming widely used. No pump is required; the method is reliable, versatile and accurate. Analysis is normally done in a specialist laboratory, so personal sampling generally suffers from delay in obtaining the results compared with the use of stain detector tubes.

Dust sampling – The most widely used technique is simple dust filtration, involving the use of a small pump to suck a measured quantity of air through a filtering membrane over a period of time. The sampling head containing the membrane is then removed from the pump and analysed in a laboratory.

Direct monitoring – Some instruments available commercially, and also for hire, can produce an immediate quantitative analysis of the level of a particular contaminant, or even a qualitative analysis of the air sample as a whole. The results are available on a meter or chart recorder. Infrared gas analysers are the most common type. This method allows the detection of the presence of short-term peak concentrations of contaminant during the work period, which is useful for working out control methods.

Hygrometers – These are instruments used for the measurement of water vapour in air. Although everyone is aware when the atmosphere becomes humid, people's ability to estimate humidity accurately is not good. Hygrometers can be very useful in measuring the humidity and comfort of the working environment.

Measurement of other environmental hazards which may be encountered involves similar considerations and techniques, although the equipment required is usually more sophisticated. **Radiation** is measured by grab sampling using a Geiger counter; **microwave** energy can be measured by a meter as it is radiated, or its presence can be detected by the fluorescing of a rare earth in a vacuum tube – an ordinary fluorescent lighting tube can be used for the purpose if small enough. **Sound** energy is measured using a proprietary meter, as is **light** for quantity and colour temperature.

Measurement of wave and particle energy, including noise, normally requires special equipment, and special training to operate it correctly and to produce reliable and useful results. Simple equipment can be very useful in identifying the presence of a hazard, but should not be relied on totally in the development of controls.

An example is in the selection of personal hearing protection (see Part 3 Section 6), where although a simple meter can identify the broad extent of the hazard, an octave band analysis would be required in order to match the characteristics of the sound source with the attenuating capabilities of different hearing protection. Note that this matching of the protective equipment to the risk it will control and to the wearer is a requirement of the Personal Protective Equipment at Work Regulations 1992 (see Part 4 Section 12).

Interpreting the results
Interpretation of the results is a skilled task – it involves making judgements about the results and the norms and standards laid down. The interpretation will determine the control strategy (see the next Section).

Summary of techniques
For **chronic hazards** – continuous personal dose measurement, continuous measurement of average background levels, spot readings of contaminant levels at selected positions and times.

For **acute hazards** – continuous personal monitoring with rapid response, continuous background monitoring with rapid response, spot readings of background contaminant levels at selected positions and times.

For **analysis of whether an area is safe to enter** – direct-reading instruments. Qualitative and quantitative particle analysis can be carried out by direct-reading instruments, which are expensive.

Revision
3 main forms of environmental monitoring in the workplace:
* grab sampling
* long-term sampling
* direct monitoring

Selected references
Legal
Control of Substances Hazardous to Health Regulations 2002 (as amended)
Ionising Radiations Regulations 1999
Control of Asbestos at Work Regulations 2002
Control of Lead at Work Regulations 2002
and associated Approved Codes of Practice

Guidance (HSE)
HSG173   *Monitoring strategies for toxic substances*

Self-assessment questions

1 Explain the advantages of monitoring air quality standards using stain detector tubes.

2 Previously, you found exposure limit values for two substances present in your workplace air. Find out what methods are, or can be, used to measure how much of each is present.

4 Environmental engineering controls

Introduction

Having established a potential for injury in the workplace, selection of one or more control measures is necessary. An integral part of the effectiveness of the control is the monitoring of the controls once in place. It is important to remember that engineering and designing the problems out must be the primary consideration. The use of PPE is low in the list of control measures to be considered as a single solution. The **safety precedence sequence** applies to these control measures, placing the need for human intervention lower on the scale of acceptability than 'hardware' solutions.

The following control measures can be used, in descending order of efficacy and priority:

- substitution
- isolation
- enclosure of the process
- local exhaust ventilation
- general (dilution) ventilation
- good housekeeping
- reduced exposure time
- training
- personal protective clothing and equipment
- welfare facilities
- health surveillance.

The use of warnings (such as signs) may be regarded as an aid to these controls, not as a substitute. As they depend on the correct action being taken in response to the warning, they are not effective unless combined with other measures. It may be necessary to include warnings in order to comply with national or local regulatory requirements.

Types of control

Substitution of safer alternatives in procedures or materials is the first stage in the review of existing processes and procedures.

Isolation and enclosure of the process can be achieved by the use of physical barriers, or by relocation of processes and/or facilities.

Local exhaust ventilation (LEV) is achieved by trapping the contaminant close to its source, and removing it directly by purpose-built ventilation prior to its entry into the breathing zone of the operator or the atmosphere.

LEV systems have four major parts, all of which must be efficiently maintained:

- hood – the collection point
- ducting – to transport the contaminant away
- air purifying device – eg charcoal filters to prevent further pollution
- fan – the means of moving air through the system.

The efficiency of LEV systems is affected by draughts, capture hood design and dimensions, air velocity achieved, and distance of capture point from the source. A major design consideration is that sucking air is very inefficient as an alternative to blowing it into a capture hood.

General or dilution ventilation uses natural air movement through open doors or assisted ventilation by roof fans or blowers to dilute the contaminant. It should only be considered if:

- there is a small quantity of contaminant
- the contaminant is produced uniformly in the area
- the contaminant material is of low toxicity.

Good housekeeping lessens the likelihood of accidental contact with a contaminant. It includes measures to anticipate and handle spillages and leaks of materials, and minimising quantities in open use.

Reduced exposure time to a contaminant may be appropriate, provided that the possible harmful effect of the dose rate is taken into account, ie high levels of exposure for short periods of time may be damaging.

Training should emphasise the importance of using the control measures provided, and give an explanation of the nature of the hazard which may be present together with the precautions which individuals need to take.

Personal protective clothing and equipment may be used where it is not possible to reduce the risk of injury sufficiently using the above control strategies. In that case, suitable PPE must be used (see Part 3 Section 6).

Welfare facilities allow workers to maintain good standards of personal hygiene, including regular washing and showering, and using appropriate clean protective clothing and equipment. The presence of adequate first aid and emergency facilities minimises the effects of exposure to hazards.

Health surveillance may detect early signs of ill health. In some cases, this can be carried out by supervisors trained to recognise the effects of exposure to workplace materials, or otherwise by the use of trained nursing and medical staff and facilities.

Failure of controls

Once a control strategy has been devised and introduced, it is tempting to assume that there will be no further problems. A brief study of the possible ways in which things can go wrong will show that there are two broad areas where problems can occur – in the introduction phase and as a result of change.

Inadequate initial design because of inappropriate choice of the type of control system, lack of consultations between designers, users and workers, failures to foresee future demands on the system, and failure to

consider the possible consequences of introducing the system (eg increased noise).

Inadequate installation because of incompetence, or lack of adequate instructions or specifications.

Incorrect use may be due to lack of training, inadequate supervision or poor ergonomic design – or all of these.

Inadequate maintenance resulting in blocked, damaged or removed parts, filters clogged or badly fitted, or worn fans.

Failure to anticipate changes, which may include:
* the process itself
* the materials used
* the workers and supervisors concerned
* work methods
* the local environment including operator adjustments and additions
* regulatory changes such as exposure limits and LEV changes.

Legal requirements
The reader's attention is drawn to the comprehensive requirements for risk assessments to establish the need for environmental control measures. These include the Management of Health and Safety at Work Regulations 1999, the Workplace (Health, Safety and Welfare) Regulations 1992 and the Control of Substances Hazardous to Health Regulations 2002 (COSHH), as well as the Health and Safety at Work etc Act 1974. Summaries of these can be found in Part 4 of this book.

Revision

11 **control strategies:**
- substitution
- isolation
- enclosure of the process
- local exhaust ventilation
- general (dilution) ventilation
- good housekeeping
- reduced exposure time
- training
- PPE
- welfare facilities
- medical surveillance

5 **causes of failure of controls:**
- inadequate initial design
- inadequate installation
- incorrect usage
- inadequate maintenance
- unanticipated change

Selected references

Legal

Control of Substances Hazardous to Health Regulations 2002 (as amended)
Ionising Radiations Regulations 1999
Control of Asbestos at Work Regulations 2002
Control of Lead at Work Regulations 2002
and associated Approved Codes of Practice

Guidance (HSE)

HSG37	*An introduction to local exhaust ventilation*
HSG54	*The maintenance, examination and testing of local exhaust ventilation*

Self-assessment questions

1 Differentiate between local exhaust ventilation and dilution ventilation.

2 Identify the various environmental control strategies employed in your workplace.

5 Noise and vibration

Introduction

Noise enables us to communicate, and can create pleasure in the form of music and speech. However, exposure to excessive noise can damage hearing. Noise is usually defined as 'unwanted sound', but in strict terms noise and sound are the same. Noise at work can be measured using a sound-level meter. Sound is transmitted as waves in the air, travelling between the source and the hearer. The frequency of the waves is the **pitch** of the sound, and the amount of energy in the sound wave is the **amplitude**.

How the ear works

Sound waves are collected by the outer ear and pass along the **auditory canal** for about 2.5cm to the **ear drum**. Changes in sound pressure cause the ear drum to move in proportion to the sound's intensity. On the inner side of the ear drum is the middle ear which is completely enclosed in bone. Sound is transmitted across the middle ear by three linked bones, the **ossicles**, to the **oval window** of the **cochlea**, the organ of hearing which forms part of the **inner ear**. This is a spirally wound tube, filled with fluid which vibrates in sympathy with the ossicles. Movement of the fluid causes stimulation of very small, sensitive cells with hairs protruding from them and rubbing on a plate above them. The rubbing motion produces electrical impulses in the **hair cells** which are transmitted along the auditory nerve to the **brain** which then interprets the electrical impulses as perceived sound. Hair cells sited nearest the middle ear are stimulated by high frequency sounds, and those sited at the tip of the cochlea are excited by low frequency. **Noise-induced hearing loss** occurs when the hair cells in a particular area become worn and no longer make contact with the plate above them. This process is not reversible, as the hair cells do not grow again once damaged.

How hearing damage occurs

Excessive noise energy entering the system invokes a protection reflex, causing the flow of nerve impulses to be damped and as a result making the system less sensitive to low noise levels. This is known as threshold shift. From a single or short duration exposure, the resulting temporary threshold shift can affect hearing ability for some hours, but recovery then takes place. Repeated exposure can result in irreversible permanent threshold shift. The following damage can occur as a result of exposure to noise:

Acute effects

- **acute acoustic trauma** from gunfire, explosions; usually reversible, affects the ear drum, ossicles
- **temporary threshold shift** from short exposures, affecting the cochlea
- **tinnitus** (ringing in the ears) results from intense stimulation of the auditory nerves, usually wears off within 24 hours.

Chronic effects

- **permanent threshold shift** from long duration exposure; affects the cochlea and is irreversible
- **noise-induced hearing loss** from (typically) long duration exposure; affects ability to hear human speech, irreversible, compensatable. It involves reduced hearing capability at the frequency of the noises that have caused the losses
- **tinnitus** may become chronic without warning, often irreversibly.

Presbycusis is the term for hearing losses in older people. These are thought to be due to changes as a result of ageing in the middle ear ossicles, which causes a reduction in their ability to transmit higher frequency vibrations.

Measurement of noise

The range of human hearing from the quietest detectable sound to engine noise at the pain threshold is enormous, involving a linear scale of more than 100,000,000,000 units. Measuring sound intensity on such a scale would be clumsy, and so a method of compressing it is used internationally. The sound intensity or pressure is expressed on a logarithmic scale and measured in **bels**, although as a bel is too large for most purposes, the unit of measurement is the **decibel** (dB). The logarithmic decibel scale runs from 0 to 160dB.

A consequence of using the logarithm scale is that an increase of 3dB represents a doubling of the noise level. If two machines are measured when running separately at 90dB each, the sound pressure level when they are both running together will not be 180dB, but 93dB. To establish noise levels on this scale, several different types of measurement are used.

Three weighting filter networks (A, B and C) are incorporated into sound-level meters. They each adjust the reading given for different purposes, and the one most commonly used is the **A-weighted dB**. This filter recognises the fact that the human ear is less sensitive to low frequencies, and the circuit attenuates or reduces very low frequencies to mimic the response of the human ear, and attaches greater importance to the values obtained in the sensitive frequencies. Measurements taken using the A circuit are expressed in dB(A).

In most workplaces and most types of work, noise levels vary continuously. A measurement taken at a single moment in time is unlikely to be representative of exposure throughout the work period, yet this needs to be known as the damage done to hearing is related to the total amount of noise energy to which the ear is exposed.

A measure called L_{Eq} – the **Continuous Equivalent Noise Level** – is used to indicate an average value over a period which represents the same noise energy as the total output of the fluctuating real levels. L_{Eq} can be obtained directly from a sound-level meter with an integrating circuit, which captures noise information at frequent timed intervals and recalculates the average value over a standard period, usually eight hours. It can also be calculated from a series of individual readings coupled with timings of the duration of each sound level, but this is laborious and relatively inaccurate.

Noise dose is a measure which expresses the amount of noise measured as a percentage, where eight hours at a continuous noise level of 90dB(A) is taken as 100 per cent. If the work method and noise output is uniform, and the dose measured after four hours is 40 per cent, then the likely 8-hour exposure will be less than 100 per cent. However, if the dose reading after two hours is 60 per cent, this will be an indication of an unacceptably high exposure.

$L_{EP,d}$ measures a worker's daily personal exposure to noise, expressed in dB(A). $L_{EP,w}$ is the measure of the worker's weekly average of the daily personal noise exposure, again expressed in dB(A).

Peak pressure is the highest pressure level reached by the sound wave, and assessments of this will be needed where there is exposure to impact or explosive noise. A meter capable of carrying out the measurement must be specially selected, because of the damping event of needle-based measuring which will consistently produce underreading. A similar effect can be found in 'standard' electronic circuitry.

Controlling noises

This can be achieved by:

Engineering controls – purchasing equipment which has low vibration and noise characteristics, and achieving designed solutions to noise problems including using quieter processes (eg presses instead of hammers), design dampers, making mountings and couplings flexible, and keeping sudden direction and velocity changes in pipework and ducts to a minimum. Operate rotating and reciprocating equipment as slowly as practicable.

Location and orientation – moving the noise source away from the work area, or turning the machine around.

Enclosure – surrounding the machine or other noise source with sound-absorbing material, but the effect is limited unless total enclosure is achieved.

Use of silencers – can suppress noise generated when air, gas or steam flow in pipes, or are exhausted to atmosphere.

Lagging – can be used on pipes carrying steam or hot fluids as an alternative to enclosure.

Damping – can be achieved by fitting proprietary damping pads, stiffening ribs or by using double skin construction techniques.

Screens – are effective in reducing direct noise transmission.

Absorption treatment in the form of wall applications or ceiling panels; these must be designed for acoustic purposes to have a significant effect.

Isolation of workers – in acoustically insulated booths or control areas properly enclosed, coupled with scheduling of work periods to reduce dose will only be effective where there is little or no need for constant entry into areas with high noise levels. This is because even a short duration exposure to high sound pressure levels will exceed the permitted daily dose.

Personal protection – by the provision and wearing of earmuffs or earplugs. This must be regarded as the last line of defence, and engineering controls should be considered in all cases. Areas where personal protective devices must be worn should be identified by signs, and adequate training should be given in the selection, fitting and use of the equipment, as well as the reasons for its use.

Choice of hearing protection

Hearing protection should be chosen to reduce the noise level at the wearer's ear to below the recommended limit for unprotected exposure. Selection cannot be based on A-weighted measurement alone, because effective protection will depend on the ability of the protective device to attenuate (reduce) the sound energy actually arriving at the head position. Sound is a combination of many frequencies (unless it consists of a pure tone), and it can happen that a particular noise against which protection is

required has a frequency component which is not well handled by the 'usual' protection equipment. Therefore, a more detailed picture of the sound spectrum in question should be made before selection, checking the results obtained by **octave band analysis** against the sound-absorbing (attenuation) data supplied by the manufacturers of the products under consideration.

Vibration

Vibration, or oscillation, can be thought of as movement about a central reference point. The human body is exposed to many sources of vibration, some involving the whole body – when driving a vehicle, for example. Of most common concern to the health and safety practitioner is **hand–arm vibration** (HAV), which is vibration from work processes and equipment transmitted to the hands and arms of users. Common examples of HAV causes include operating road drills and hand polishers, and holding materials being worked on by machines.

Regular, long-term exposure to HAV can cause damage to body parts including the bones, nerves and circulatory system, in turn resulting in disabling injury. The commonest consequence of exposure to HAV is **vibration white finger** (VWF). Symptoms of VWF may include painful episodes of numbness and whitening of the fingers, especially in cold weather, coupled with loss of grip strength and manual dexterity. HAV syndrome is a condition reportable under RIDDOR as a prescribed industrial disease.

Controlling vibration

This can be achieved by:
- eliminating the vibrating equipment by finding alternative work systems and methods
- substitution with equipment having better vibration performance, and adopting a procurement policy with performance specifications
- limiting exposure by better job design and job rotation
- maintenance procedures which ensure that worn parts and mountings are replaced before they introduce vibration, and which keep tools sharp and properly adjusted

- information and training given to employees to make them aware of the problem, the risk factors, how to report injury, health surveillance arrangements, the maintenance of good circulation, and the proper use of tools to as to reduce the forces applied to hold them
- health surveillance – for those identified as at risk, a supervised health surveillance programme is essential, including pre-employment checks.

Legal requirements

Specific legislation on noise in all places of work is contained in the Noise at Work Regulations 1989. At the time of writing, it is anticipated that these Regulations will be revoked and replaced with an updated version in early 2006. Part 4 contains a summary of the current law, and also a summary of the Personal Protective Equipment at Work Regulations 1992, which require PPE to be selected according to criteria established in the risk assessment. New equipment provided must also conform to EC standards as a result of the activation of the PPE Product Directive.

Specific duties concerning the control of vibration at work were introduced in July 2005. Part 4 Section 18 contains a summary of the new Regulations.

Revision

2 **types of noise-related damage:**
- acute – acute acoustic trauma, temporary threshold shift, tinnitus
- chronic – permanent threshold shift, noise-induced hearing loss, tinnitus

11 **noise control techniques:**
- engineering
- orientation/location
- maintenance
- enclosure
- silencers
- lagging

- damping
- screens
- absorption treatment
- isolation of workers
- use of PPE

6 **vibration control techniques:**
- elimination
- substitution
- limiting exposure
- maintenance procedures
- information and training
- health surveillance

Selected references

Legal

Noise at Work Regulations 1989
Control of Vibration at Work Regulations 2005
Reporting of Injuries, Diseases and Dangerous Occurrences Regulations 1995

Guidance (HSE)

L108	*Guidance on the Noise at Work Regulations 1989*
HSG88	*Hand–arm vibration*
HSG138	*Sound solutions: techniques to reduce noise at work*
INDG175(2)	*Control the risks from hand–arm vibration*

Self-assessment questions

1 Explain how noise-induced hearing loss can be caused by noise at work.

2 What general controls are used in your workplace to reduce noise exposure?

6 Personal protective equipment

Introduction

This Section discusses the methodology and practicalities of selecting and using PPE. For a review of the legal requirements, see Part 4 Section 12.

PPE has two serious general limitations. It does not eliminate a hazard at source, and it cannot be guaranteed to work for 100 per cent of wearers for 100 per cent of the time. If the PPE fails and the failure is not detected, the risk increases greatly. Where used, this equipment must be appropriately selected, and its use and condition monitored. Workers required to use it must be trained.

For a PPE scheme to be effective, three elements must be considered:

Nature of the hazard – details are required before adequate selection can be made, such as the type of contaminant and its concentration.

Performance data for the PPE – the manufacturer's information will be required concerning the ability of the PPE to protect against a particular hazard.

The acceptable level of exposure to the hazard – for some hazards, the only acceptable exposure level is zero. Examples are work with carcinogens and the protection of eyes against flying particles. Occupational exposure limits can be used, bearing in mind their limitations.

Factors affecting use

There are three interrelated topics to consider before an informed choice of adequate PPE can be made.

The workplace

What sorts of hazards remain to be controlled? How big are the risks which remain? What is an acceptable level of exposure or contamination? What machinery or processes are involved? What movement of objects or people will be required?

The work environment

What are the physical constraints? They can include temperature, humidity, ventilation, size, and movement requirements for people and plant.

The PPE wearer

Points to consider include:

Training – users (and supervisors) must know why the PPE is necessary, any limitations it has, the correct use, how to achieve a good fit, and the necessary maintenance and storage for the equipment.

Fit – a good fit for the individual wearer is required to ensure full protection. Some PPE is available only in a limited range of sizes and designs.

Acceptability – how long will the PPE have to be worn by individuals? Giving some choice of the equipment to the wearer without compromising on protection standards will improve the chances of its correct use.

Wearing pattern – are there any adverse health and safety consequences which need to be anticipated? For example, any need for frequent removal of PPE, which may be dictated by the nature of the work, may, in turn, affect the choice of design or type of PPE.

Interference – consideration is needed for the practicability of the item of equipment in the work environment. Some eye protection interferes with peripheral vision; other types cannot easily be used with respirators. Correct selection can alleviate the problem, but full consideration must be given to the overall protection needs when selecting individual items, so that combined items of equipment may be employed. For example, the 'Airstream' helmet gives respiratory protection, and has fitted eye protection incorporated into the design.

Management commitment – the *sine qua non* of any safety programme, required especially in relation to PPE because it constitutes the last defence against hazards. Failure to comply with instructions concerning the wearing of it raises issues of industrial relations and corporate policy.

Types of PPE

The types of PPE have different functions, including eye protection, hearing protection, respiratory protection, protection of the skin, and general protection in the form of protective clothing and safety harnesses and lifelines.

Hearing protection

There are two main forms of hearing protection – objects placed in the ear canals to impede the passage of sound energy, and objects placed around the outer ear to restrict access of sound energy to the outer ear as well as the ear drum and middle and inner ear. It should be noted that neither of these forms of protection will prevent a certain amount of sound energy reaching the organ of hearing by means of bone conduction effects in the skull.

Earplugs fit into the ear canal. They may be made from glass down, polyurethane foam or rubber, and are disposable. Some forms of re-usable plugs are available, but these are subject to hygiene problems unless great care is taken to clean them after use, and unless they are cast into the individual ear canal, a good fit is unlikely to be achieved in every case. Even though some plugs are available in different sizes, the correct size should only be determined by a qualified person. One difficulty is that a reasonable proportion of people have ear canals of different sizes.

Earmuffs consist of rigid cups which fit over the ears and are held in place by a head band. The cups generally have acoustic seals of polyurethane foam or a liquid-filled annular bag to obtain a tight fit. The cups are filled with sound-absorbing material. The fit is a function of the design of the cups, the type of seal and the tightness of the head band. The protective value of earmuffs may be lost almost entirely if objects such as hats or spectacles intrude under or past the annular seals.

Respiratory protective equipment (RPE)

There are two broad categories: respirators, which purify the air by drawing it through a filter to remove contaminants, and breathing apparatus, which supplies clean air to the wearer from an uncontaminated external source. Most equipment will not provide total protection; a small amount of contaminant entry into the breathing zone is inevitable.

Four main types of respirator are available:

Filtering half-mask – a facepiece covers the nose and mouth, and is made of a filtering medium which removes the contaminant; generally used for up to an 8-hour shift and then discarded.

Half-mask respirator – which has a rubber or plastic facepiece covering the nose and mouth, and which carries one or more replaceable filter cartridges.

Full-face respirator – covering the eyes, nose and mouth, and having replaceable filter canisters.

Powered respirator – supplies clean, filtered air to a range of facepieces, including full, half and quarter-masks, hoods and helmets via a battery-operated motor fan unit.

NB: Respirators do not provide **any** protection in oxygen-deficient atmospheres.

There are three main types of breathing apparatus which provide continuous air flow (in all cases the delivered air must be of respirable quality):

Fresh air hose apparatus – which supplies clean air from an uncontaminated source, pumped in by the breathing action of the wearer, by bellows or an electric fan.

Compressed airline breathing apparatus (CABA) – using flexible hosing delivering air to the wearer from a compressed air source. Filters in the airline are required to remove oil mist and other contaminants. Positive pressure continuous flow full-face masks, half-masks, hoods, helmets and visors are used.

Self-contained breathing apparatus (SCBA) – in which air is delivered to the wearer from a cylinder via a demand valve into a full-face mask. The complete unit is usually worn by the operative, although cylinders can be remote and connected by a hose.

To make proper selection of any type of RPE, an indication is needed of its likely efficiency, when used correctly, in relation to the hazard guarded against. The technical term used in respiratory protection standards to define the equipment's capability is the **nominal protection factor** (NPF). For each class of equipment, it is the total inward leakage requirement set for that class, and therefore does not vary between products meeting the class standards. Manufacturers will supply information on the NPF for their product range – it is the simple ratio between the contaminant outside the respirator (in the ambient air) and the acceptable amount inside the facepiece. Examples of typical NPFs are:

Disposable filtering half-mask respirator	4.5 to 50
Positive pressure-powered respirator	20 to 2,000
Ventilated visor and helmet	10 to 500
SCBA	2,000
Mouthpiece SCBA	10,000

The NPF can be used to help decide which type of RPE will be required. It is necessary to ensure that the concentration of a contaminant inside the facepiece is as far below the OEL as can reasonably be achieved. The required protection factor to be provided by the RPE will be expressed as:

$$\frac{\text{measured ambient concentration}}{\text{occupational exposure limit}}$$

and this value must be less than the NPF for the respirator type under consideration. Other limitations may apply, including restrictions on the maximum ambient concentration of the contaminant based on its chemical toxicity, and its physical properties such as the lower explosive limit.

In 1997, the revised BS 4275 presented a new approach to the efficiency issue, aimed at dispensing with the theoretical calculations of NPF. Instead, the term **assigned protection factor** (APF) was introduced to describe the level of protection that can reasonably be achieved in the workplace, given that workers have been appropriately trained and the RPE is properly fitted and working. Used in the same way as the NPF, the APF provides a higher margin of safety. A full discussion of the use of the APF is beyond the scope of this book. Guidance on best practice and the setting up of an RPE programme can be found in BS 4275, to which the reader is referred.

Eye protection

Assessment of potential hazards to the eyes and the extent of the risks should be made in order to select equipment effectively. There are three types of eye protection commonly available:

Safety spectacles/glasses – which provide protection against low-energy projectiles, such as metal swarf, but do not assist against dusts, are easily displaced, and have no protective effect against high-energy impacts.

Safety goggles – to protect against high-energy projectiles and dusts. They are also available as protection against chemical and metal splashes with additional treatment. Disadvantages include a tendency to mist up inside (despite much design effort by manufacturers), lenses which scratch easily, limited vision for the wearer, lack of protection for the whole face, and high unit cost. Filters will be required for use against non-ionising radiation (see Part 3 Section 8).

Face shields – offering high-energy projectile protection, also full-face protection and a range of special tints and filters to handle various types of radiation. The wearer's field of vision may be restricted. The high initial cost of this equipment is a disadvantage, although some visors allow easy and cheap replacement of shields. Weight can be a disadvantage, but this is compensated by relative freedom from misting up.

Protective clothing

A range of protective clothing provides body protection against a range of hazards, including heat and cold, radiation, impact damage and abrasions, water damage and chemical attack.

Head protection – is provided by two types of protectors: the safety helmet, and the scalp protector (also known as the 'bump cap'), which is usually brimless. Their function is to provide protection against sun and rain, and against impact damage to the head. The ability of the scalp protector to protect against impacts is very limited, and its use is mainly to protect against bruising and bumps in confined spaces. It is not suitable for use as a substitute for a conventional safety helmet. Safety helmets have a useful life of about three years, which can be shortened by prolonged exposure to ultraviolet light and by repeated minor or major impact damage.

Protective outer garments – are normally made of PVC material and often of high-visibility material to alert approaching traffic. PVC clothing can be uncomfortable to wear because of condensation, and vents are present in good designs. Alternatively, non-PVC fabric can be used, which allows water vapour to escape, but garments made from this material are significantly more expensive.

Protective indoor garments – such as overalls and coats are made of polycotton, and some makes are disposable. If overalls are supplied, arrangements for cleaning must be made to prevent unhygienic conditions developing if the clothing is worn in circumstances where oils or chemicals are handled. Failure to keep the clothing clean and changed regularly may result in dermatitis or skin cancer formation. Aprons and over-trousers should be fire-resistant, and trousers worn during cutting operations require protection in the form of ballistic nylon or similar material. Clothing may limit movement, and become entangled in machinery – careful

selection of type and manufacturer is required, together with necessary training about its proper use. This may involve rules concerning the buttoning of coats in the vicinity of rotating machinery. Wearing anti-static clothing is of major importance in reducing static electricity effects – local rules should be strictly followed.

Gloves – must be carefully selected, taking account of use requirements such as comfort, degree of dexterity required, temperature protection offered, and ability to grip in all conditions likely to be encountered – these factors being weighed against considerations of cost and the hazards likely to be encountered by the wearer. Resistance to wearing gloves can be found; it is often claimed that gloves impede work by reducing sensitivity in the fingers, and some users find that excessive sweating causes the gloves to become damp. It is important to change gloves frequently where solvents and fuels are handled because they permeate many protective materials quickly. Gloves should not be used for longer than half the indicated breakthrough time.

The main types of material and their features are shown in Table 1 overleaf.

Footwear – is designed to provide protection for the feet, especially for the toes, if material should drop or fall to the ground. It should also protect against penetration from beneath the sole of the foot, be reasonably waterproof, provide a good grip, and be designed with reference to comfort. Steel toecaps are inflexible, and it is important to purchase the right size of footwear. Electrical insulation can be assisted by the correct footwear for the circumstances, and anti-static conducting shoes are essential where static effects need to be eliminated.

Skin protection

Where protective clothing is not a practicable solution to a hazard, barrier creams may be used together with a hygiene routine before and after work periods. There are three types of barrier cream commonly found: water miscible, water repellent and special applications.

Table 1: Good glove guide

Material	Good for
Leather	Abrasion protection, heat resistance
PVC	Abrasion protection, water and limited chemical resistance, good for people with latex allergy
Rubber	Degreasing, paint spraying
Cloth/nylon, latex coated	Hand grip
Natural rubber latex	Electrical insulation work (but this material is a recognised allergy risk)
Nitrile rubber, 0.4mm or thicker	Resistance to liquid fuels and solvents, most paints (but not toluene and acetate-based products), rust inhibitors
Chain mail	Cut protection

Safety harnesses

These are not replacements for effective fall prevention practices. Only where the use of platforms, nets or other access and personal suspension equipment is impracticable is their use permissible. The functions of belts and harnesses are to limit the height of any fall, and to assist in rescues from confined spaces. In addition to comfort and freedom of movement, selection of this equipment must take into account the need to provide protection to the enclosed body against energy transfer in the event of a fall. Because of this, harnesses are preferable to belts except for a very limited number of applications where belts are required because of the movement needs of the work.

Harness attachments to strong fixing points must be able to withstand the snatch load of any fall. A basic principle is to attach the securing lanyard to a fixing point as high as possible over the area of the work, so as to limit the fall distance. Similarly, a short lanyard should be provided. Equipment which has been involved in arresting a fall should be thoroughly examined before further use, according to the manufacturer's instructions.

The CE mark

One of the definitions of 'suitable' PPE is that it must conform to legal requirements. Workwear and other PPE which meets the relevant strict rules for products must be sold carrying the mark 'CE'. This stands for 'Conformité Européenne', and the letters show the item complies with necessary design and manufacturing standards based on several levels and classes of protection. The same rules apply throughout the European Union, so that PPE which complies and bears the 'CE' mark can be sold in any member state, as part of the harmonisation of standards. There are three categories or levels of 'conformance' within the 'CE' mark system – basically, equipment is designated as being suitable for protection against low risks, medium risks and high risks. In some cases, PPE in the higher protection categories is subject to examination by an approved body before the mark can be displayed, and the category details will be displayed alongside the mark.

Workwear in the first **simple** category will protect against, for example, dilute detergents. Gloves capable of handling items at less than 50 degrees Celsius will also be in this category, and equipment capable of protecting against minor impacts. The manufacturer of products in this category can assess his/her own product and award himself the 'CE' mark to be shown on the equipment, but must keep a technical file.

In the second **intermediate** category, the manufacturer must submit a technical file for review, which holds details of the product's design, and claims made for it, and details of the quality system used by the manufacturer. The product itself is liable to be tested to ensure it can do what is claimed for it.

The third high risk or **'complex design'** category mostly covers hazards where death is a potential outcome of exposure. Products which can protect against limited chemical attack, ionising radiation protection, and respiratory equipment protecting against asbestos are examples of products in this group. Manufacturers must comply with all the requirements for the second category, and also submit to an EC examination annually or have a recognised quality system in place, such as ISO 9000.

Legal requirements
The general duties sections of the Health and Safety at Work etc Act 1974 require a safe place of work to be provided together with safe systems of work, and these may involve use of PPE. General requirements for the provision, use, maintenance and storage of PPE are contained in the Personal Protective Equipment at Work Regulations 1992. Specific Acts, regulations and orders also contain requirements for PPE, and should be consulted for particular applications.

Revision

7 **usage factors:**
  * fit
  * period of use
  * comfort
  * maintenance
  * training
  * interference
  * management commitment

6 **types of PPE:**
  * hearing protection
  * RPE
  * eye/face protection
  * protective clothing
  * skin protection
  * safety harnesses

Selected references

Legal

Personal Protective Equipment at Work Regulations 1992
Control of Substances Hazardous to Health Regulations 2002 (as amended)
Noise at Work Regulations 1989
Construction (Head Protection) Regulations 1989
Control of Asbestos at Work Regulations 2002
Control of Lead at Work Regulations 2002

Guidance (HSE)
L25   *Personal protective equipment – guidance on Regulations*
HSG53   *RPE – selection, use and maintenance*

British Standards
BS EN 360–363:1993   PPE against falls from a height – retractable type fall arresters
BS EN 397:1995   Specification for industrial safety helmets
BS 4275:1997   Recommendations for selection, use and maintenance of RPE
BS EN 352   Hearing protectors
BS 7028:1999   Eye protection for industrial and other uses
BS EN 340:2003   Protective clothing. General requirements

Self-assessment questions

1 An oxygen-deficient atmosphere has been established. Discuss the forms of RPE which should be used for (a) continuous work in the area and (b) short-term work only.

2 List the PPE used in your workplace. Can you identify its limitations and suggest alternative controls?

7 Asbestos

Introduction

Asbestos is the name given to a group of naturally occurring mineral silicates, which are grouped because of their physical and chemical similarity and their consequent general properties. Asbestos is strong, inert, resilient and flexible – and, therefore, almost indestructible. It has been used in a wide range of products requiring heat resistance and insulation properties. In humans it has been claimed with varying degrees of certainty to be a factor in the causation of asbestosis, lung cancer, cancers of the stomach, intestines and larynx, and mesothelioma. There is no safe level of exposure to any form of asbestos; in most cases, long periods of time pass before any effect on an individual can be diagnosed.

Asbestos produces its effects more because of the size, strength, sharpness and jagged shape of the very small fibres it releases than as a result of its chemical constituents, although these do also have relevance. The health hazards arise when these small fibres become airborne and enter the body, and when they are swallowed. The body's natural defence mechanisms can reject large (visible) dust particles and fibres, but the small fibres reaching inner tissues are those that are both difficult to remove and the most damaging. They are particularly dangerous because they cannot be seen by the naked eye under normal conditions, and they are too small (less than 5 microns in length) to be trapped by conventional dust-filter masks.

Asbestos is normally encountered in the demolition or refurbishment processes, but even simple jobs such as drilling partitions or removing ceiling tiles can disturb it. It is important to be aware that asbestos is normally present in a mixture containing a low percentage of asbestos in combination with filling material. Therefore, neither the colour nor the fibrous look of a substance is a good guide. The only reliable identification of the presence of asbestos is by microscopic analysis in a laboratory. The alternative to this is to adopt a default position of assuming that blue asbestos is present (which requires the strictest controls) and proceeding accordingly.

There has been much debate over whether it is better to remove asbestos wherever it is found as a matter of course, or to leave that which is in good condition in place, recorded and monitored. Arguments other than safety, health and environmental ones have a place in the risk debate. Cost is an issue, as is the fact that not all the fibres can ever be controlled when asbestos is removed and the background level of asbestos in air would rise.

Common forms of asbestos

Crocidolite – usually known as blue asbestos. The colour of asbestos can only be an approximate guide, because of colour washes and coatings following application, and also because the colour changes due to heat, either in the manufacturing process or in use, or following application of a product or chemical action on it. This is the type of asbestos considered by some to be the most dangerous, linked with a high cancer risk and particularly with mesothelioma – a specific form of cancer affecting the lining of the lungs and occasionally the stomach. The condition can take up to 25 years to develop following exposure. As with other forms of asbestos-induced cancer, even a minimal exposure may trigger the disease, which is invariably fatal.

Crocidolite has been banned in practice in the UK since 1969 but is still found in boiler lagging, piping, older insulation boards and sheeting. Asbestos installed before 1940 has a high chance of containing a measurable percentage.

Amosite – brownish in colour unless subjected to high temperatures, this form of asbestos is now considered as dangerous as crocidolite. In 1984, it was made subject to the same control limits.

Chrysotile – white asbestos and the most common form, found in cement sheets, older 'Artex' ceiling compounds, lagging and many other products, and, with all other forms of asbestos, now banned from use. Prolonged exposure causes asbestosis, which is a gradual and irreversible clogging of the lungs with insoluble asbestos fibres and the scar tissue produced by the body's defence mechanisms trying to isolate the fibres. Even short periods of exposure have been known to trigger lung cancer, especially in smokers.

Less common substances which fall within the legal definition of asbestos are fibrous actinolite, fibrous anthophyllite and fibrous tremolite (and any mixture containing any of the above).

Legal requirements

The information provided here is intended only as a summary and guide to the requirement of the Regulations, which must be studied together with the relevant ACoPs to obtain an authoritative interpretation of the requirements.

The importation, supply and use (and re-use) of all types of asbestos has now been banned. The main direct legislation dealing with work with asbestos is the Control of Asbestos at Work Regulations 2002 (CAWR). Anyone who has to handle asbestos to any extent should obtain a copy of the current edition of the ACoP. Work on land contaminated by asbestos is not covered by an ACoP.

In simple terms, most work involving asbestos-containing material requires the contractor doing the work to hold a **licence** issued by the HSE under the Asbestos Licensing Regulations 1983 (as amended). Any mixture containing any of the above substances, regardless of the quantity or proportion, falls within the definition of 'asbestos' and, therefore, within the scope of the Regulations. The purpose of the mixture is not relevant, so that, for example, insulating board containing asbestos which is not insulating anything currently is covered.

Almost all of the huge range of asbestos insulation and coating products is covered – anything with any asbestos in it. There are some exceptions to the licensing requirements – articles made of rubber, plastic, resin or bitumen which also contain asbestos, such as vinyl floor tiles, roofing felts and electric cables, and other asbestos products which have no insulation purposes, such as gaskets and ropes.

The major common exception to the Licensing Regulations is for asbestos-cement mixture products such as roofing sheets, gutters and pipes. The defining point is the bulk density of the product, which to be classed as asbestos cement must exceed one tonne per cubic metre. Even though the risk from asbestos is less with these products because the fibres are firmly bound by the cement, the work is still covered by CAWR.

The Licensing Regulations apply to work with asbestos insulation, coating or insulating board, and 'work' means removing, repairing or disturbing the material. Painting when the surface is in sound condition is not included in their scope, but sealing or painting damaged boards, insulation or coating is.

A licence is required for work with asbestos insulation, coating or insulating board except:
- where any individual worker does not spend more than a total of one hour on the work in any period of seven consecutive days, and the total time spent by all those working on that work does not exceed two hours (this exception allows minor repair work to be done without a licence, and the time refers only to the time spent removing, repairing or disturbing the asbestos, not the length of the whole job)
- where the work is being done at the employer's own premises, and the employer has given at least 14 days' notice of the work to the HSE and takes all the appropriate precautions
- where the work is solely air monitoring, clearance inspections, or the collection of samples for identification.

Licences may be granted with conditions, and may refer to insulation, coating or board in any combination. All licences are issued by the HSE's Asbestos Licensing Unit, and can be refused to applicants:

- with previous convictions for health and safety offences whether or not related to asbestos work
- who have failed HSE inspections of their work previously
- who cannot demonstrate adequate competence
- who have had two enforcement notices issued against them within a two-year period.

There is an appeal process against licence decisions, which unsuccessful applicants will be advised about.

The techniques of asbestos stripping, building and testing of enclosures, employee protection and other important matters relevant to those employers working with licences are beyond the scope of this book.

Before starting any work with asbestos, CAWR requires a specific assessment to be made of likely exposure of employees to asbestos. This need not be repeated for repetitive work, but where there is significant variance between jobs, as in demolition, a new assessment will be required. There may also be a need for other assessments under the COSHH Regulations. The ACoP for CAWR gives details of the scope of the assessment as:

- description of the work and its expected duration
- mention of the type of asbestos and the results of analysis
- controls to be applied to prevent or control exposure
- the reason for the chosen work method
- details of expected exposures, including whether they are likely to exceed the action level or control limit (see below) and the number of people affected
- steps to be taken to reduce exposure to the lowest level reasonably practicable
- steps to be taken to reduce release of asbestos into the environment and for the removal of waste
- procedures for provision and use of PPE
- procedures for dealing with emergencies where appropriate
- any other relevant information.

The assessment will need to be reviewed as circumstances change, such as fibre control or work methods.

CAWR Regulation 4

An important new concept introduced in the 2002 revision of CAWR is the duty to manage asbestos in non-domestic premises. The new duty came into effect on 21 May 2004, and is believed to affect around half a million commercial, industrial and public buildings in the UK.

The duty extends to all those who have any contractual or tenant obligations in relation to maintaining or repairing of non-domestic premises, including their means of access and egress, and also to those who have any control over these areas, whether or not there is a contract in place.

There is also a duty for those who control premises to co-operate with whoever manages the asbestos risk.

The duty holder is required to ensure that a suitable and sufficient assessment is made of whether asbestos is present in the premises, or is liable to be present. The assessment must involve an inspection of the reasonably accessible parts of the premises, and take account of the condition of the asbestos. The results must be recorded, and where asbestos is shown to be liable to be or actually present, the level of risk must be determined and a written action plan prepared to identify and control the risks.

The plan must include details of future monitoring of the condition of the asbestos, and be kept up to date. It must be made available to the emergency services on demand, and be provided to anyone liable to disturb the asbestos, such as contractors carrying out maintenance or alterations to the premises. The duty holder is responsible for the plan and for carrying it out, although the survey, assessment and plan preparation can be contracted out. There is no requirement to remove any asbestos found, but to control the risks appropriately (which may include removal in appropriate cases).

Table 1: Control limits and action levels for asbestos

Asbestos type	4-hour control limit (fibres/ml)	10-minute control limit (fibres/ml)	Action level (fibre-hours/ml)
Chrysotile (white asbestos) alone	0.3	0.9	72
Any other form, alone or in mixtures, including chrysotile mixtures with any other form of asbestos	0.2	0.6	48

Control limits and action levels

A **control limit** is that concentration of asbestos in the air, averaged over any continuous four-hour or 10-minute period, to which employees must not be exposed unless they are wearing suitable RPE. Control limits are expressed in fibres per millilitre of air averaged over the period.

Action levels are cumulative exposures calculated over a longer time – any continuous 12-week period. The period chosen should be a 'worst case', not a 'best case' period. If any employee's exposure could or does exceed an action level, then the CAWR requirements for notification, designated areas and medical surveillance apply (see below). Action levels are expressed in fibre-hours per millilitre, calculated by multiplying the airborne exposure in fibres/ml by the time in hours for which the exposure lasts. The cumulative exposure is arrived at by adding all the individual exposures over the chosen 12-week period together.

The likely concentration of asbestos fibres to be encountered can be estimated and where necessary confirmed by measuring and monitoring. The latter is done by personal sampling followed by analysis. Monitoring results must be retained to supplement the health records, and for the same length of time (see below).

If the assumption is made that the material is not chrysotile alone, then the more stringent limits and level must be used and there will not be a need to identify the type of asbestos.

Notification

CAWR requires notification of work where exposure to employees above the action level is likely to occur (Regulation 6). Licence holders only have to make a single one-off notification, others have to notify each time the work is planned. The enforcing authority depends on the main activity at the premises where the work is planned to be done.

Designated areas

The intention of marking asbestos areas (Regulation 14) is to make sure workers do not enter them unknowingly. A work area can be designated as an asbestos area (because the workers' exposure may exceed the action level) or a respirator zone (because either of the control limits may be exceeded), or both.

Medical surveillance

CAWR's Regulation 16 requires that anyone who is exposed above the action level must have been medically examined within the previous two years, and employers will need to check with the medical examiner or previous employer that the certificates are genuine. Employers must keep a health record for 40 years, containing prescribed information.

Work with asbestos cement

The general principles to be followed are:
- reduce the risk of breathing fibres by avoiding the need to do the work where reasonably practicable
- segregate the area, considering use of a physical barrier such as an enclosure
- post appropriate warning notices
- keep the material wet while working on it
- avoid using abrasive power and pneumatic tools, choosing hand tools instead
- where power tools are used, provide local exhaust ventilation and use at low speed
- wear suitable PPE and RPE

- organise essential high-risk tasks such as cutting and drilling at a central point
- minimise dust disturbance by choosing appropriate cleaning methods (not sweeping)
- provide training and information for employees.

Asbestos cement sheets

Cleaning by dry scraping or wire brushing should never be allowed. Water jetting can produce slurry which contaminates the area, and the process can cause the sheets to break. Use of surface biocides followed by hand brushing from a safe working platform is the method of choice.

Removal of asbestos cement sheets causes problems because of their potential height and fragility, and weathering which produces fibrous surface dust. Basic principles to follow are:

- do not allow anyone to walk or stand on sheets or their fixing bolts or purlins – they should be removed from underneath
- take asbestos cement sheets off before any other demolition is done
- do not break the sheets further
- keep the material wet and lower it onto a clean and hard surface
- remove waste and debris quickly to avoid further breakdown
- do not dry sweep debris
- follow safe waste disposal practice
- consider background air sampling at the site perimeter to reassure neighbours as well as workers

- conduct a final inspection to confirm removal of all debris and that a clean area has been left behind.

Disposal of asbestos waste

Asbestos waste is subject to the Special Waste Regulations 1996, which require it to be consigned to a site authorised to accept it. Whether put into plastic sacks or another type of container, the requirement is for the container to be strong enough not to puncture and to contain the waste, be capable of being decontaminated before leaving the work area, and be properly labelled and kept secure on site until sent for disposal. The label design is specified.

Large pieces or whole sheets should not be broken up, but transferred direct to covered trucks or skips, or wrapped in polythene sheeting before disposal. Smaller pieces can be collected and put into a suitable container. This should be double wrapped or bagged, and labelled to show it contains asbestos. Dust deposits are removed preferably using a Type H vacuum cleaner. Outer surfaces of waste containers should be cleaned off before removal.

Acknowledgement: *The publishers acknowledge with thanks the permission and co-operation given by Blackwell Science Limited in allowing reproduction of selected material from the author's* Principles of construction safety.

Selected references

Legal
Control of Asbestos at Work Regulations 2002
Asbestos Licensing Regulations 1983 (as amended)
Asbestos (Prohibitions) Regulations 1992 (as amended)
Construction (Design and Management) Regulations 1994

Guidance (HSE)

L11	*Guide to the Asbestos Licensing Regulations*
L27	*Control of Asbestos at Work Regulations – ACoP*
L28	*Work with asbestos insulation, asbestos coating and asbestos insulating board*
HSG 189/1	*Controlled asbestos stripping techniques for work requiring a licence*

HSG213	*Introduction to asbestos essentials*
HSG210	*Asbestos essentials: task manual*
HSG189/2	*Working with asbestos cement*
MDHS100	*Surveying and sampling asbestos-containing material*
INDG223	*Managing asbestos in premises*
INDG288	*Selection of suitable respiratory protective equipment for work with asbestos*
INDG289	*Working with asbestos in buildings*

Self-assessment questions

1 What are the main arguments for and against the precautionary removal of asbestos-containing material from the workplace?

2 Is there a register of asbestos-containing material in your workplace? Have all potential asbestos-containing materials been identified?

8 Radiation

Introduction

Energy which is transmitted, emitted or absorbed as particles or in wave form is called radiation. Radiation is emitted by a variety of sources and appliances used in industry. It is also a natural feature of the environment. Transmission of radiation is the way in which radios, radar and microwaves work.

The human body absorbs radiation readily from a wide variety of sources, mostly with adverse effects. All types of electromagnetic radiation are similar in that they travel at the speed of light. Visible light is itself a form of radiation, having component wavelengths which fall between the infrared and the ultraviolet portions of the spectrum.

Essentially, there are two forms of radiation, ionising and non-ionising, which can be further subdivided.

Ionisation and radiation

All matter is made up of **elements**, which consist of similar **atoms**. These, the basic building blocks of nature, are made up of a **nucleus** containing **protons** and orbiting **electrons**.

Protons have a mass and a positive charge.

Electrons have a negligible mass and a negative charge.

Neutrons have a mass but no charge.

If the number of electrons in an atom at a point in time is not equal to the number of protons, the atom has a net positive or negative charge, and becomes **ionised**.

Ionising radiation is that which can produce ions by interacting with matter, including human cells, which leads to functional changes in body tissues. The energy of the radiation dislodges electrons from the cell's atoms, producing ion pairs (see below), chemical-free radicals and oxidation products. As body tissues are different in composition and form as well as in function, their response to ionisation is different. Some cells can repair radiation damage, others cannot. The cell's sensitivity to radiation is directly proportional to its reproductive capability.

Ionising radiation

Ionising radiations found in industry are alpha, beta, gamma and X-rays. Alpha and beta particles are emitted from **radioactive** material at high speed and energy. Radioactive material is unstable, and changes its atomic arrangement so as to emit a steady but slowly diminishing stream of energy.

Alpha particles are helium nuclei with two positive charges (protons), and thus are comparatively large and attractive to electrons. They have short ranges in dense materials and can only just penetrate the skin. However, ingestion or inhalation of a source of alpha

Table 1: Radiation sources and hazards

Radiation	Emitted from	Examples of hazards
Radio-frequency and microwaves	Communications equipment, catering equipment, plastics welding	Heating of exposed body parts
Infrared	Any hot material	Skin reddening, burns, cataracts
Visible radiation	Visible light sources, laser beams	Heating, tissue destruction
Ultraviolet	Welding, some lasers, carbon arcs	Sunburn, skin cancer, ozone
X-rays and other ionising radiations	Sources, radiography and X-ray machines	Burns, dermatitis, cancer, body cell damage

particles can place it close to vulnerable tissue, so essential organs can be destroyed. **Beta particles** are fast-moving electrons, smaller in mass than alpha particles but having longer range, so they can damage the body from outside it. They have greater penetrating power, but are less ionising.

Gamma rays have great penetrating power and are the result of excess energy leaving a disintegrating nucleus. Gamma radiation passing through a normal atom will sometimes force the loss of an electron, leaving the atom positively charged – an **ion**. This and the expelled electron are called an ion pair. Gamma rays are very similar in their effects to **X-rays**, which are produced by sudden acceleration or deceleration of a charged particle, usually when high-speed electrons strike a suitable target under controlled conditions. The electrical potential required to accelerate electrons to speeds where X-ray production will occur is a minimum of 15,000 volts. Equipment operating at voltages below this will not, therefore, be a source of X-rays. Conversely, there is a possibility of this form of radiation hazard being present at voltages higher than this. X-rays and gamma rays have high energy, and high penetration power through fairly dense material. In low density substances, including air, they may have long ranges.

Common sources in industry of ionising radiation are X-ray machines and isotopes used for non-destructive testing. They can also be found in laboratory work and in communications equipment.

Non-ionising radiation
Generally, non-ionising radiations do not cause the ionisation of matter. Radiation of this type includes that in the electromagnetic spectrum between ultraviolet and radio waves, and also artificially produced laser beams.

Ultraviolet radiation comes from the sun, and is also generated by equipment such as welding torches. Much of the natural ultraviolet in the atmosphere is filtered out by the ozone layer, but a sufficient amount penetrates to cause sunburn and even blindness. Its effect is thermal and photochemical, producing burns and skin thickening, and eventually skin cancers. Electric arcs and ultraviolet lamps can produce a photochemical effect by absorption on the conjunctiva of the eyes, resulting in 'arc eye' and cataract formation.

Infrared radiation is easily converted into heat, and exposure results in a thermal effect such as skin burning, and loss of body fluids. The eyes can be damaged in the cornea and lens, which may become opaque (cataract). Retinal damage may also occur if the radiation is focused, as in lasers. These are concentrated beams of radiation, having principally thermal damage effects on the body.

Radio-frequency radiation is emitted by microwave transmitters including ovens and radar installations. The body tries to cool exposed parts by blood circulation. Organs where this is not effective are at risk, as for infrared radiation. These include the eyes and reproductive organs. Where the heat of the absorbed microwave energy cannot be dispersed, the temperature will rise unless controlled by blood flow and sweating to produce heat loss by evaporation, convection and radiation. Induction heating of metals can cause burns when touched.

Controls for ionising radiation
The intensity of radiation depends on the strength of the source, the distance from it and the presence and type of shielding. Intensity will also depend on the type of radiation emitted by the source. Radiation intensity is subject to the **inverse square law** – it is inversely proportional to the square of the distance from the source to the target. The dose received will also depend on the duration of the exposure. These factors must be taken into account when devising the controls. **Elimination of exposure** is the priority, to be achieved by restricting use and access, use of shielded enclosures, and written procedures to cover:
- use, operation, handling, transport, storage and disposal of known sources
- identification of potential radiation sources
- training of operators
- identification of operating areas

- monitoring of radiation levels around shielding
- monitoring of personal exposure of individuals, by dosimeters
- medical examinations for workers at prescribed intervals
- hygienic practices in working areas
- wearing of disposable protective clothing during work periods
- clean-up practice
- limiting of work periods when possible exposure could occur.

Controls for non-ionising radiation

Protection against **ultraviolet radiation** is relatively simple (sunbathers have long known that anything opaque will absorb ultraviolet light); that emitted from industrial processes can be isolated by shielding and partitions, although plastic materials differ in their absorption abilities. Users of emitting equipment, such as welders, can protect themselves by the use of goggles and protective clothing – the latter to avoid 'sunburn'. Assistants often fail to appreciate the extent of their own exposure, and require similar protection.

Visible light can, of course, be detected by the eye, which has two protective control mechanisms of its own – the eyelids and the iris. These are normally sufficient, as the eyelid blink reflex has a reaction time of 150 milliseconds. There are numerous sources of high-intensity light which could produce damage or damaging distraction, and sustained glare may also cause eye fatigue and headaches. Basic precautions include confinement of high-intensity sources, matt finishes to paintwork, and provision of optically correct protective glasses for outdoor workers in snow, sand or near large bodies of water.

Problems from **infrared** radiation derive from thermal effects and include skin burning, sweating, and loss of body salts leading to cramps, exhaustion and heat stroke. Clothing and gloves will protect the skin, but the hazard should be recognised so that effects can be minimised without recourse to PPE.

Controls for **laser** operations depend on prevention of the beam from striking persons directly or by reflection. Effects will depend on the power output of the laser, but even the smallest is a potential hazard if the beam is permitted to strike the body and especially the eye. Workers with lasers should know the potential for harm of the equipment they work with, and should be trained to use it and authorised to do so. If the beam cannot be totally enclosed in a firing tube, eye protection should be worn which is suitable for the class of laser being operated (classification is based on wavelength and intensity). Work areas should be marked so that inadvertent entry is not possible during operations. Laser targets require non-reflecting surfaces, and much care should be taken to ensure that this also applies to objects nearby which may reflect the laser beam. Toxic gases may be emitted by the target, so arrangements for ventilation should be considered. It is also necessary to ensure that the beam cannot be swung unintentionally during use, and that lasers are not left unattended while in use.

Equipment which produces **microwave** radiation can usually be shielded to protect the users. If size and function prohibits this, restrictions on entry and working near an energised microwave device will be needed. Metals, tools, and flammable or explosive materials should not be left in the electromagnetic field generated by microwave equipment. Appropriate warning devices should be part of the controls for each such appliance. Commercially available kitchen equipment is now subject to power restrictions and controls over the standard of seals to doors, but regular inspection and maintenance by manufacturers is required to ensure that it does not deteriorate with use and over time.

General strategy for the control of exposure to radiation

In addition to the previous specific controls, the following general principles must be observed:
- radiation should only be introduced to the workplace if there is a positive benefit
- safety information must be obtained from suppliers about the type(s) of radiation emitted or likely to be emitted by their equipment

- there is a requirement under the Management of Health and Safety at Work Regulations 1999 for written assessments of risks to be made, noting the control measures in force. All those affected, including employees of other employers, the general public and the self-employed must be considered, and risks to them evaluated. They are then to be given necessary information about the risks and the controls. Relevant Sections of this book contain reviews of the requirements
- all sources of radiation must be clearly identified and marked
- protective equipment (see below) must be supplied and worn so as to protect routes of entry of radiation into the body
- safety procedures must be reviewed regularly
- protective equipment provided must be suitable and appropriate, as required by relevant regulations. It must be checked and maintained regularly
- a radiation protection adviser must be appointed with specific responsibilities to monitor and advise on use, precautions, controls and exposure

- emergency plans must cover the potential radiation emergency, as well as providing a control strategy for other emergencies which may threaten existing controls for radiation protection
- written authorisation by permit should be used to account for all purchase, use, storage, transport and disposal of radioactive substances
- workers should be classified by training and exposure period. Those potentially exposed to ionising radiation should be classed as 'persons especially at risk' under the Management of Health and Safety at Work Regulations 1999 (see Part 4).

Legal requirements

General duties sections of the Health and Safety at Work etc Act 1974 apply to exposure to all forms of radiation, with its potential for harm. Specifically, the Ionising Radiations Regulations 1999 are the main control measure, together with duties concerning information provision contained in the Management of Health and Safety at Work Regulations 1999.

Revision

2 forms of radiation:
- ionising – alpha and beta particles, gamma rays, X-rays
- non-ionising – ultraviolet, infrared, visible light, lasers, radio-frequency, microwaves

Selected references

Legal
Ionising Radiations Regulations 1999
Radiological Protection Act 1970

Guidance (HSE)
L49   *Protection of outside workers against ionising radiations*
L121   *Work with ionising radiation*
HSG53   *The selection, use and maintenance of respiratory protective equipment*
HSG94   *Safety in the design and use of gamma and electron irradiation facilities*

Self-assessment questions

1 Explain the difference between ionising and non-ionising radiations.

2 What control measures would be appropriate to ensure the health and safety of employees using or working near a microwave oven?

9 Ergonomics

Introduction

Ergonomics is the applied study of the interaction between people and the objects and environment around them. In the work environment, the objects include chairs, tables, machines and workstations. Ergonomics looks at more than just the design of chairs, though. A complete approach to the work environment is the aim, including making it easier to receive information from machines and interpret it correctly.

Careful ergonomic design improves the 'fit', and promotes occupational wellbeing. It also encourages employee satisfaction and efficiency. Ergonomics is concerned with applying scientific data on human mental and physical capabilities and performance to the design of workplaces, hardware and systems. Usually, the ergonomic design emphasis is on designing tools, equipment and workplaces so that they and the job fit the person rather than the other way around.

A combination of techniques is normally used. These include:

Work design – incorporating ergonomics into the design of tools, machines, workplaces and work methods. These topics are not mutually exclusive.

Organisational arrangements – aimed at limiting the potentially harmful effects of physically demanding jobs on individuals. They may be concerned with selection and training, matching individual skills to job demands, job rotation methods and work breaks.

Studies carried out by the Swedish construction employment and insurance organisation Bygghälsan with employees and manufacturers have shown that considerable gains in productivity can be associated with optimising working conditions using ergonomic solutions. Bygghälsan investigated musculoskeletal disorders in the necks and shoulders of construction workers. A survey showed that almost half of all construction workers work more than 10 hours a week with their arms above shoulder level, leading to increased risk of problems in neck and shoulders. Sickness absences due to these problems increase for individuals over 30 in all trade categories in the industry, indicating that work-related problems in the neck and shoulders manifest themselves after 10 to 15 years of exposure.

It is impossible to eliminate totally the need to work with the hands above shoulder height, and so Bygghälsan looked at alterations in the way the work was organised, work methods, improved equipment and techniques to make the work easier. One of the discoveries made was that the use of micropauses – very short breaks – reduced muscular load and resulted in faster work. Screws were tightened into a beam at eye level, and those who took a 10-second pause after every other tightening did the work 12 per cent faster than those who did not. Interestingly, the workers themselves thought it took longer to carry out the work with micropauses than without.

Designing work and work equipment to suit the worker can reduce errors and ill health – and accidents. Examples of problems which can benefit from ergonomic solutions are:
- workstations which are uncomfortable for the operator
- hand tools which impose strain on users
- control switches and gauges which cannot be easily reached or read
- jobs reported to be found excessively tiring.

An ergonomic approach is used in the Manual Handling Operations Regulations 1992. It begins by forcing a review of the need to handle loads manually at all, and permits handling only following specific assessment of the risks associated with the task. The contrast with the former approach could not be more marked – we no longer train people to lift heavy weights as the sole solution to the problem.

Anthropometry

Anthropometry is the measurement of the physical characteristics of the human body. People vary enormously in basic characteristics such as weight, height and physical strength. A car built for the 'average' user may require tall people to bend at uncomfortable angles, while smaller people may not be able to reach the controls.

Designers use information on variations in size, reach and other physical dimensions to produce cars and other objects which most people can handle comfortably and conveniently.

This has particular application in the design of workplaces and work equipment. Including these measurements in system design assists in the process of making man and machine more compatible, and produces considerable benefits to industry because of improvements in efficiency, quality and safety.

Anthropometry is concerned with measurements of body movement as well as static dimensions, and includes the discipline of **biomechanics**, which studies the forces involved in movements. Knowledge of all these factors is combined to make improvements in workplace design – seating, for example, and work equipment, where the knowledge assists in the design of appropriate machine guards.

The 'average' person

Ergonomic designers usually try to meet the physical needs of the majority of a population. For any dimension, usually 10 per cent of the population will fall outside the range used. For example, the height of a workstation will only be wrong for the shortest 5 per cent and the tallest 5 per cent – the range is 5 to 95 per cent.

The designer will have to make deliberate compromises of this kind. It is not possible to produce a single size to fit everyone. Information about the design will enable purchasers to make informed decisions about equipment so that it meets the physical needs of the workforce.

Ergonomic issues

Display screen equipment

Over the past 20 years, display screen equipment (DSE) has been claimed to be responsible for a wide variety of adverse health effects, including radiation damage to pregnant women. Medical research shows, however, that radiation levels from the equipment do not pose significant risks to health. The range of symptoms which is positively linked relates to the visual system and working posture, together with general increased levels of stress and fatigue in some cases.

Repetitive strain injury (RSI) arising from work activities is now known with other musculoskeletal problems as work-related upper limb disorders (see below). One difficulty with 'RSI' has been that ill health is not necessarily due solely to repetitive actions, and is not limited to strains. The physical effects of DSE work range from temporary cramps to chronic soft tissue disorders, such as carpal tunnel syndrome in the wrist. It is likely that a combination of factors produces them. They can be prevented by action following an analysis of the workstation as required by the Regulations, together with an appreciation of the role of training, job design and work planning.

Eye and eyesight defects do not result from use of DSE, and it does not make existing defects worse. Temporary fatigue, sore eyes and headaches can be produced by poor positioning of the DSE, poor legibility of screen or source documents, poor lighting and screen flicker. Staying in the same position relative to the screen for long periods can have the same effect.

Fatigue and stress are more likely to result from poor job design, work organisation, lack of user control over the system, social isolation and high-speed working than from physical aspects of the workstation. These factors will also be identified in an assessment, and through consultation with the workforce.

Epilepsy is not known to have been induced by DSE. Even photosensitive epileptics (who react to flickering light patterns) can work safely with display screens.

Facial dermatitis has been reported by some DSE users, but this is quite rare. The symptoms may be due to workplace environmental factors including low humidity and static electricity near the equipment.

Pregnancy is not put at risk by DSE work, but those worried about the dangers should be encouraged to talk with someone aware of current advice – this is that DSE radiation emissions do not put unborn children, or anyone else, at risk.

Work-related upper limb disorders

'Upper limb disorders' (ULDs) is a label that refers to a wide range of conditions caused or aggravated by work. In the USA, the term 'cumulative trauma disorders' covers the same phenomena. From the neck to the fingertips, underlying changes in anatomy produce symptoms including pain, restriction of movement, and reduction in strength or sensation. Not all ULDs are related to work. Jobs with a high level of manual work and restrictions on posture are likely to have a significant percentage of sufferers from ULDs. These jobs include assembly and processing line working, cleaners and hairdressers.

An ergonomic approach encourages taking into account all relevant parts of the system of work, and encourages participation. Work over long periods that is repetitive, and involves awkward or static postures and/or high levels of applied manual force should be specially assessed. The publication HSG60 contains an excellent set of worksheets that can be used to do this.

Hand–arm vibration

Some pieces of work equipment have characteristics that can have lasting and damaging effects on users. Regular exposure to high levels of vibration from tools can lead to permanent injury and functional impairment. Hand-guided and hand-fed powered equipment and hand-held power tools used as a regular part of a job are potential causes of hand–arm vibration – examples are road-breaking hammers, lawnmowers and grinders. The permanent injuries that can result from the use of this equipment range from numbness and tingling in the fingers through pain and loss of grip strength, and loss of the sense of touch. These effects are known collectively as hand–arm vibration syndrome (HAV). The most well-known form of HAV is vibration white finger (VWF).

The most effective control to prevent HAV is to specify low vibration equipment (if its use at all is justified). Manufacturers are required to provide information on vibration levels, but this needs to be interpreted and may not be reliable in terms of actual exposure under field conditions. Poor maintenance is also a factor; machines with worn parts are likely to vibrate. The use of special gloves is rarely a substitute for low-vibration equipment – they are not likely to reduce significantly the amount of vibration reaching the hands. Changes in work practices can be effective, such as the use of jigs and holders at grinding machines.

Where HAV risks cannot be eliminated, health surveillance should be provided. Using a questionnaire to gather information can be productive. It should establish the daily amount of use for each type of equipment, and ask people to report symptoms of fingers going white on exposure to cold, tingling or numbness following equipment use, hand and arm joint and muscle problems, and difficulties with fine gripping. Health surveillance should include regular health checks, and should itself be supervised by a medical practitioner.

Selected references

Legal

Manual Handling Operations Regulations 1992
Health and Safety (Display Screen Equipment) Regulations 1992
Workplace (Health, Safety and Welfare) Regulations 1992
Provision and Use of Work Equipment Regulations 1998
Personal Protective Equipment at Work Regulations 1992

Guidance (HSE)

L22	*Safe use of work equipment – guidance on Regulations*
L23	*Manual handling – guidance on Regulations*
L25	*Personal protective equipment at work – guidance on Regulations*
L26	*Display screen equipment – guidance on Regulations*
HSG38	*Lighting at work*
HSG48	*Reducing error and influencing behaviour*
HSG57	*Seating at work*
HSG60(rev)	*Work-related upper limb disorders*
HSG121	*A pain in your workplace – ergonomic problems and solutions*
INDG171(rev1)	*Upper limb disorders – assessing the risk*
INDG175(rev2)	*Health risks from hand–arm vibration: advice for employers*
HSG88	*Hand–arm vibration*
HSG170	*Vibration solutions: practical ways to reduce the risk of hand–arm vibration injury*

Other sources

McKeown, C, and Twiss, M (2001), *Workplace ergonomics: a practical guide*, IOSH Services Ltd
Putz-Anderson, V (1988), *Cumulative trauma disorders: a manual for musculoskeletal diseases of the upper limbs*, Taylor & Francis, London
Pheasant, S (1996), *Bodyspace – anthropometry, ergonomics and the design of work*, 2nd edition, Taylor & Francis, London

Self-assessment questions

1 Identify and list examples where ergonomic principles could be applied with benefit to your workplace. What improvements would you expect to result?

2 How can a knowledge of ergonomics help in assessing manual handling risks?

Part 4
Law

1 Introduction to health and safety law

Much of this Part covers the legal framework and requirements in England and Wales. There is no fundamental difference between the legal systems of these countries. The systems in Scotland and Northern Ireland are different, although not such as to make significant differences in the standards of health and safety required. Differences in the organisation and administration of the law, in procedures and terminology, will, however, be significant for the health and safety practitioners in those countries.

On matters of health and safety law, the Scottish system is essentially the same as the English system. Acts of Parliament state whether or not they apply in Scotland – for example, Section 182 of the Factories Act 1961 states that the Act applies in Scotland. Where a statute is not applicable, legislative harmonisation is necessary. Northern Ireland also has its own legal system, ultimately controlled by the United Kingdom Parliament. In most circumstances, additional legislation is required by Acts, regulations or orders to introduce health and safety provisions into Northern Ireland.

The selection of statutes to be included in this Part was made on the basis that most affect everyone and should be included, and there should also be discussion of legislation which is either important or can serve as an example of a type.

Some of the regulations discussed include technical terms either in the regulations themselves or in the accompanying ACoPs. Where these terms are defined in other Sections of the book, they are not defined again in the following Sections.

The reader will probably be aware that regulations are made under the Health and Safety at Work etc Act 1974 or the European Communities Act 1972 to implement European Union directives. These require similar laws to be passed in each member state on the matters covered by the directives, which must at least meet the requirements of the directives, and may exceed them. In general, UK laws made in response to directives on health and safety at work exceed the directives because they apply to the self-employed as well as employees, employers and other duty holders.

As an EU member state, the UK is required to activate laws implementing directives by no later than the deadline given in the directive concerned. For the six sets of health and safety regulations described in later Sections which were first laid before Parliament in 1992, the activation date was 1 January 1993, to coincide with the beginning of the Single Market in Europe. Several have been amended (and redated) since then, in response to other directives.

On the other hand, the activation date for the Construction (Design and Management) Regulations was 15 months later than the relevant directive required. There is considerable variance in response to directives throughout the member states.

This book does not cover the regulations made to comply with directives on product and machinery safety, which impose standards on products and equipment rather than on people and systems. There are considerable links between the two types of regulations, which are discussed briefly in this Part. For a detailed treatment, the reader is directed to the various ACoPs and guidance published with the regulations.

Self-assessment questions have not been provided for many of the Sections in this Part. Decisions about the need for questions were taken on the basis of the need to self-test understanding of basic requirements.

2 The English legal system

Introduction

The English legal system uses different types of courts to hear different types of cases. Some hear only criminal matters, where guilt is the matter at issue, and others hear only civil cases, where the aim (for health and safety purposes) is to obtain compensation for a person who has suffered injury or damage. It should be said that there is a financial limit on much of the jurisdiction of the civil courts, which is otherwise wide-ranging. There are numerous remedies available in civil courts, of which damages are but one. In some instances it is possible for courts to hear both types of case. The senior courts hear the more important cases and appeals from the lower courts. Appeals can generally only be made by the defendant, although points of law can be raised as grounds for appeal by either side. The prosecution cannot appeal against a sentence in a magistrates' court, but there are now limited rights of appeal for it in the Crown court. In civil cases, the right of appeal and the grounds on which appeal may lie are very much wider.

The framework of the court system is shown in the diagram on page 191.

Magistrates' courts

Magistrates' courts hear both civil and criminal cases, and most criminal prosecutions begin in these courts (the majority also ending there as well). The civil jurisdiction of magistrates' courts extends to the recovery of certain debts, such as income tax and electricity, gas and water charges. Care proceedings are dealt with in the Family Proceedings Court. Magistrates' courts also deal with cases brought by the enforcing authorities under health and safety legislation, although as the court's powers are relatively limited these are sometimes committed to the Crown court for trial and/or sentence.

Cases are heard by members of the public who are appointed as magistrates (Justices of the Peace). They need have no legal qualifications, acting only as decision-makers on both fact and law, and take legal advice from the clerk to the justices for each Bench (the magistrates). Each 'commission area' – usually a county in the shires – is divided into petty sessional divisions. A Bench serves the petty sessional division. Each individual court normally sits with three magistrates, and a qualified legal adviser. Lay magistrates are unpaid and carry out their duties in addition to their normal employment. There are about 34,000 in England and Wales. There are also some professional magistrates with legal qualifications, formerly known as 'stipendiaries' but now called district judges (magistrates' courts), who are not required to sit

with other magistrates and adjudicate by themselves on longer cases or where complex legal arguments may be heard.

Some matters are **triable summarily** (only at the magistrates' courts), some are **triable either way** (the defendant and the prosecution can make representations about whether they think a jury trial and higher sentencing powers are appropriate, and the Bench decides whether it is prepared to hear the case in its court). The defendant is then given an opportunity to select the court of his or her choice, having been told that the magistrates may still decide to send the case to the higher court for sentence if they feel their own powers of sentencing are insufficient.

Some offences are triable only on **indictment** at the Crown court (murder, for example). For these most serious cases, the functions of the examining magistrates are set out in Sections 4–8 of the Magistrates' Courts Act 1980. In short, the function is to hold committal proceedings in order to enquire into the evidence of the prosecution, and to examine the prosecution case to decide whether there is sufficient evidence to put the defendant on trial by jury. The test is whether the prosecution has brought forward sufficient evidence to satisfy the examining magistrates that there is a triable issue to be put before the jury.

Penalties

The powers of magistrates' courts are limited to fines of up to £5,000 and/or up to six months' imprisonment for breaches of health and safety law generally. It is anticipated that this maximum will be raised shortly. For breaches of the 'general duties' sections of the Health and Safety at Work etc Act 1974 (Sections 2–6), and for some other regulations, maximum fines were increased in 1992 to £20,000. Fines at this level can also be imposed on those ignoring prohibition notices.

Prohibition and improvement notices are themselves penalties, as they are a matter of public record and almost always result in disruption of the work activity to which they refer. Occupying unplanned management time and business delays can be more costly than a fine, which can be imposed in addition to the notices. The notices are discussed in more detail in Section 4.

If magistrates decline to hear a case, because they feel that it is so serious that their powers of punishment are likely to be insufficient, they can direct that the case be heard in the Crown court before a judge and jury. At this level, the maximum penalty is an unlimited fine, and up to two years' imprisonment for a limited range of offences. Imprisonment can only follow breach of a very small number of requirements – contravening a licensing requirement, disobeying a prohibition notice, or the unauthorised acquisition, possession or use of explosives.

The size of financial penalties imposed by the courts for health and safety offences is rising. Each case is to be decided on its merits, and there is no set tariff. In 1999, a fine of £1.2 million was imposed against Balfour Beatty Civil Engineering Ltd. The prosecution followed the collapse, during construction, of tunnels at London Heathrow Airport in October 1994. The Austrian tunnel specialist Geoconsult was fined £500,000 after conviction on similar charges – breaches of Sections 2(1) and 3(1) of the Act. Costs of £200,000 were also directed to be paid by the companies. At the Old Bailey in July

1999, the Great Western Railway Co. was fined £1.5 million following the Southall rail crash.

Charges can be brought under the Act and specific regulations simultaneously. The lack of a guard to a conveyor belt's drive roller resulted in multiple arm fractures to an agency worker. There had been a similar accident at another of the company's depots, from which appropriate lessons had not been learned. A magistrates' court fined the depot owner £20,000 under Section 3 of the Act, £5,000 under Regulation 11 of the Provision and Use of Work Equipment Regulations 1998 for failure to prevent access to dangerous machinery, and £5,000 for failing to make risk assessments in relation to people not directly employed. Enforcing authorities commonly link allegations of failure to make suitable and sufficient risk assessments with other offences. Also, costs are usually awarded to the enforcing authority to cover the costs of investigation and prosecution.

In November 1998 the Court of Appeal issued important sentencing guidelines in the course of an appeal judgment (R -v- F Howe & Son (Engineers) Ltd, Court of Appeal case 97/101/Y3). Lower courts take this into account when deciding on the level of penalty and costs.

Howe & Son had pleaded guilty to four offences at Bristol Crown Court in 1997 and the company was made to pay costs of £7,500 and fines totalling £48,000 – £40,000 was for a breach of Section 2(1) of the Act, and a further £2,000 was for failure to make a suitable and sufficient risk assessment of a cleaning operation. The company appealed against the level of fines and costs. The Court of Appeal reduced the fines to £15,000 on the Section 2(1) charge, imposing no separate penalty on the others. But it upheld the earlier decision on costs.

Howe & Son's appeal was based on the company's financial position. When the fatal accident which gave rise to the charges and conviction occurred, Howe & Son had employed 12 people. In that trading year it had made a profit of almost £27,000 on a turnover of

£355,000. The Court of Appeal felt that insufficient weight had been given to the company's ability to pay a fine. But it did not accept as mitigating factors that there was a low level of risk in the task (which had led to a fatal electrocution), the small size of the company and its inability to have specialist in-company advisors, or that the general level of fines for this type of incident was lower than that imposed at the Crown court. It did accept in mitigation the company's previous good record, its guilty plea and its means.

In the judgement, the Court of Appeal said that the annual accounts should be supplied to both court and prosecution in good time before the hearing, and where this is not done a court will be entitled to assume that the defendant is in a position to pay any financial penalty which may be imposed.

The sentencing criteria which were summarised for the first time in this judgement are:
* how far short of the appropriate standard did the defendant fall (in failing to meet the test of reasonable practicability)?
* what happened? The penalty should reflect public disquiet at unnecessary loss of life. Where death is a consequence of a criminal act, it is an aggravating feature of an offence
* was there a deliberate breach of legislation? Was it with a view to profit, which seriously aggravates an offence, as does running a risk deliberately to save money
* was attention paid to warnings given previously?

The appeal judges commented that the standard of care imposed by legislation is the same regardless of the size of the organisation offending or its financial strength. It said that mitigation may include:
* the degree of risk and extent of the danger
* the extent and duration of the breach
* prompt admission of responsibility and a timely guilty plea
* a good safety record
* active steps to remedy the breach after being notified of it.

Tribunals

Tribunals are creatures of statute, and provide a complex system, with different tribunals being regulated by different statutory provisions. They are not courts – they are bodies established to take decisions in particular areas of the law, and they are regulated by the legal instrument which brings them into being. Examples of tribunals are those covering employment disputes including some health and safety matters (employment tribunals), and rent disputes (rent tribunals). Tribunal members are appointed and are not necessarily legally qualified.

The growth and success of the tribunal system may be due to the fact that they provide a cheap and less formal way of resolving disputes. Activities of tribunals are controlled ultimately by the conventional court system. The Divisional Court of the Queen's Bench Division hears appeals which have exhausted the full tribunal mechanism. Appeals against tribunal decisions are allowed on questions of law; decisions will be referred back if the tribunal has acted improperly, exceeded its jurisdiction or refused to hear a case.

County courts

The county court is the junior of the two main civil courts, the other being the High Court. It only hears civil cases, which range from landlord/tenant disputes to hire purchase repossessions. All cases in this court are heard by a judge or a registrar (the junior judge), the latter normally hearing the less important cases. Its jurisdiction is extensive, and this brief summary cannot hope to do more than indicate its range.

Crown courts

The Crown court is responsible for hearing the more serious criminal cases, particularly those not tried, and those which cannot be tried (indictable offences) by the magistrates' court. It also deals with cases where the defendant has elected trial by jury (if the offence is 'triable either way' – ie in either court), or because the magistrates have determined that a particular offence should be tried by a jury.

Crown courts also hear appeals against convictions by magistrates' courts, as well as other matters. While no jury is present, lay magistrates can be.

Crown courts are widespread geographically and are graded according to the seriousness of the offence which, in turn, determines the seniority of the judge who will hear the case. The Old Bailey is the most famous and senior of the Crown courts. Appeals from these courts are heard by the Court of Appeal (Criminal Division).

The High Court

All the important civil cases are heard in the High Court, which also has some jurisdiction over criminal appeals. Its work is divided into three main divisions:

The **Queen's Bench Division** primarily administers the common law. It hears cases dealing with contracts, or with civil wrongs (known as **torts**). The tort of negligence, for example, is the means by which compensation may be obtained following an accident at work. The Division also hears those cases which cannot be brought to the county court, such as those where claims involve more than £5,000. Within the Division there are two specialist courts: Commercial (hearing major commercial disputes in private where the judge acts as arbitrator) and the Admiralty Court (hearing maritime disputes).

The **Chancery Division** primarily administers equity. It specialises in hearing cases involving money matters, such as tax, trusts, bankruptcy, and disputes over wills or with the Inland Revenue.

The **Family Division** deals with family disputes which cannot be handled by the magistrates' courts. These include defended divorces, separation, child custody, adoption and distribution of family assets on marriage breakdown.

Within each High Court Division there is a Divisional Court, presided over by three judges at each sitting, as opposed to the single judge in the High Court.

The **Divisional Court of the Queen's Bench Division** hears appeals from magistrates' courts, or from the Crown court on appeal from a magistrates' court on criminal matters, and can undertake judicial reviews. It also oversees the activities of the junior courts and tribunals. Further appeal lies from the Divisional Court to the House of Lords.

The **Divisional Court of the Chancery Division** deals with appeals from county courts on bankruptcy matters.

The **Divisional Court of the Family Division** hears appeals from magistrates' courts in domestic and matrimonial cases.

The Court of Appeal

The Court of Appeal is divided into two distinct courts: one for hearing appeals arising from civil proceedings and one for hearing criminal appeals.

The **Civil Division of the Court of Appeal** hears appeals from county court and High Court decisions. The court rehears the case, listening only to legal argument from counsel and relying on transcripts taken from the previous hearings. Most appeals are based on points of law, but some are based on the claim that the original judge may have drawn wrong conclusions from the evidence. Three Lord Justices of Appeal constitute the court, and the decisions they reach can be made on a majority vote. The Master of the Rolls is the most senior of the Lords Justices of Appeal.

The **Criminal Division of the Court of Appeal** hears appeals from the Crown court. This court can overturn a conviction, vary the sentence imposed or substitute a conviction for one for another offence. Appeals can be made by the defendant, arising out of points of law. Also, where a defendant is acquitted, the Attorney General may refer a point of law involved in the case to the Court of Appeal in order to have a ruling for future cases. There is now an important provision enabling the prosecution to refer to the Court of Appeal a sentence imposed in the Crown court which it regards as being unduly lenient. Generally, appeals in this Division are heard by two judges; if they cannot agree, the appeal is reheard in front of three. The Lord Chief Justice is the head of the Division.

Appeals against the decisions of the two Divisions of the Court of Appeal are made to the House of Lords.

The House of Lords

This is the highest appeal body (short of the European Court) for both criminal and civil cases. All cases heard at this level require approval from the Court of Appeal, or from the Appeals Committee of the House of Lords. Permission is only given if the appeal is of significant legal importance, or in the wide public interest (such as the 'Spycatcher affair'). In exceptional cases, an appeal can 'leap-frog' the Court of Appeal to be heard immediately by the House of Lords. For this to happen, the High Court judge and the Lords must certify that there is a legal point of public importance involved which may alter previous legal decisions.

Judges in this court are known as the Law Lords, and all are professional judges who have been awarded a life peerage. Also, the Lord Chancellor and peers who have held high judicial office can sit as judges in the court. Technically, any Lord can be involved, but in practice this never happens. A minimum of three is required to make a decision; in practice there are always five. Due to the high cost involved in bringing cases to the Lords, very few criminal cases reach the court. Most cases involve tax, commercial and property disputes.

The Judicial Committee of the Privy Council

Although this committee does not form a part of the English legal system, it exerts a considerable influence over it. It is the highest appeal forum for cases within Commonwealth countries, but it also hears cases from the Isle of Man and the Channel Islands. Decisions of the Judicial Committee are respected by countries which have based their legal system on English law. Essentially, judges involved in the activities of the Committee are those who sit in the House of Lords, but Commonwealth judges can also be involved.

European courts

The **European Court of Justice**, sitting in Luxembourg, adjudicates on the law of the European Union, and its decisions are binding on British courts by reason of the European Communities Act 1972. It gives rulings regarding the interpretation and application of provisions of European Union law referred to it by member states. A single judgement is given, from which there is no right of appeal. Any decisions it makes are enforceable throughout the network of courts and tribunals within each of the member states.

The **European Court of Human Rights** enforces the agreed standards on the protection of human rights and fundamental freedoms.

The European Union system

Administration of the EU is carried out through a network of four bodies:

The **Commission** is led by a Board of Commissioners of representatives from each member state. The Commission administers the law throughout the EU. Proposals for future legislation originate from the Commission in the form of directives, which are sent for approval to the Council of Ministers.

The **Council of Ministers** is composed of members of government from each member state and is responsible for decision-making within the EU.

The **European Parliament** is formed of representatives elected from constituencies from within member states and consults on and debates all proposed legislation.

The **European Courts** – see above.

Making health and safety legislation in the EU

Proposals for health and safety legislation are made either under Article 100A (now Article 95) or Article 118A (now Article 138) of the Treaty of Rome. The original Article 118A has been completely reworded as Article 138. Following consultation in international committees and Council working groups, proposals are generated by the Directorates General, of which DG V is the one most often associated with health and safety matters. These are submitted to the Commission, which submits its proposals to the Council of Ministers. Consultation follows with the European Parliament and the Economic and Social Committee, both of which issue opinions on the proposals after voting.

The Commission receives the opinions and may modify its proposals as a result, although it is not required to do so. The proposals then go to the Council of Ministers. Health and safety matters are normally proposed under Article 138 of the Treaty, and in these cases the relevant council is the Labour and Social Affairs Council. The Council discusses the proposal, and votes using the qualified majority vote system to adopt a Common Position.

The Common Position is the official starting point for the European Parliamentary process. The Parliament can reject a Common Position. Where this happens, it can only be adopted after a unanimous vote by the Council. The European Parliament can approve a Common Position either directly, or indirectly by taking no decision on it. In that case the proposal is automatically adopted by the Council of Ministers. The third possibility is that the European Parliament proposes amendments to the Common Position. This forces a resubmission, starting back at the Commission and ending at the Council where a unanimous vote is required in order to insert amendments proposed by the European Parliament and not supported by the Commission.

Elements of European law

Apart from the general provisions of the Treaty of Rome, which is signed by all member states, there are three elements or instruments of European law:

European Regulations, passed by the Council of Ministers, are adopted immediately into the legal framework of each member state.

Directives, on the other hand, set out objective standards to be achieved but allow member states to decide on how their own individual legal frameworks will be altered to adopt them. In the UK, the most common method of implementation of directives is by statutory instrument.

Decisions are reached in cases which are taken to the European courts, and affect only the company or individual concerned.

Revision

The English legal system is summarised in the diagram on the opposite page.

Self-assessment questions

1 Write notes on the structure and functions of magistrates' courts, and explain the extent of the powers of the court in relation to health and safety legislation.

2 How do European directives affect health and safety in the UK?

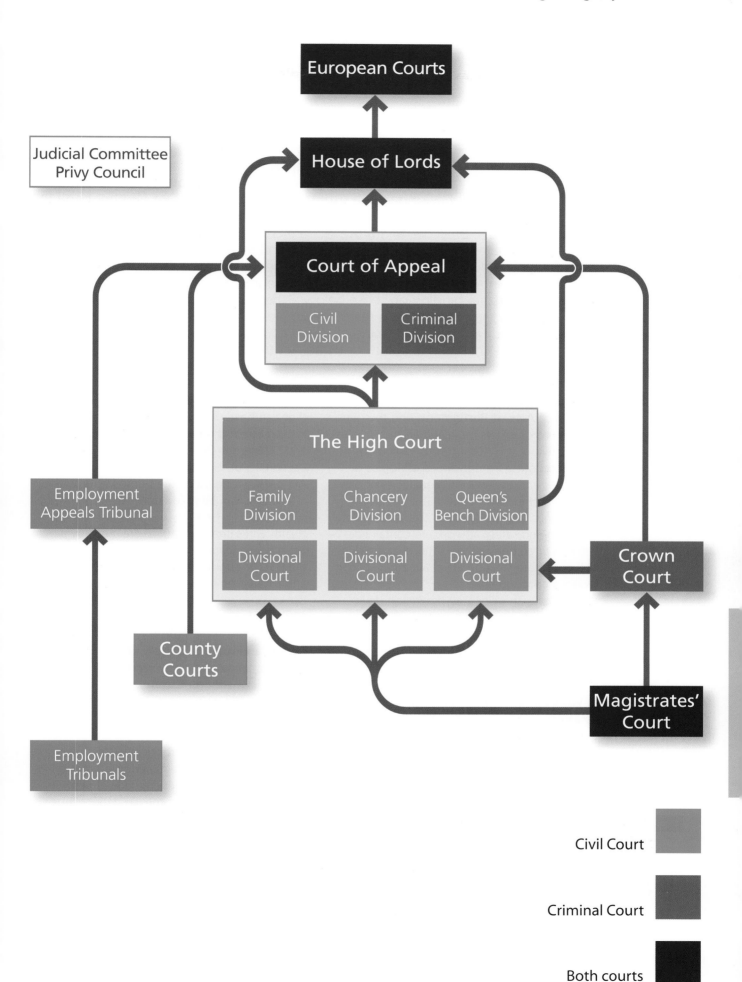

Judicial Committee Privy Council

European Courts

House of Lords

Court of Appeal
Civil Division
Criminal Division

The High Court
Family Division
Chancery Division
Queen's Bench Division
Divisional Court
Divisional Court
Divisional Court

Employment Appeals Tribunal

County Courts

Employment Tribunals

Crown Court

Magistrates' Court

Civil Court

Criminal Court

Both courts

3 Statute and common law

Introduction
'The law' concerning health and safety matters is a mixture of criminal law set out in statutes, and known as statute law, and common law, which provides a means of compensation for injuries or damage suffered because of failure of another party to comply with statute law or to carry out any of the duties established by common law over the years. Breaches of statute law are criminal offences for which financial penalties, and even prison sentences in special and relatively uncommon situations, can be imposed. Individual cases are always judged on their particular facts. 'The law' stands as interpreted by the courts and announced in court decisions. Decisions of higher courts are binding on those below – these are called **precedents**. Lower court decisions may be helpful in reaching decisions; these are called **persuasive**. Decisions of the House of Lords bind all lower courts, and can only be overridden by the House itself changing its mind, or by an Act of Parliament.

Statute law
Statute law is the written law of the land and consists of **Acts of Parliament**, and the rules, regulations or orders made within the parameters of the Acts. Acts of Parliament usually set out a framework of principles or objectives and use specific regulations or orders to achieve these, sometimes after a delay written into the Act. Most of the laws relating to health and safety at work are contained within a large body of regulations (some rules and orders are still operative) aimed at specific requirements and enforced by the HSE or local authorities with delegated powers to do so.

The Act of Parliament is, therefore, the primary or principal legislation; regulations made under Acts are secondary, subordinate, delegated legislation. For example, the Health and Safety at Work etc Act 1974 is a primary Act of Parliament, and regulations made under it have the same terms of reference and applicability – the Noise at Work Regulations 1989 and the Electricity at Work Regulations 1989 being only two of many.

Regulations are made by ministers of the Crown where they are enabled to do so by an Act of Parliament. Regulations have to be laid before Parliament, but most do not need a vote and become law on the date specified in the regulations. They can still be vetoed by a vote within 40 days of being laid before Parliament. Regulations can apply to employment conditions generally, eg the Noise at Work Regulations; they can control specific hazards in specific industries, such as the Construction (Health, Safety and Welfare) Regulations; and they can have the most general application, as in the case of the Reporting of Injuries, Diseases and Dangerous Occurrences Regulations. Section 15 of the Health and Safety at Work etc Act gives a long list of the purposes for which regulations can be made under it.

Health and safety regulations are made by a Secretary of State, acting on proposals made by the HSC or after consultation with the Health and Safety Commission (HSC). Normally, the HSC issues consultative documents to all interested parties before making such recommendations.

Orders and **rules** technically give the force of law to an executive action. Thus, commencement orders can be made by the Secretary of State which activate parts of Acts or regulations.

Approved Codes of Practice (ACoPs) can supplement Acts and regulations in order to give guidance on the general requirements which may be set out in the legislation (thus effectively enabling the legislation to be kept up to date by revising the ACoP rather than the law). The HSC has the power to approve codes of practice of its own, or of others such as the British Standards Institution. Approval cannot be given without consultation, and consent of the Secretary of State. Failure to comply with an ACoP is not an offence of itself, but failure is held to be proof of contravention of a requirement to which a code applies unless a defendant can show that compliance was achieved in some equally good way (Health and Safety at Work etc Act, Section 17).

Guidance Notes are documents issued by the HSC or HSE as opinions on good practice. They have no legal force, but because of their origin and the experience employed in their production, they will be 'persuasive' in practice to the lower courts and useful in civil cases to establish reasonable standards prevailing in an industry. ACoPs and Guidance Notes can therefore be said to have a 'quasi-legal' status, rather like the Highway Code.

Breaches of statute law can be used in civil claims to establish negligence, unless this is prohibited by the Act or regulation itself. The Health and Safety at Work etc Act contains such a provision, so breaches of it cannot be used to support civil claims for compensation.

Types of statutory duty

There are three levels or types of duty imposed by statute, which allow different responses to hazards. These are: absolute duty, duty to do what is practicable, and duty to take steps that are reasonably practicable. Over the years, a body of case law has built up which gives guidance on the meaning of these duties in practice. Mostly, the cases have been decided in common law, in the course of actions for personal injuries which were based on alleged breach of statutory duty.

Recently the attention of the criminal courts has been directed to the interpretation and application of Section 3 of the Health and Safety at Work etc Act 1974 in relation to the responsibility of the occupier of premises to contractors and others.

Absolute duty (or strict liability)

There are circumstances when the risk of injury is very high unless certain steps are taken, and some Acts and regulations have recognised these by placing an absolute duty on the employer to take specific steps to control the hazard. The words 'must' or 'shall' appear in the section or regulation to indicate this – there is no choice or evaluation of risk or feasibility to be made by the employer. The best-known example is probably the former Section 12 of the Factories Act 1961, which required 'the fencing of every moving part of every prime mover'. The absolute nature of this duty was upheld by courts, even in circumstances where a machine had become practically or financially unusable because of the strict guarding requirement. However, an 'easement' is sometimes given where such absolute requirements are subject to a 'defence' clause where, in any proceedings for an offence consisting of a contravention of an absolute requirement, it shall be a defence for any person to prove that he/she took all reasonable steps and exercised all due diligence to avoid the commission of that offence. This is known as the 'due diligence' defence, and an example can be found in Regulation 29 of the Electricity at Work Regulations 1989.

'Practicable'

Some regulations specify that steps must be taken 'so far as is practicable'; for example, Regulation 11(2)(a) of PUWER 98 requires the provision of 'fixed guards enclosing any dangerous part or rotating stock-bar where and to the extent that it is practicable to do so'. 'Practicable' means something less than physically possible. To decide on whether the requirement can be achieved, the employer would have to consider current technological knowledge and feasibility, not just difficulty of the task, its inconvenience or cost. Some degree of reason can be applied and current practice can be taken into account.

'Reasonably practicable'

The phrase 'so far as is reasonably practicable' qualifies almost all the general duties imposed on employers by the Health and Safety at Work etc Act. Its use allows the employer to balance the cost of taking action (in terms of time and inconvenience as well as money) against the risk being considered. If the risk is insignificant when weighed against the cost, then the steps need not be taken. This analysis of cost versus benefit must be done prior to any complaint by an enforcing authority. An employer would need to keep up to date with health and safety developments affecting his/her activities, as steps might become reasonably practicable over a time. Also, a single analysis of precautions could be insufficient in the sense that other less effective but cheaper alternatives might be available in addition to the one being evaluated. 'Reasonably practicable' steps are, simply, those which are the best under the circumstances.

Development of health and safety law

Since the early part of the 19th century, when more and more people began to earn their living in factories and workshops instead of agricultural and handicraft work, Parliament has become correspondingly active in passing laws to promote health and safety at work. The first law was passed in 1802 and was promoted by pioneer reformers attempting to improve the conditions of child labour. This was "An Act for the preservation of the health and morals of apprentices and others employed in cotton and other mills and other factories." This Act was important, not only for the sanitary improvements it introduced, but also because it influenced further legislation and its direction. From then onwards until 1974, the law was aimed at types of **premises** (eg factories, shops and mines), and where necessary at types of work activity or **processes** going on there – grinding of metals, the use of special types of machinery, etc.

No further factory safety legislation was passed until 1819, but between then and 1859 there was a steady stream of statutes and subordinate regulations covering areas such as safety, hours of work, the employment of women and children, and other important issues. By 1875, factory safety law was scattered between numerous documents with no real pattern of development. Then, under a review by a Royal Commission, the Factory and Workshop Act was passed in 1878, which attempted to provide the first comprehensive piece of factory legislation. However, even this new system required reviews at intervals to cope with changing technology and developing legal precedents, and further extensions of this Act

were made in 1883, 1889, 1891 and 1897. In 1901, the Factory and Workshop Act became the principal statute controlling safety in factories, and remained so until its repeal by the Factories Act of 1937.

Many nations within the former British Empire and the Commonwealth adopted British safety law with little or no modification, and small or large pieces of the original Factories Act can still be found on the statute books of countries that now have little remaining cultural contact with Britain. Thus, British influence on world safety standards has been profound.

The 1937 Factories Act provided for the first time a comprehensive code for safety, health and welfare requirements, applicable to all factories. It included many new requirements and processes including building activities (not previously covered by any safety laws) and work repairing ships in harbour. Further amendments and alterations were made in 1948 and 1959, with a consolidating measure in 1961 to produce the final version.

Since then it has been progressively whittled away by newer measures until, by the time of writing, there is almost nothing left. One of the difficulties with the Factories Act was that it only applied to premises defined within it (notably factories, of course) and not to other workplaces. This was one of the reasons behind the change of course begun in 1974 by the Health and Safety at Work etc Act and later EU-driven regulations, with their obligations based more on relationships between organisations and employees than on narrow definitions of types of premises, and removing the need to interpret what is meant by a 'factory' for new premises.

The 'opposite number' of the Factories Act 1961 was the Offices, Shops and Railway Premises Act 1963, which has also virtually disappeared. All that is left of the Factories Act applies to premises within its scope which were first taken into use before 1993.

Common law

Common law has evolved over hundreds of years as a result of the decisions of courts and judges. Common principles or accepted standards fill the gaps where statute law has not supplied specific requirements. The accumulation of common law cases has resulted in a system of **precedents**, or decisions in previous cases, which are binding on future similar cases unless overruled by a higher court or by statute.

Case law on health and safety did not appear until the 19th century, when the rapid introduction of industrial technology led to an equally rapid rise in work injuries. In 1840 a young female mill worker successfully sued her employer and was awarded damages (Cottrell -v- Stocks (1840), Liverpool Assizes). Claims at common law were often resisted successfully by employers, especially if there had been any contribution to the accident by the worker's own actions or if the worker had claimed additionally for compensation under the Workman's Compensation Act. All these restrictions on claims were removed by reforms (but not until after the Second World War), which were completed by the Law Reform (Personal Injuries) Act 1948.

As a result of the cases which have been decided over the years, there is now a body of precedents which defines the duties of both employers and employees at common law. The employer must take reasonable care to protect his/her employees from the risk of foreseeable injury, disease or death at work by the provision and maintenance of a safe place of work, a safe system of work, safe plant and equipment, and reasonably competent fellow employees. The basis of the employer's duty to employees derives from the existence of a contract of employment, but the duty to take reasonable care may also extend to matters which affect others but are within the employer's control. To avoid common law claims (and to avoid the accidents which prompt them), the employer has to do what is reasonable under the particular circumstances. This will include providing a safe system of work, instruction on necessary precautions and indicating the risks if the precautions are not followed.

The employer can also be held liable for the actions of his/her employees causing injury, death or damage to others, providing the actions were committed in the course of employment. This liability of the employer is known as **vicarious liability**.

Employees have a general duty of care in common law towards themselves and to other people – so they can also be sued by anyone injured as a result of lack of reasonable care on their part. In practice this happens rarely, and the liability is assumed by the employer under vicarious liability.

The burden of proof in common law cases rests with the claimant – the injured party. He/she must show that he/she was owed a duty by whoever he/she is suing, that there was a breach of the duty, and that as a result of that breach he/she suffered damage. The claimant can be helped by procedural rules and rules of evidence, one of which saves him/her having to show exactly how an accident happened if all the circumstances show that there would not have been an accident but for the lack of reasonable care on the part of the person being sued. This principle is known as **res ipsa loquitur** – 'let the facts speak for themselves'.

Defences can be raised against common law claims: there was no negligence, there was no duty owed, the accident was the sole fault of the employee, the accident did not result from the lack of care, contributory negligence (very rarely a complete defence), and **volenti non fit injuria** – the employee knowingly accepted the risk.

Civil actions have to be started within three years of the date when the accident happened, or the date when the injured person became aware that there was an injury suffered which could be the fault of the employer. The court has discretion to extend the three-year period in appropriate cases.

The Woolf reforms

Reforms to the civil claims procedures proposed by Lord Woolf were adopted on 26 April 1999. These were designed to ensure that courts deal with claims justly, and the resulting Civil Procedure Rules constitute probably the most far-reaching single set of changes in the administration of civil justice. With the virtual disappearance of legal aid for injury claims, the courts deal with cases to ensure as far as practicable that the parties are on an equal footing without excessive cost, that each case is dealt with in proportion to its claim value, complexity and importance, and that each case is dealt with promptly and fairly. The larger the size of the claim, the more the courts will direct management of the proceedings and appoint a target trial date. Small claims up to £15,000 enter a fast-track system ensuring that trials will not last more than one day and will take place within about 33 weeks of the commencement of proceedings.

To assist in this, the Personal Injury Protocol gives a mechanism for early exchange of documentation so that each side can establish what their position will be. A detailed initial letter written by a claimant's lawyer must be acknowledged by the employer within 21 days. A maximum of three months is given for exchange of documents and investigation. If liability is denied, written reasons for that denial are to be provided to the claimant. Contributory negligence can be argued, but again the reasons and any documents to be used for relying on that argument must be given. Failure to respond or otherwise not to work within the set time periods is likely to prejudice both claimant and defence in a variety of ways.

The system requires an active response from lawyers and employers, rather than allowing lengthy delays to develop and costs to mount up. A further small but important change is that the injured party is now known as the claimant, not as the plaintiff.

Damages awarded are difficult to quantify but they depend on loss of a faculty, permanent nature of the injury and its effect on the ability to earn a living, the expenses incurred (including estimated and agreed loss of earnings), and a range of other factors. There is a 'tariff' of awards used informally by judges, but this is often not followed. English courts do not use the jury system (except in libel cases), which decides the amount of damages in other jurisdictions. Awards may be reduced by a percentage if there is found to be a contribution to the injury caused by the claimant's own negligence, but the trend of the courts is to minimise this as a significant factor.

Revision

2 types of applicable law:
- statute law
- common law

Statute law: made by Acts, regulations, orders, rules, supplemented by Approved Codes of Practice, Guidance Notes

3 types of statutory duty:
- absolute (strict liability)
- practicable
- reasonably practicable

Common law:
- based on judicial precedent
- involves a general duty of care
- breach of statutory duty can usually establish breach of general duty
- defences can be based on:
  - lack of a duty
  - lack of negligence
  - sole fault of injured person
  - contributory negligence
  - voluntary acceptance of the risk
- system controlled by Lord Woolf's Civil Procedure Rules

Self-assessment questions

1 Outline the main differences between common law and statute law.

2 Explain the meaning of 'reasonably practicable'.

4 The Health and Safety at Work etc Act 1974

Introduction

This Act is the major piece of health and safety legislation in Great Britain. It provides the legal framework to promote, stimulate and encourage high standards. Previous Acts had concentrated on prescription of solutions within the law; a series of Factories Acts in particular had raised standards progressively since 1833 by stipulating what measures needed to be taken in factories, and although the legislative base was gradually widened to include other kinds of premises, modern work practices and businesses were no longer fully covered by legislation by the third quarter of the 20th century. Consultations carried out by the Robens Committee between 1970 and 1972 produced the basis of a new type of law – one which placed responsibility on employers and employees together to produce their own solutions to health and safety problems, subject to the test of reasonable practicability (see Part 4 Section 3).

The Act introduced for the first time a comprehensive and integrated system dealing with workplace health and safety and the protection of the public from work activities. By placing duties of a general character on employers, employees, the self-employed, manufacturers, designers and importers of work equipment and materials, the protection of the law, rights and responsibilities are available and given to all at work. As regulations made under the Act have the same scope, there now exists the potential to achieve clear and uniform standards.

As an 'enabling' Act, much of the text is devoted to the legal machinery for creating administrative bodies, combining others and detailing new powers of inspection and enforcement. The HSC – see below – carries responsibility for policy-making and enforcement, and is answerable to a Secretary of State. Its executive arm is the HSE, whose functions range from enforcement to research and European liaison on standards.

The gradual replacement of previous piecemeal health and safety requirements by revised and updated measures applicable to the whole of the workforce of the country has been a feature of the past three decades. This overhaul has been done by the repeal of statutes and their replacement with regulations and ACoPs prepared in consultation with industry and workers. One of the key features of the report of the Robens Committee, echoed by European directives, is the principle of consultation at all levels in order to achieve consensus and combat apathy. This consultative process starts within the HSC, and continues to the workplace, where employers are required to consider the views of workers in the setting of health and safety standards.

The Act consists of four parts:

Part 1 contains provisions on:
- health and safety of people at work
- protection of others against health and safety risks from work activities
- control of danger from articles and substances used at work
- controlling certain atmospheric emissions.

Part II establishes the Employment Medical Advisory Service (EMAS).

Part III amends previous laws relating to safety aspects of the Building Regulations.

Part IV contains a number of general and miscellaneous provisions.

General duties of employers

These are contained in **Sections 2, 3, 4** and **9**. By **Section 40**, the burden of proof is transferred from the prosecution to the defence in prosecutions where it is alleged that the accused person or employer failed to do what was practicable or reasonably practicable as required in the particular circumstances. **Sections 36** and **37** provide for the personal prosecution of members of management in certain circumstances, notably in Section 37 where they can be charged as well as, or instead of, the employer if the offence in question was due to their consent, connivance or neglect.

Employers must, as far as is reasonably practicable, safeguard the health, safety and welfare of employees (**Section 2**). In particular, this extends to the provision and maintaining of:

- safe plant and safe systems of work
- safe handling, storage, maintenance and transport of (work) articles and substances
- necessary information, instruction, training and supervision
- a safe place of work, with safe access and egress
- a safe working environment with adequate welfare facilities.

There is an absolute duty on employers with five or more employees to prepare and revise as necessary a written statement of safety policy, which details the general policy and the particular organisation and arrangements for carrying it out. The policy must be brought to the notice of all the employees (**Section 2(3)**). This topic is discussed in Part 1 Section 4.

Employers must consult with employees on health and safety matters. They must also set up a statutory health and safety committee at the request of trade union-appointed safety representatives, with the main function of keeping under review the measures taken to ensure the health and safety at work of employees (**Section 2(4–7)**). Regulations made in 1977 expand employers' duties to consult with trade union appointees. These are the Safety Representatives and Safety Committees Regulations, which describe the functions of safety representatives but do not impose any duties on them. The functions are: investigations of potential hazards and dangerous occurrences; examining the causes of accidents; investigation of complaints by employees; making representations to the employer on those matters; carrying out inspections of several kinds; representing

employees in consultations with enforcing authorities; receiving information from inspectors and attending meetings of safety committees. A safety committee must be set up at the request of two or more safety representatives. Safety representatives have the right to training to carry out their functions; their entitlements are covered in an ACoP and Guidance Notes.

In 1996 a wider entitlement was introduced which gave formal consultation rights to all employees, regardless of trade union membership or representation. It offers employers the choice of consulting with the whole of the workforce or of allowing (and funding) elections of representatives with whom the employer must consult. These representatives have not been given the detailed functions of safety representatives, or their immunity from prosecution while carrying out their functions. The topic of employee consultation is covered in detail in Part 4 Section 28, where both sets of Regulations are summarised.

The self-employed, other employees and the public must not be exposed to danger or risks to health and safety from work activities (**Sections 3** and **4**).

Harmful emissions into the atmosphere must be prevented, from prescribed operations (**Section 5**).

General duties of the self-employed
Similar duties to the above rest on the self-employed, with the exception of the safety policy requirement (unless they, in turn, employ others) (**Section 3(2)**).

General duties of employees
By **Section 7**, employees must take reasonable care of their own health and safety and that of others who may be affected by their acts or omissions. They must also co-operate with their employer so far as is necessary to enable the employer to comply with his/her duties under the Act. By **Section 8**, it is an offence for anyone to intentionally or recklessly interfere

with or misuse anything provided in the interests of health, safety or welfare. Members of management who are also employees are vulnerable to prosecution under Section 7 if they fail to carry out their health and safety responsibilities (as defined in the safety policy statement), in addition to their liability under Sections 36 and 37 as noted above.

General duties of manufacturers and suppliers

Section 6 of the Act focuses attention on the role of the producer of products used at work, and includes designers, importers and hirers-out of plant and equipment in the list of those who have duties under it. The Section refers to articles and substances for use at work, and requires those mentioned to ensure so far as is reasonably practicable that articles and substances are safe when being 'used' in the widest sense. They must be tested for safety in use, or tests are to be arranged and done by a competent authority. Information about the use for which an article or substance was designed, including any necessary conditions of use to ensure health and safety, must be supplied with the article or substance.

Since the introduction of the Act, Section 6 has been modified so as to apply to fairground equipment, to provide a more complete description of the kinds of use an article may be put to, and to ensure that necessary information is actually provided and not merely 'made available' as was originally required.

Hire purchase companies are not regarded as suppliers for the purposes of Section 6. An interesting provision requires anyone installing an article for use at work to ensure, so far as is reasonably practicable, that nothing about the way in which it is installed or erected makes it unsafe or a risk to health when properly used.

Charges

Section 9 forbids the employer to charge his/her employees for any measures which he/she is required by statute to provide in the interests of health and safety.

The Health and Safety Commission and the Health and Safety Executive

These bodies were established by the Act, in **Section 10**. Their chief functions are as follows:

The **HSC** takes responsibility for developing policies in health and safety away from government departments. Its nine members are appointed by the government from industry, trade unions, local authorities and other interest groups (currently consumers). It can also call on government departments to act on its behalf.

The **HSE** is the enforcement and advisory body to the HSC. For lower risk activities and premises such as offices, the powers of enforcement are delegated to local authorities' environmental health departments, whose officers have the same powers as HSE inspectors when carrying out these functions.

Powers of inspectors

These are contained in **Sections 20–22** and **25** of the Act. An appointed inspector can:
- gain access without a warrant to a workplace at any time
- employ the police to assist in the execution of his or her duty
- take equipment or materials onto premises to assist the investigations
- carry out examinations and investigations as seen fit
- direct that locations remain undisturbed for as long as is seen fit
- take measurements, photographs and samples
- order the removal and testing of equipment
- take articles or equipment away for examination or testing
- take statements, records and documents
- require the provision of facilities needed to assist him or her with the enquiries
- do anything else necessary to enable him or her to carry out these duties.

Enforcement

If an inspector discovers what is believed to be a contravention of any Act or regulation, he or she can:

- issue a **prohibition notice**, with deferred or, more usually, immediate effect, which prohibits the work described in it if the inspector is of the opinion that the circumstances present a risk of serious personal injury. The notice is effective until any steps which may be specified in it have been taken to remedy the situation. Appeal can be made to an employment tribunal within 21 days, but the notice remains in effect until the appeal is heard
- issue an **improvement notice**, which specifies a time period for the rectification of the contravention of a statutory requirement. Appeal can be made to an employment tribunal within 21 days; this has the effect of postponing the notice until its terms have been confirmed or altered by the tribunal
- prosecute any person who contravenes a requirement or fails to comply with a notice as above
- seize, render harmless or destroy any article or substance which is considered to be the cause of imminent danger or serious personal injury.

The financial consequences of receiving prohibition and improvement notices are often severe because of work delays and disruption. Rights for businesses under the HSC's Enforcement Policy include the right to a letter of explanation from an inspector stating what needs to be done and why, and in the case of an improvement notice representation can be made to the inspector's manager if it is felt that the notice should not be issued or its terms should be changed. Two weeks are allowed for this process to take place, which is in addition to appeal rights through employment tribunals. Details of rights of appeal to tribunals are given in writing with notices as they are issued.

Complaints procedures have been formalised; the first line of complaint is always to the inspector's manager. Beyond that, the Director-General of the HSE oversees the investigation of complaints concerning HSE inspectors. Complaints about local authority inspectors which cannot be resolved by the appropriate manager can be directed to the authority's chief executive. The HSE's Local Authority Unit will see that complaints are followed up if necessary.

The detailed Management Regulations

The detailed provisions of the Management of Health and Safety at Work Regulations 1999 are discussed in Section 6. It has been claimed that their effect is solely to provide more detailed descriptions of the general duties owed under the Act, and as such they make little difference to employers already complying with the Act. However, it should be noted that there are several aspects of the Regulations which impose additional duties on employers and others. Examples include mandatory induction training, measures to deal with serious and imminent danger, information required to be given to others, and the general requirement for most employers to make risk assessments and record in writing their significant findings.

Revision

The Health and Safety at Work etc Act 1974 is an enabling Act in four parts:

Part I:
- sets out general duties of employers, employees, the self-employed and those involved in the supply process of anything used at work
- established the HSC and HSE and their enforcement functions
- provides for future regulations and ACoPs

Part II establishes the Employment Medical Advisory Service (EMAS)

Part III amends the Building Regulations

Part IV contains miscellaneous and general provisions

Self-assessment questions

1 What are the powers of an HSE inspector? What enforcement action can be taken by an inspector to prevent dangerous acts taking place?

2 Summarise the employer's duties under the Health and Safety at Work etc Act 1974.

5 The Framework Directive (89/391/EEC)

Introduction

The European Directive No 89/391/EEC, known as the 'Framework Directive', is the measure dealing generally with 'the introduction of measures to encourage improvements in the safety and health of workers at work'. The European Parliament adopted four resolutions in February 1988 which specifically invited the Commission to draw up a framework to serve as a basis for more specific directives covering all the risks connected with health and safety at the workplace. This Directive was the result of that invitation, and its structure is typical of directives in general.

This Section summarises the Framework Directive (FD) and outlines its structure. The next Section follows its implementation within the United Kingdom as an example of how the general requirements were taken up by a member state and passed into specific laws within the allowed time period for implementation.

Preliminary

The reasons for the presence of the FD are set out in a series of paragraphs, each beginning with the word 'Whereas', summarised as follows:

- the FD is founded on the requirement of the former Article 118A of the Treaty of Rome, by which the Council must use the directive system to adopt minimum requirements for encouraging improvements, especially in the working environment, to guarantee a better level of protection of the health and safety of workers
- the FD does not justify any reduction in existing levels of protection already in place in member states
- workers can be exposed to the effects of dangerous environmental factors at the workplace during the course of their working lives
- the former Article 118A of the Treaty of Rome required that directives must not impose constraints holding back the creation and development of small and medium-sized undertakings
- member states have a responsibility to encourage improvements in worker safety on their territory, and taking these measures can also help in preserving the health and safety of others
- member states have different legislative systems on health and safety at work which differ widely and need to be improved, and national provisions in this field may result in different levels of protection and allow competition at the expense of health and safety
- the incidence of accidents at work and occupational disease is still too high and measures ensuring a higher level of protection are to be introduced without delay
- ensuring improved protection requires the contribution and involvement of workers and/or their representatives by means of balanced participation to see that necessary protective measures are taken
- information, dialogue and balanced participation must be developed between employers and workers and/or their representatives
- improvement of workers' safety, hygiene and health at work is an objective which should not be subordinated to purely economic considerations
- employers must keep themselves informed of technological advances and scientific findings on workplace design and inform workers accordingly
- the FD applies to all risks, especially those arising from chemical, physical and biological agents
- the Advisory Committee on Safety, Hygiene and Health Protection at Work is consulted by the Commission on the drafting of proposals in this field.

General provisions

Article 1 states the object of the FD as the introduction of measures to encourage improvements in the safety and health of workers at work, to which end it contains general principles and implementation guidelines and is without prejudice to existing or future Community/Union or national provisions which are more stringent.

Article 2 states the scope of the FD applies to all sectors of activity, both public and private, but not to specified public service activities, such as the armed forces or police, where the objectives of the Directive must be followed as far as possible.

Article 3 contains definitions of the following terms: worker, employer, workers' representative, prevention.

Article 4 requires member states to take the necessary steps to ensure their laws extend the Directive's provisions to employers, workers and their representatives, with adequate controls and supervision.

Employers' obligations

Article 5 describes general duties of employers to ensure the safety and health of workers in every aspect related to their work. These are duties which cannot be delegated to workers or third parties. An option is given to member states to exclude or limit employers' responsibility where occurrences are due to unusual and unforeseeable circumstances beyond the employers' control, and to exceptional events where the consequences could not have been avoided despite exercise of all due care.

Article 6 sets out the general obligations of employers. These include the general principles of prevention, the need for co-operation with other employers sharing a workplace, not involving workers in the financial cost of any measures taken, worker instruction and capability assessment, and the assessment of risks.

In **Article 7** the employer is required to designate one or more workers to carry out protection and prevention activities, who must be allowed time

to do this and not be placed at any disadvantage because of these activities. If there are no competent persons in the undertaking, the employer must enlist competent external services or persons. These persons must be given access to information required to be made available by Article 10. All designated workers must be sufficient in number and ability to deal with the organising of health and safety measures appropriate to the undertaking, the hazards there and their distribution throughout the entire undertaking. Member states must define the necessary abilities required to do this, and how many such people will be designated. They can also define those undertaking(s) where size and activities indicate that the employer who is competent can take personal responsibility for carrying out the activities referred to.

Article 8 covers first aid, firefighting and evacuation of workers, and the concept of serious and imminent danger. The employer must designate personnel required to implement the chosen measures in these areas. Workers who are assessed as being potentially exposed to serious and imminent danger of the risks involved must be identified and given information about the risks and the steps to be taken concerning stopping work and proceeding to a place of safety. In general, workers must not be asked to resume work where a risk of serious and imminent danger remains. When, because of such a risk, workers leave a dangerous area, they must be protected against any harmful or unjustified consequences. Actions of workers in this context do not place them at any disadvantage unless they acted carelessly or there was negligence on their part.

Various additional obligations are given in **Article 9**, including a requirement for the employer to be in possession of risk assessments including those for risks facing particular groups of workers, and a record of accidents resulting in a worker being unfit for work for more than three working days.

Article 10 obliges employers to provide a wide range of information on identified risks and preventive and protective measures to workers, outside employers of workers sharing or using common workplaces, persons appointed under Article 7, and workers' representatives. They are also entitled to see accident reports and summaries, and information provided by regulatory agencies.

Worker participation and consultation is to be guaranteed under **Article 11**. Consultation is to be in advance and in good time. Such consultation is to be held on any measure which substantially affects safety and health, designation of workers under Articles 7 and 8, information provision under Articles 9 and 10, appointing of external services or persons under Article 7, and planning and organising of training referred to in Article 12. Workers must be given the right to ask for and propose appropriate measures to mitigate hazards, must be allowed time off with pay to exercise their rights and functions and to liaise during inspection visits by the enforcing authority, and may not be placed at a disadvantage because of their activities in relation to this Article.

Article 12 deals with training of workers. This is to be provided by the employer on recruitment, on changing jobs and following introduction of new technology or new equipment. Training takes the form of information and instructions specific to the job or workstation, and must be repeated where necessary, especially when changed or new risks are identified. The employer is also to ensure that workers from outside undertakings have received appropriate instructions on the risks to be found during activities in the employer's undertaking. Worker representatives with a specific role in protecting health and safety of workers are entitled to appropriate training.

Generally, no training under this Article is to be at the workers' own expense, and it must take place during working hours.

Workers' obligations

Workers' own obligations under **Article 13** are to, as far as possible, take care of their own health and safety and that of others who might be affected by their acts at work in accordance with training and instructions given by the employer. The Article details that workers must make correct use of the means of production and the PPE supplied to them, use supplied safety devices correctly and not change or remove any devices fitted to structures or equipment, and co-operate with their employer and other employees with health and safety responsibilities. They must also inform their employer immediately about any work situation which they reasonably consider to be dangerous and about any shortcomings in protection arrangements.

By **Article 14**, health surveillance may be received by any worker wishing it, appropriate to the risks incurred at work. Such surveillance may be provided as part of a national health system. Particularly sensitive risk groups are to be specifically protected (**Article 15**).

Individual directives

Article 16 binds the Council to adopt individual directives in the areas listed in the FD Annex, which without prejudice provide more stringent or specific requirements. It also allows for amendment of the FD and these other 'daughter' directives. The areas specified in the Annex are:
- workplaces
- work equipment
- PPE
- work with visual display units
- handling of heavy loads involving risk of back injury
- temporary or mobile construction sites
- fisheries and agriculture.

Oversight

A mechanism is prescribed in **Article 17** for taking account of needs for technical adjustment, harmonisation, new findings and technological progress, and international regulatory changes as far as they affect the FD or the 'daughter' directives. In this, the Commission will be assisted by a representative committee.

Action

The final **Article 18** details the date by which the member states must bring into force their provisions necessary to comply with the Directive. This was 31 December 1992. The texts of national laws in place to cover the requirements must be lodged with the Commission. Every five years, member states are to report their consulted comments on the practical implementation of the provisions of the Directive. The Commission then submits a report to the Council, Parliament and the Economic and Social Committee.

Despite initial argument by the UK that the Health and Safety at Work etc Act 1974 itself satisfied the requirements of the Directive, more detailed legislation was demanded. As a result the Management of Health and Safety at Work Regulations 1992 were introduced – see the next Section.

Revision

The Framework Directive:
* is based on the former Article 118A (now Article 138) of the Treaty of Rome
* encourages improvements in the safety and health of workers
* is binding on member states
* is activated largely by the Management of Health and Safety at Work Regulations in the UK

6 The Management of Health and Safety at Work Regulations 1999

Introduction

These Regulations implement most of European Directive 89/391/EEC of 29 May 1990 (the 'Framework Directive'), Council Directive 91/383/EEC, dealing with the health and safety of those who are employed on a fixed term or other temporary basis, and Council Directive 94/33/EC, on the protection of young people. The Regulations are in addition to the requirements of the Health and Safety at Work etc Act 1974, and they extend the employer's general safety obligations by requiring additional specific actions on the employer's part to enhance control measures.

Originally one of the set of six issued at the end of 1992 and widely known as the 'six-pack', the Management of Health and Safety at Work Regulations (MHSWR) were enlarged and extended in 1994, and were renumbered, updated and reissued at the end of 1999.

The MHSWR do not cover all the matters mentioned in the Directives, partly because some of them are dealt with elsewhere and partly because existing provisions are considered to cover the points adequately. Examples include first aid, use of PPE and firefighting.

Generally, a breach of a duty imposed by the MHSWR does not confer a right of action in any civil proceedings. Major exceptions to this are the Regulations covering risks to new and expectant mothers and young persons (Regulations 16(1) and 19). The Regulations do not apply to the master or crew of a seagoing ship, or to their employer in respect of the normal shipboard activities of the crew.

There are some overlaps between these more general Regulations and specific regulations such as the Manual Handling Operations Regulations 1992. Where the MHSWR general duties are similar to specific ones elsewhere, as in the case of general risk assessments, the legal requirement is to comply with both the general and the specific duty. However, specific assessments required elsewhere will also satisfy the general requirement under the MHSWR for risk assessment (**as far as those operations alone are concerned**). The employer will still have to apply the general duties in all work areas.

Because they provide the detail lacking in the Health and Safety at Work etc Act, a knowledge of the MHSWR is important for every employer. The Regulations contain specific requirements on risk assessments, training and dealing with other employers – to name only three of the significant areas requiring action by employers. They also include the development of arrangements which need to be written into the health and safety policy.

1999 modifications to the 1992 Regulations

This section has been introduced to provide a quick update for readers already familiar with the main provisions of the original 1992 Regulations. The new MHSWR came into force on 29 December 1999. The main effect of the changes was to introduce four new requirements, to incorporate duties and references from the Fire Precautions (Workplace) Regulations 1997, and to integrate the previous (1994) amendments as full numbered regulations in their own right.

Four new Regulations were introduced:

Regulation 4 requires employers to put in place preventive and protective measures as specified in the originating Framework Directive. These specifics are now included as Schedule 1 to the MHSWR.

Regulation 7(8) requires that competent persons appointed by the employer be chosen from among the employer's employees in preference to competent persons not in the employer's employment. This requirement is said by the HSE to be a close transposition from the Directive, rather than being 'anti-consultant'.

Regulation 9 requires employers to arrange any 'necessary' contacts with external services, citing first aid, emergency medical care and rescue work as possible examples. This is especially significant for the construction industry.

Regulation 21 removes a previously available line of defence for an employer to take against prosecution. The employer cannot now rely on a breach of a regulation by an employee, or a competent person employed by the employer, as a reason for not being prosecuted him/herself.

The following Regulations have been revoked:
* Management of Health and Safety at Work Regulations 1992
* Management of Health and Safety at Work (Amendment) Regulations 1994
* Health and Safety (Young Persons) Regulations 1997
* Fire Precautions (Workplace) Regulations 1997, Part III.

Objective of the Regulations
The objective of the Regulations is to implement the Directives referred to above.

Requirements of the Regulations
Regulation 1 contains the date of commencement and relevant definitions. The latter include:
* **child** – in simple terms, a person who is not over compulsory school age
* **young person** – a person who has not reached the age of 18
* **new or expectant mother** – an employee who is pregnant, who has given birth within the previous six months, or who is breast-feeding.

Regulation 2 disapplies the MHSWR to the master and crew of a seagoing ship, and most of the young persons' requirements in respect of 'occasional or short-term working involving

either domestic service in a private household', or work 'not regarded as harmful, damaging or dangerous to young people in a family undertaking'. No definition is offered of 'family undertaking'.

Regulation 3 requires every employer and self-employed person to make a suitable and sufficient assessment of the health and safety risks to employees and others not in his/her employment to which his/her undertakings give rise, in order to put in place appropriate control measures. It also requires review of the assessments as appropriate, and for the significant findings to be recorded (electronically if desired) if five or more are employed. Details are also to be recorded of any group of employees identified by an assessment as being especially at risk.

With regard to young persons, an employer must not employ a young person unless a risk assessment has been made or reviewed in relation to the risks to health and safety of young people, taking into account in particular:
* the inexperience, immaturity and lack of awareness of risks of young people
* the fitting-out and layout of the workstation and workplace
* the nature, degree and duration of exposure to physical, chemical and biological agents
* the form, range and use of work equipment and the way in which it is handled
* the organisation of processes and activities
* risks from special processes, which are listed.

Regulation 4 requires any preventive and protective measures to be implemented on the basis of Schedule 1 to the Regulations. In practice, it can be assumed that existing measures, particularly those required by other regulations, will be regarded as complying with the Schedule, which is couched in rather general terms overall.

Regulation 5 requires employers to make appropriate arrangements, given the nature and size of their operations, for effective planning, organising, controlling, monitoring and review of preventive and protective measures. If there are more than five employees, the arrangements are to be recorded.

Regulation 6 requires the provision by the employer of appropriate health surveillance, identified by the assessments as being necessary.

Adequate numbers of 'competent persons' must be appointed by employers under **Regulation 7**. They are to assist the employer in complying with obligations under all the health and safety legislation, unless (in the case only of sole traders or partnerships) the employer already has sufficient competence to comply without assistance. This Regulation also requires the employer to make arrangements between competent persons to ensure adequate co-operation between them, to provide the facilities they need to carry out their functions, and to give specified health and safety information (on temporary workers, assessment results and risks notified by other employers under other regulations within the MHSWR). Regulation 7 repays study because it sets out clearly what is required of the competent person and the appointment arrangements, including the Regulation 7(8) 'preference' for internal appointment(s).

Regulation 8 requires employers to establish and give effect to procedures to be followed in the event of serious and imminent danger to persons working in their undertakings, to nominate competent persons to implement any evacuation procedures, and to restrict access to danger areas. Persons will be classed as competent when trained and able to implement these procedures. One effect of this Regulation is to require fire evacuation procedures with nominated fire marshals in every premises.

Regulation 9 directs that any necessary contacts with external agencies are arranged, as stated above.

Regulation 10 requires employers to give employees comprehensible and relevant information on the health and safety risks

identified by the assessment, protective and preventive measures, any procedures under Regulation 8(1), the identities of competent persons appointed under that Regulation, and risks notified to the employer by those in shared facilities and other third parties as required by Regulation 11. Regulation 10 also contains two paragraphs on the employment of children and young persons as defined above.

The first requirement is that, before employing a child, an employer must provide the parent of that child with comprehensible and relevant information on assessed health and safety risks, any preventive or protective measures employed, and details of any third-party risks notified to the employer under Regulation 11. The second paragraph defines the word 'parent' in terms of people who have parental rights (in Scotland) and parental responsibility (in England and Wales).

Regulation 11 deals with the case of two or more employers or self-employed persons **sharing a workplace temporarily or permanently**. Each employer or self-employed person in this position is to co-operate and co-ordinate with any others as far as necessary to enable statutory duties to be complied with, reasonably co-ordinate his/her own measures with those of the others, and take all reasonable steps to inform the other employers of risks arising out of his/her undertaking.

Regulation 12 covers sharing of information on workplace risks. It requires employers and the self-employed to provide employers of any employees, every employee from outside undertakings, and every self-employed person who is working in their undertakings, with appropriate instructions and comprehensible information concerning any risks from the undertaking. Outside employers, their employees and the self-employed must also be told how to identify any person nominated by the employer to implement evacuation procedures.

Regulation 13 requires employers to take capabilities regarding health and safety into account when allocating tasks. Adequate health and safety training must be provided to all employees on recruitment, and on their exposure to new or increased risks because of:

- a job or responsibility change
- the introduction of new work equipment or a change in use of existing work equipment
- the introduction of new technology
- the introduction of a new system of work or a change to an existing one.

The training must be repeated where appropriate, take account of new or changed risks to the employees concerned, and take place during working hours.

Regulation 14 introduces employee duties. They must use all machinery, equipment, dangerous substances, means of production, transport equipment and safety devices in accordance with any relevant training and instructions, and inform their employer or specified fellow employees of dangerous situations and shortcomings in the employer's health and safety arrangements.

Regulation 15 covers temporary workers. Employers and the self-employed have to provide any person they employ on a fixed-term contract or through an employment agency with information on any special skills required for safe working (and any health surveillance required) before work starts. The employment agency must be given information about special skills required for safe working, and specific features of jobs to be filled by agency workers where these are likely to affect their health and safety. The agency employer is required to provide that information to the employees concerned.

Regulation 16 requires specific risk assessment of the work of new and expectant mothers and the taking of appropriate measures as a result, including variation of working hours or conditions where reasonable to do so and effective against the risks, up to suspension from work where necessary. **Regulation 17** requires the employer to suspend new and expectant mothers from work 'for as long as is necessary' for their health and safety, either

when night work is involved, or when a medical certificate indicates that this should be done. In both cases, the employee's rights to alternative work and remuneration are protected by the Employment Rights Act 1996 and earlier legislation.

Regulation 18 permits the employer to apply his/her knowledge of the medical condition of those who may be affected by Regulations 16 and 17. For Regulation 16 to apply, the employer must have received written notice from the employee that she is pregnant, breast-feeding or has given birth within the preceding six months. Failure to produce a medical certificate where one is required, or other specified failure to show that the provisions apply, may disapply them. Cases where the employer either knows the person is no longer a new or expectant mother, or cannot establish whether she remains so, also disapply the provisions.

Regulation 19 is set in the most general terms and contains a duty on employers to ensure that employed young persons are protected while at work from any risks to their health or safety which result from their lack of experience and awareness of risks or their immaturity (paragraph 1). In particular (paragraph 2), young persons cannot be employed to do work beyond their physical or psychological capacity, or which involves:

- harmful exposure to toxic, carcinogenic or other chronic agents of harm to human health
- harmful exposure to radiation
- risks of accidents which, it can reasonably be assumed, cannot be recognised or avoided by young persons because of lack of experience or training, or because of their insufficient attention to safety
- a risk to health from extreme heat or cold, noise or vibration.

In deciding whether the work will involve such harm or risks, employers must pay attention to the results of their risk assessments (which are required by Regulation 3).

Paragraph 3 does not prevent the employment of a young person who is not a child, where this is necessary for training, or where the young person will be supervised by a competent person and where any risk will be reduced to the lowest level that is reasonably practicable. These provisions are without prejudice to any other provisions or restrictions on employment arising elsewhere.

The Home Secretary can invoke national security where necessary to exempt various armed forces from the MHSWR by **Regulation 20**.

Regulation 21 denies the employer a defence in criminal proceedings because of any act or default by an employee or a competent person appointed by him/her. Breach of the MHSWR cannot be used to give a right of civil action (**Regulation 22**), with the important exceptions of Regulations 16(1) and 19.

Regulations 23–30 deal with procedural matters including application outside Great Britain, amendments and revocations. The consequential amendments are listed in an extensive Schedule 2.

Readers should note that the foregoing attempts only to summarise a complex set of Regulations, and that they should consult the Regulations and the ACoP for full details and an interpretation to ensure accuracy.

Revision

12 requirements of the Management of Health and Safety at Work Regulations 1999:
- formal risk assessments
- formal management control systems
- specific protective and preventive measures on the part of employers (in Schedule 1)
- health surveillance
- competent person appointments
- arrangement of 'necessary contacts' with external services
- procedures for serious and imminent danger
- information for workers
- inter-employer co-operation
- job-specific training
- capability assessment
- further and detailed employee duties

Self-assessment questions

1 Review your organisation's health and safety policy. Does it set out the details of the operation of all the arrangements required by Regulation 5?

2 Summarise the main ways in which these Regulations expand the duties of the employer under the Health and Safety at Work etc Act 1974.

7 The Workplace (Health, Safety and Welfare) Regulations 1992

Introduction

The Regulations implement most of the requirements of the Workplace Directive (89/654/EEC) concerning minimum standards for workplace health and safety. Most of the provisions have long been part of British law, which was, however, not comprehensive in application. As a result of them, most of the pre-1974 law in this field is repealed or revoked.

The Regulations have applied in full to all workplaces since 1 January 1996 when the transitional provisions ended. Some workplaces not previously subject to specific requirements are within their scope, including schools and hospitals.

The major **repeals and revocations** are Sections 1–7, 18, 28, 29, 57–60 and 69 of the Factories Act 1961, and Sections 4–16 of the Offices, Shops and Railway Premises Act 1963. Twenty-five sets of regulations and orders and parts of a further 17 disappear.

Objective of the Regulations

The objective of the Regulations is to place obligations on employers and others in control of workplaces to reduce the risks associated with work in or near buildings. Opportunity was taken to rationalise the different, wide-ranging and detailed requirements in pre-1993 UK health and safety law, and to combine them all in one comprehensive set of regulations.

Some significant definitions

Traffic route means a route for pedestrians, vehicles, or both – and includes stairs, fixed ladders, doors, ramps and loading bays.

Workplace means any non-domestic premises or part of premises made available to any person as a place of work. It includes any place within the premises to which a person has access while at work, and means of access to or egress from the premises other than a public road. A modification, conversion or extension is part of a workplace only when completed.

Premises means any place, including an outdoor place. Domestic premises are not within the scope of the Regulations, which as a result do not cover homeworkers.

Requirements of the Regulations

The application of these Regulations is very general, in contrast to the Factories Act and other legislation which they largely replace. They do not apply to a workplace inside a

means of transport, workplaces where the only activities are construction activities as defined, mineral resource extraction or ancillary activities. Fishing boats are also excluded. By **Regulation 3(2)**, the provisions of Regulations 20–25 as they affect temporary work sites are to be qualified by 'reasonably practicable'. **Regulation 3(3)** removes application of Regulations 5–12 and 14–25 from workplaces in or on most means of transport. Regulation 13 only applies when the means of transport is stationary inside a workplace or, for licensed vehicles, off the road.

By **Regulation 3(4)**, Regulations 5–19 and 23–25 do not apply to places of work in woods, fields and other agricultural land not inside buildings, and in the same way Regulations 20–22 are subject in those circumstances to the test of reasonable practicability.

Where construction work is in progress within a workplace, it can be treated as a construction site and not subject to these Regulations only if it is fenced off, otherwise these Regulations and the various Construction Regulations apply. For temporary work sites, Regulations 20–25 apply so far as is reasonably practicable only. These include work sites used infrequently or only for short periods, and fairs and similar structures only present for short periods.

Generally, **Regulation 4** obliges employers and others to ensure that every workplace under their control and where any of their employees work complies with any applicable requirement. The same duty is placed on any person having control of a workplace to any extent in connection with trade, business or other undertaking (whether or not for profit), in relation to matters within his/her control. The self-employed are not under a requirement by Regulation 4 in respect of their own work or that of a partner. Every person deemed to be a factory occupier by Section 175(5) of the Factories Act 1961 shall ensure that those premises comply with these Regulations.

Regulation 5 covers maintenance of workplaces, equipment, devices and systems, which are to be cleaned, and maintained in an efficient state, working order and good repair. If a fault in any equipment or device is liable to result in a breach of any of these Regulations, it must be subject to a suitable system of maintenance. For example, this would apply to emergency lighting and ventilation and would extend to arrangements to ensure its continuous and effective operation. These and other examples would be expected to be identified by the risk assessments made under the Management of Health and Safety at Work Regulations 1999.

Ventilation is addressed specifically by **Regulation 6**, which requires every enclosed workplace to be ventilated by a sufficient quantity of fresh or purified air, and plant to achieve this must be properly maintained and include an effective device to give visible or audible warning of any failure where this is necessary for health or safety reasons. The only exception to this is any confined space, where use of breathing apparatus may be necessary.

A reasonable temperature during working hours in all workplaces inside buildings replaces specified temperatures in previous legislation (**Regulation 7**), together with a ban on heating or cooling methods which allow injurious or offensive fumes to enter a workplace. Suitable thermometers in sufficient numbers are to be provided and maintained for workers to establish the temperature in any workplace

inside a building. The ACoP gives a minimum temperature of 16 degrees Celsius for sedentary work, and 13 degrees Celsius where there is severe physical effort, unless lower maximum room temperatures are required by law. (Note, however, that general food hygiene regulations do not specify maximum room temperatures.)

Suitable and sufficient lighting is required by **Regulation 8** for every workplace, by natural light as far as is reasonably practicable. Additionally, where there is potential danger due to failure of any artificial light source, suitable and sufficient emergency lighting is required.

Cleaning and decoration requirements are given in **Regulation 9**. Floor, wall and ceiling surfaces inside buildings must be capable of being kept sufficiently clean. All workplaces and their furniture, furnishings and fittings must be kept clean, which includes keeping them free from any effluvia from drains or toilets, and keeping them free of waste materials (apart from what is kept in suitable receptacles) so far as is reasonably practicable.

Every room where persons work must have sufficient floor area, height and unoccupied space for purposes of health, safety or welfare (**Regulation 10**). Personal space is defined in the ACoP as $11m^3$. There is a notional maximum ceiling height of 3m to be used in this calculation. For some existing workplaces (factories, subject to the provisions of the Factories Act 1961), compliance with Part 1 of Schedule 1 to the Regulations (see below) will be deemed to satisfy the requirement.

By **Regulation 11**, workstations must be arranged to be suitable for any person who is likely to work there, and for any work likely to be done there. For those workstations outside buildings, there must be protection from adverse weather so far as is reasonably practicable, and so arranged that persons at the workstation are not likely to slip or fall. Also, the arrangement must enable anyone to leave it swiftly or receive assistance in an emergency. For each person at work in the workplace where a substantial part of the work must be done sitting, a suitable seat must be provided. This means suitable for the person as well as the task, with a suitable footrest where necessary.

Floors and traffic route surfaces must be constructed so that they are suitable for the purpose for which they are used (**Regulation 12**). It is specified that this means having no hole or slope, or being uneven or slippery, so as to expose persons to risks to health or safety. Floors must have effective means of drainage where necessary. For holes, no account is to be taken of one where adequate measures have been provided to prevent falls; for slopes, account is to be taken of any handrails provided, in deciding if there is exposure to risk to health or safety. Except where a traffic route would be obstructed, suitable and sufficient handrails and guards must be provided.

So far as is reasonably practicable, suitable and effective safeguards are required against any person falling a distance likely to cause personal injury, or being struck by a falling object likely to cause personal injury. **Regulation 13** adds that any area where there is a risk to health and safety from either of these must be clearly indicated as appropriate. So far as is reasonably practicable, every tank, pit or structure shall be securely covered or fenced where there is a risk of falling into a dangerous substance in it. Also, every traffic route over these must be securely fenced. 'Dangerous substances' are defined as those which are likely to burn or scald, and those which are poisonous, corrosive or as fumes, gases or vapours are likely to overcome a person. Also included are any granular or free-flowing solid, or viscous substance, of a nature or quantity likely to cause danger to anyone.

Windows, doors, gates, walls and partitions glazed wholly or partially so that they are transparent or translucent must, where necessary for health or safety, be of safe material and appropriately marked (**Regulation 14**). The ACoP makes it clear that a risk assessment is needed before a decision is taken on replacement. The Regulation does not mean that

all glass everywhere in a workplace must be replaced if it is not safety glass or other suitable material. An alteration to the ACoP was made by the HSC in October 1995 to make this issue more clear.

By **Regulation 15**, windows, skylights and ventilators which can be opened are required to be designed so that they cannot be opened, closed or adjusted in a way which causes danger to anyone. They are also not to be capable of remaining open in a dangerous position. Windows and skylights are to be designed or constructed so that they can be cleaned safely, taking account of equipment or devices fitted to the window, skylight or building (**Regulation 16**).

The safe circulation of pedestrians and vehicles in the workplace must be organised, to comply with **Regulation 17**. Traffic routes are to be suitable for the use made of them – sufficient, in the right places, and big enough. This means that traffic routes must be so arranged that vehicles using them do not endanger those at work nearby, there is sufficient separation of vehicles and pedestrians at doors and gates, and where both use the same route there is sufficient separation between them. All traffic routes are to be signed where necessary for health and safety.

Doors and gates (**Regulation 18**) must be suitably constructed, to include fitting with any necessary safety device. This means particularly that sliding and powered doors must not cause injury by falling on or trapping people. Powered doors and gates require suitable features to prevent injury by trapping anyone, and should be operable manually unless they open automatically if the power fails. Any door or gate which can be pushed open from either side must provide a clear view of both sides of the space when closed.

Regulation 19 requires travelators and escalators to function safely, be equipped with any necessary safety devices and have one or more identifiable and accessible emergency stop controls.

Sanitary conveniences must be suitable, sufficient, and in readily accessible places for all persons at work. The former means, according to **Regulation 20**, that the rooms containing them must be adequately lit and ventilated, rooms and conveniences are to be kept clean and orderly, and separate rooms containing conveniences are to be provided for men and women. The last is **not** required where a convenience is in a room intended for use by one person at a time and which has a door that can be secured from inside.

Washing facilities which are suitable and sufficient are required for all persons at work in any workplace, and are to be readily accessible. This also includes showers if required for health or work reasons. **Regulation 21** defines 'suitable' as those which:

- are provided in the immediate vicinity of every sanitary convenience (whether or not provided elsewhere as well)
- are provided in the vicinity of any changing rooms required by these Regulations (whether or not provided elsewhere as well)
- include a supply of clean hot and cold, or warm, water which is running water as far as practicable
- include soap or other suitable cleanser
- include towels or other suitable means of drying
- are contained in rooms sufficiently ventilated and lit
- are together with their rooms kept clean and orderly
- provide separate facilities for men and women, unless they are in a room which can only be used by one person at a time and where the door can be secured from the inside.

The final requirement, for separate facilities, does not apply where they are used only for washing the face, hands and forearms.

An adequate supply of wholesome drinking water is to be provided and maintained for all persons at work in the workplace, which is to be readily accessible in suitable places and conspicuously marked by a sign where this is necessary for health and safety. An example would be where people could otherwise drink from a contaminated water supply. A sufficient number of drinking vessels is also required by **Regulation 22**, unless a water jet is installed where persons can drink easily.

Accommodation for clothing which is suitable and sufficient is required for personal clothing not worn during work hours, and for special clothing worn at work but not taken home (**Regulation 23**). Accommodation is 'suitable' if:

- where changing facilities are required by Regulation 24, it is secure for clothes not worn
- where necessary to avoid risks or damage to the clothing, it includes separate facilities for work clothes and for other clothes
- it allows or includes drying facilities so far as is reasonably practicable
- it is in a suitable location.

Regulation 24 requires suitable and sufficient changing facilities in all cases where special work clothing has to be worn and a person cannot be expected to change elsewhere for reasons of health or propriety. 'Suitable' facilities will include separate facilities or separate use of facilities by men and women where necessary for propriety.

Suitable and sufficient rest facilities are to be provided at readily accessible places (**Regulation 25**), and the facilities must include one or more rest rooms. They must also include facilities to eat meals where food eaten in the workplace would otherwise become contaminated. Rest rooms must include suitable arrangements to protect non-smokers from discomfort caused by tobacco smoke. Suitable facilities must be provided for any working pregnant women or nursing mothers to rest. Suitable and sufficient facilities are to be provided for persons at work to eat meals, where meals are regularly eaten in the workplace.

Exemption certificates may be issued by the Secretary of State for Defence for the armed forces (**Regulation 26**).

Revision

14 major topics in the Workplace (Health, Safety and Welfare) Regulations 1992:
- maintenance of workplace and equipment servicing it
- ventilation, temperature and lighting
- cleanliness
- workspace allocation
- workstation design and arrangement
- traffic routes and floors
- fall protection
- glazing
- doors and gates
- travelators and escalators
- sanitary and washing facilities
- drinking water supply
- accommodation for clothing
- facilities for changing, rest and meals

Self-assessment questions

1 Consult the ACoP to establish the minimum number of toilets for each sex required in your workplace.

2 Write notes on the application of the Regulations to a forestry worker in a wood.

8 The Provision and Use of Work Equipment Regulations 1998

Introduction
These Regulations (PUWER 98) replace an earlier 1992 set of the same name, which implemented the first EC directive on work equipment – the Use of Work Equipment Directive (known as UWED). The European Council of Ministers agreed a further directive – the Amending Directive to the Use of Work Equipment (AUWED) – in 1995, and PUWER 98 (SI 1998 No 2306) implements that Directive in the UK.

The **main changes** between the 1992 and the 1998 Regulations are that:
- there are new requirements for mobile work equipment
- they cover lifting equipment which is also subject to the Lifting Operations and Lifting Equipment Regulations 1998 (LOLER)
- previous schemes for inspections, thorough examinations and tests are changed
- regulations for power presses and woodworking have been incorporated into PUWER 98 (although both topics have their own separate ACoPs explaining in detail how the Regulations and their principles are to be applied).

The 1998 Regulations came into effect on 5 December 1998. The requirements in respect of mobile work equipment in Regulations 25–30 have applied to such equipment provided for use in the establishment or undertaking before 5 December 1998 since 5 December 2002.

The Regulations revoke the remaining parts of the Power Presses Regulations 1965, the Abrasive Wheels Regulations 1970, the Woodworking Machines Regulations 1974, and Regulation 27 of the Construction (Health, Safety and Welfare) Regulations 1996 (CHSWR). They are accompanied by an ACoP, the study of which is essential to an understanding of the application of the Regulations. In this commentary, details from the ACoP have been added where appropriate to allow a full understanding of some requirements.

Objectives of the Regulations
The Regulations set objectives to be achieved, rather than establish prescriptive requirements. They simplify and clarify previous legislation using a clear set of requirements to ensure the provision of safe work equipment and its safe use, irrespective of its age or place of origin. As a result some equipment, notably that used for lifting, is subject to at least two sets of Regulations, and these should be read in conjunction to determine particular obligations.

Some significant definitions
Work equipment – any machinery, appliance, apparatus, tool or installation for use at work, whether exclusively or not. This definition covers almost any equipment used at work and any assembly of components which, in order to achieve a common end, are arranged and controlled so that they function as a whole. Examples include:
- 'tool-box tools' such as hand tools, including hammers, knives, handsaws
- single machines including drills, circular saws, photocopiers, dumpers
- laboratory apparatus
- lifting equipment, such as hoists, fork-lift trucks, elevating work platforms, lifting slings (all of which are also covered by LOLER)
- other equipment including ladders and pressure cleaners
- installations such as a series of machines connected together, as in production lines, and scaffolding (except where the Construction (Health, Safety and Welfare) Regulations 1996 impose more detailed requirements).

Not included in the definition are:
- livestock
- substances such as acids, alkalis, slurry, cement, water
- structural items such as walls, stairs, roofs, fences, private cars.

Inspection – such visual or more rigorous inspection by a competent person as is appropriate for the purpose.

Thorough examination – a thorough examination by a competent person, including testing, the nature and extent of which are appropriate for the purpose described in Regulation 32.

Use – any activity involving work equipment. This includes starting, stopping, programming, setting, transporting, repairing, modifying, maintaining, servicing and cleaning.

Danger zone – means any zone in or around machinery in which a person is exposed to a risk to health and safety from contact with a dangerous part of machinery or a rotating stock-bar.

Requirements of the Regulations

Regulations 1–3 contain citation, definitions and application. Almost all employment relationships and the places of work involved, where the Health and Safety at Work etc Act 1974 applies, are within scope. This includes all industrial sectors, offshore operations and service occupations. All activities involving work equipment are covered, as is any work equipment made available for use in non-domestic premises. All the requirements apply to employers, the self-employed in respect of personal work equipment, and persons holding obligations under Section 4 of the Health and Safety at Work etc Act 1974 (control of premises) in connection with the carrying on of a trade, business or other undertaking.

Regulation 4 requires employers and others to ensure that work equipment is constructed or adapted so as to be suitable for the purpose for which it is used or provided. In selecting it, employers must have regard to the working conditions and the risks to health and safety which exist where the equipment is used, as well as any additional risks posed by its use. The equipment is to be used only for the operations for which, and under conditions for which, it is 'suitable'. 'Suitable' is defined as being suitable in any respect which it is reasonably foreseeable will affect the health or safety of any person.

Regulation 5 requires the equipment to be maintained in an efficient state, in efficient working order, and in good repair. Where any machinery has a maintenance log, that log is to be kept up to date. There is no requirement to keep such a log, but the guidance to the Regulations recommends that one be kept in respect of high-risk equipment.

Regulation 6 covers the general inspection requirements. The employer must ensure that, where the safety of work equipment depends on the installation conditions, it is inspected by a competent person to ensure that it has been installed correctly and is safe to operate:

- after installation and before being put into service for the first time
- after assembly at a new location.

Work equipment that is exposed to conditions causing deterioration liable to result in dangerous situations must be inspected to ensure that health and safety conditions are maintained, and that any deterioration can be detected and remedied in good time. This has to be done at suitable intervals and each time that exceptional circumstances liable to jeopardise the safety of the work equipment have occurred.

The 'competent person' must have the necessary knowledge and experience. The level of competence will vary according to the type of equipment and how and where it is used.

Employers must ensure that the results of inspections are recorded and kept until the next inspection is recorded.

No work equipment is to leave an employer's undertaking or, if it is obtained from another person, is to be used in his/her undertaking, unless it is accompanied by physical evidence that the last inspection which was required to be carried out has been carried out. This evidence could be a copy of the record of inspection, or for smaller items, a tag, colour code or label attached to the item.

The minimum inspection regime should be set by a competent person on behalf of the owner/supplier of the equipment, based on the manufacturer's information and other statutory obligations. The need, if any, for additional inspections will be identified by the user. Factors to be taken into account by the user include the work being carried out, any specific risks on site that may affect the condition of the equipment, and the intensity of use. An inspection may include visual checks, a strip-down of the equipment and functional tests. Advice should be sought from the manufacturer's instructions and a competent person for guidance on what an inspection should consist of for each piece of equipment.

A number of parties have responsibilities for inspections. Hire companies must ensure that the equipment they hire out complies with PUWER 98. Those using hired equipment must agree with the plant hire company on who is to carry out the inspections and when they will be carried out. The employer or self-employed person has a duty to ensure that equipment they use or provide complies with PUWER 98, including ensuring that inspections are carried out by a competent person. Those using equipment provided by another employer have a duty to ensure that the equipment is safe for use.

If the equipment is provided for common use on site, such as a scaffold, it must be established who is to take responsibility for ensuring that it complies with the Regulations. Each employer must still ensure that the equipment is suitable and safe for use by his/her employees.

Equipment that poses a significant risk will need to be considered by a person who is competent to determine a suitable regime. These inspections are in addition to the daily checks by the operator and must be carried out by a competent person. For the majority of equipment the formal inspection will be undertaken weekly (as was the case with certain items of construction plant under the previous regulations). Some equipment will need more frequent inspections, eg items used in confined spaces may require to be inspected before each shift.

There are some exceptions to these inspection requirements where other legislation details a particular inspection regime, eg CHSWR, LOLER, and for certain power presses and mine winding equipment.

Low-risk equipment used for low-risk activities will not require a formal inspection. A visual check may be required by the user before each use to ensure that it is in good condition, and the person carrying out these checks must be competent. There is no requirement for the operator to record the results of the visual check by the operator, but this may be an appropriate practice in some circumstances. In circumstances where additional hazards exist, low-risk equipment may need a more detailed check.

Equipment of a higher risk, and that which has moving parts, should have a visual check before each use and may require a more formal check at specified intervals. A competent person should ascertain how often these formal checks should take place.

Specific risks are dealt with by **Regulation 7**, which restricts use of equipment likely to involve a specific risk to health or safety to those persons given the task of using it. Where maintenance or repairs have to be done, this work is restricted to those who have been specifically designated to do it, having received adequate training for the work.

Users and supervisors of work equipment must have available to them adequate health and safety information and, where appropriate, specific written instructions pertaining to the use of the equipment (**Regulation 8**). User and supervisor training, including work methods, risks and precautions, are covered in **Regulation 9**. Some specific training which is referred to in the ACoP includes young persons, and operators of self-propelled work equipment and chainsaws.

Employers are required by **Regulation 10** to ensure that work equipment has been designed and constructed in compliance with any 'essential safety requirements'. These are those relating to its construction or design in any of the statutory instruments listed in Schedule 1 to the Regulations (probably the most significant is the Supply of Machinery (Safety) Regulations 1992, but there are other requirements for tractors, noise, industrial trucks and simple pressure vessels, for example). This requirement applies only to work equipment provided for use in the employer's premises or undertaking for the first time after 31 December 1992. In practice, this Regulation means that employers must check that operating instructions are present, and that the equipment has been checked for obvious faults. Products should also carry the 'CE' mark.

Regulation 11 deals with specific hazards associated with the use of work equipment. Mostly, this and the remainder of the Regulations are aimed at the **provision** of safe equipment, fitted where necessary with safety devices or protected against failure; in other cases, the requirements affect the **use** of the equipment. The Regulation sets out requirements to reduce the risk to employees from dangerous parts of machinery. They include measures to prevent access to those dangerous parts and to stop the part's movement before someone (or part of their body) enters the danger zone.

In selecting preventive measures, the Regulation sets out a hierarchy of them on four levels. These are:
* fixed enclosing guards to the extent practicable to do so, *but where not*
* other guards or protection devices to the extent practicable, *but where not*
* protection appliances (jigs, push-sticks, etc) to the extent practicable, *but where not*
* provision of information, instruction, training and supervision.

Guards and protection devices are to be:
* suitable for the purpose
* of good construction, sound material and adequate strength

* maintained in an efficient state, in good repair and in efficient working order
* not the source of additional risk to health or safety
* not easily bypassed or disabled
* situated at sufficient distance from the danger zone
* not unduly restrictive of any necessary view of the machine
* constructed or adapted so that they allow maintenance or part replacement without removing them.

Exposure of a person using work equipment to specified hazards must be prevented by the employer as far as is reasonably practicable, or adequately controlled where it is not (**Regulation 12**). The protection is not to be by provision of PPE or of information, instruction, training and supervision so far as is reasonably practicable, and is to include measures to minimise the effects of the hazard as well as to reduce the risk. The specified hazards are:
* ejected or falling objects
* rupture or disintegration of parts of the work equipment
* fire or overheating of the work equipment
* unintended or premature discharge of any article or of any gas, dust, liquid, vapour or other substance produced, used or stored in the work equipment
* unintended or premature explosion of the work equipment or any article or substance produced, used or stored in it.

This Regulation does not apply where other specified regulations apply, including those on COSHH, ionising radiations, asbestos, noise, lead and head protection in construction activities.

Measures must be taken to ensure that people do not come into contact with work equipment, parts of work equipment or any article or substance produced, used or stored in it which are likely to burn, scald or sear (**Regulation 13**).

Specific requirements relate to the provision, location, use and identification of control systems and controls on work equipment (**Regulations 14–18**). They relate to controls for starting or making a significant change in operating conditions, stop controls, emergency stop controls, controls in general and control systems.

All work equipment is to have a means to isolate it from all its sources of energy (**Regulation 19**). The means is to be clearly identifiable and readily accessible, and reconnection of the equipment to any energy source must not expose any person using the equipment to any risk to health or safety. Work equipment is to be stabilised by clamping or otherwise where necessary, so as to avoid risks to health or safety (**Regulation 20**). Places where work equipment is used have to be suitably and sufficiently lit, taking account of the kind of work being done (**Regulation 21**).

So far as is reasonably practicable, maintenance operations are to be done while the work equipment is stopped or shut down. Otherwise, other protective measures are to be taken, unless maintenance people can do the work without exposure to a risk to health or safety (**Regulation 22**).

Employers must ensure that all work equipment has clearly visible markings where appropriate (**Regulation 23**) and any warnings or warning devices appropriate for health and safety (**Regulation 24**). Warnings will be inappropriate unless they are unambiguous, easily perceived and easily understood.

No employee can be carried on mobile work equipment unless it is suitable for carrying passengers and it has features for reducing to as low as is reasonably practicable risks to their safety, including risks from wheels or tracks (**Regulation 25**). Risks from moving parts other than wheels and tracks are covered by Regulation 11, while mounting and dismounting are covered by the Health and Safety at Work etc Act 1974.

Regulation 26 deals with risks arising from rollover while the equipment is travelling.

Where there is a risk to an employee riding on mobile work equipment from its rolling over, it must be minimised by:
- stabilising the work equipment
- providing a structure which ensures that the work equipment does no more than fall on its side
- providing a structure giving sufficient clearance to anyone being carried if it overturns further than on its side
- a device giving comparable protection.

Where there is a risk of anyone being carried by mobile work equipment being crushed by its rolling over, the employer must ensure that it has a suitable restraining system.

These requirements do not apply to:
- a fork-lift truck having a structure which prevents it falling more than on its side and giving sufficient clearance if it overturns further than that (see below)
- where such provisions would increase the overall risk to safety
- where it would not be reasonably practicable to operate the mobile equipment in consequence
- where, in relation to an item of work equipment provided for use in the undertaking or establishment before 5 December 1998, it would not be reasonably practicable.

Fork-lift trucks (which have vertical masts which give effective protection to seated operators, or have full rollover protection (ROPS) fitted) carrying an employee are to be adapted or equipped to reduce to as low as is reasonably practicable the risk to safety from overturning (**Regulation 27**). This means that restraint systems should be provided, following a risk assessment and if they can be fitted, to fork-lift trucks with a mast or a ROPS.

Where self-propelled work equipment in motion could involve a risk to anyone's safety, by **Regulation 28** it must have:
- facilities for preventing it being started by an unauthorised person
- facilities for minimising the consequences of a collision (where it is rail-mounted)

- physical means of braking and stopping
- appropriate aids where necessary to assist the driver's field of vision
- lighting for use in the dark
- firefighting appliances, depending on the circumstances of use.

Where remote-controlled, self-propelled equipment (such as radio-controlled machines) present a risk to safety while in motion, it must stop automatically once it leaves its control range. Where the risk is of crushing or impact, it must incorporate features to guard against such risks unless other appropriate devices are able to do so (**Regulation 29**).

Drive shafts are covered by **Regulation 30**. These are devices conveying power from mobile work equipment to any work equipment connected to it – the agricultural tractor power take-off shaft is a common example. The basic requirement here is to safeguard against the effect of any seizure and the shaft itself.

Regulations 31–35 deal specifically with power presses. Compliance with the former Power Presses Regulations will achieve compliance with PUWER 98, although there are some differences in terminology and scope. PUWER 98 does not specify any particular means of recording inspections, but its Schedule 3 does define what information must be recorded. There is a new requirement for periodic thorough examination of fixed guards. Young people working on and with power presses are within the scope of the amended Management of Health and Safety at Work Regulations 1999. The reader interested in the detailed application of Regulations 31–35 is advised to become familiar with L112 – *Safe use of power presses*, which is the ACoP containing the text of the Regulations.

Revision

10 requirements of the Provision and Use of Work Equipment Regulations 1998:
- suitability
- maintenance
- use by persons given the task
- information
- training
- CE conformity
- risk reduction
- control improvements
- isolation arrangements
- warning

4 new applications, to:
- woodworking machines
- power presses
- fork-lifts and other mobile work equipment
- drive shafts

9 The Lifting Operations and Lifting Equipment Regulations 1998

Introduction

These Regulations (LOLER) apply to all lifting equipment, whether it is existing, second-hand, leased or new, and to all premises and work situations covered by the Health and Safety at Work etc Act 1974. Other legislation, including the Construction (Lifting Operations) Regulations 1961, was replaced with effect from 5 December 1998, but the Docks Regulations 1988 have been only partially revoked. LOLER builds on the requirements of the Provision and Use of Work Equipment Regulations 1998 (PUWER 98) which came into force on the same day, and should be read in conjunction with them. Duty holders providing lifting equipment will need to comply with relevant parts of both PUWER 98 and LOLER.

The LOLER strategy is based on the requirement within the Management of Health and Safety at Work Regulations 1999 for a risk assessment to identify the nature and level of risks associated with a proposed lifting operation. The factors to be considered in the LOLER assessment process are:

- type of load being lifted – its nature, weight and shape
- risk of the load falling or striking something, and its consequences
- risk of the lifting equipment striking something or someone, and its consequences
- risk of the lifting equipment failing or falling over while in use, and the consequences.

LOLER (SI 1998 No 2307) applies to any item of equipment used for lifting or lowering loads and any operation concerned with the lifting or lowering of a load. The Regulations include equipment not previously considered to be lifting equipment (described in former regulations as 'lifting appliances' or 'lifting machines'). The definition, therefore, includes cranes, goods and/or passenger lifts, mobile elevating work platforms (MEWPs), scissor lifts, vehicle hoists, lorry loaders (such as HIABs and tail-lifts), gin wheels, ropes used for access pulleys, and fork-lift trucks and site handlers.

'Accessories for lifting' are also lifting equipment, and include chains, ropes, slings, and components kept for attaching loads to machinery for lifting, such as hooks, eyebolts, lifting beams or frames (described in former regulations as 'lifting gear' or 'lifting tackle').

Objective of the Regulations

LOLER sets out a coherent strategy for the control of lifting operations and the equipment associated with them. LOLER implements the Amending Directive to the Use of Work Equipment Directive (AUWED, 95/63/EC).

Some significant definitions

Lifting equipment – work equipment for lifting or lowering loads, including attachments used for anchoring, fixing or supporting them.

Lifting accessory – work equipment for attaching loads to machinery for lifting.

Lifting operation – an operation concerned with the lifting or lowering of a load.

Load – includes a person, material, or animal as well as the lifting accessories and hook block.

Examination scheme – a suitable scheme drawn up by a competent person for such thorough examination of lifting equipment at such intervals as may be appropriate for the purpose of Regulation 9.

Thorough examination – means a thorough examination by a competent person and includes any appropriate testing.

Work equipment – includes any machinery, appliance, apparatus, tool or installation for use (whether exclusively or not) at work.

Equipment outside the scope of LOLER include conveyor belts moving on a horizontal level and winches used to haul a load on level ground. However, PUWER 98 requires a very similar level of safety, so the distinction is academic. Escalators are covered by more specific requirements in Regulation 19 of the Workplace Regulations 1992. LOLER's ACoP is essential reading, partly because it devotes considerable space to commentary on the application to lifting equipment of the general principles of other regulations, especially PUWER 98, the Noise at Work Regulations 1989 and COSHH. It also discusses the division of responsibilities between duty holders such as owner/landlord and tenant, and hirers and users.

Requirements of the Regulations

Lifting equipment selected for a lifting operation must be suitable for the activity it is to carry out (**Regulation 3**). Factors to be considered in the risk assessment have been mentioned above.

Every part of a load and any lifting attachment must be of adequate strength and stability (**Regulation 4**). Particular attention must be paid to any mounting and fixing points. These terms include the fixing of the equipment to a supporting structure and where two parts of the equipment are joined, such as with two sections of a crane.

Where lifting equipment is used for lifting people, **Regulation 5** stipulates the operating conditions. The lifting carrier has to prevent crushing, trapping, being struck and falls from it so far as is reasonably practicable, and allow a trapped passenger to be freed. The lifting equipment must have suitable devices to prevent the carrier falling, except in the case of mines, where other enhanced safety coefficient suspension rope or chain can be used.

Regulation 6 applies to both fixed and mobile lifting equipment, requiring it to be positioned or installed to reduce the risk as low as is reasonably practicable of either the equipment or the load striking anyone, or of the load drifting, falling freely or being released unintentionally, and otherwise to be safe. There is also a general requirement to prevent falls down hoistways and shafts by means of gates.

Lifting equipment must be marked with the safe working load for each configuration except where information clearly showing this is kept with the machinery. **Regulation 7** also requires lifting accessories to be marked to show their safe use characteristics; lifting equipment designed for lifting people is to be marked and that which is not so designed is to be clearly marked to that effect.

Organisation of lifting operations is covered in **Regulation 8**. The duty of the employer is to ensure that every lifting operation under LOLER has been properly planned by a competent person, is appropriately supervised and carried out safely.

The previous regime of testing, thorough examination and inspection under various industry-specific regulations has been replaced by new thorough examination and inspection requirements contained in Regulations 9–11.

Inspection and thorough examination requirements are detailed in **Regulation 9**. A thorough examination may include visual examination, a strip-down of the equipment and functional tests. Advice should be sought from the manufacturer's instructions and a competent person for guidance on what a thorough examination should include for each piece of lifting equipment. Any need for additional inspections will be identified by the user of the equipment. Factors that must be taken into account by the user include the work being carried out, any specific risks on site that may affect the condition of the equipment, and the intensity of use of the equipment.

All inspections and thorough examinations must be undertaken by competent persons. The level of competence will depend on the type of equipment and the level of thorough examination or inspection required.

A **thorough examination** must be carried out as follows:
- when the equipment is put into service for the first time unless:

○ it is new equipment that has not been used before and an EC declaration of conformity has been received which was made not more than 12 months before the lifting equipment is put into service, *or*

○ the equipment was obtained from another undertaking, and it is accompanied by a copy of the previous report of a thorough examination

- where safety depends on the installation conditions to ensure that it has been installed correctly and is safe to operate:

  ○ after installation and before being put into service for the first time

  ○ after assembly and before being put into service at a new site or in a new location

  (**NB:** This applies to equipment erected or built on site such as tower cranes, hoists or gantry cranes, but not to equipment such as a mobile crane that is not 'installed'.)

- lifting equipment for lifting persons must be thoroughly examined at least once in every six months
- accessories for lifting must be thoroughly examined at least once in every six months
- all other lifting equipment must be thoroughly examined at least once in every 12 months
- after exceptional circumstances liable to affect the safety of the lifting equipment. (This thorough examination is required even if an appropriate examination scheme has been developed.)

These intervals can be followed as laid down in Regulation 9, or there is an option for competent persons to develop an examination scheme for different pieces of plant. These schemes will normally specify a more frequent interval rather than a less frequent one. The requirement is therefore to possess either a written examination scheme, or to follow the specified periods for examination and testing within the Regulations. The need for and the nature of a test will be a matter for the competent person to decide on, taking into account the manufacturer's information.

An **inspection** may also be required at suitable intervals for certain types of lifting equipment, following the identification of significant risk to the operator or other workers by a competent person. Unless indicated otherwise by the manufacturer, it will be appropriate to continue inspecting lifting equipment, such as hoists and excavators, at weekly intervals. For the majority of lifting equipment the driver should be competent to carry out the regular inspection.

The employer receives a notification forthwith if the competent person carrying out the thorough examination finds a defect which could in his/her opinion become a danger to anyone. The owner must then take immediate steps to ensure the equipment is not used before the defect has been rectified. Reports of thorough examinations are to be sent to the employer and any owner/hirer, and if there is found to be an imminent or existing risk of serious personal injury, the competent person is required to notify the enforcing authority. **Regulation 10** sets out inspection requirements, which include making the competent person inspecting the equipment notify the employer forthwith of any defects and to make a written record of the inspection.

Regulation 11 contains requirements for record-keeping. Any EC conformity declaration received must be kept for as long as the equipment is operated. The reports of thorough examinations must be kept as follows:

- for equipment first put into use by the user – until he/she no longer uses the lifting equipment (except for an accessory for lifting)
- for accessories for lifting first put into use by the user – for two years after the report is made
- for equipment dependent on installation conditions – until the user no longer uses the lifting equipment at that place
- for all other thorough examinations – until the next report is made or for two years, whichever is the longer.

If the intervals laid down in the Regulations are followed, all thorough examination reports will need to be kept for two years from the date that the last report was made. The initial declaration must be kept until the equipment is disposed of. Reports of inspections must be kept available until the next report is made.

Reports of thorough examinations and inspections should be kept available for inspection at the place where the lifting equipment is being used. If this is not possible, the information should be readily accessible. Reports must be readily available to the HSE or local authority inspectors if required.

No lifting equipment should leave any undertaking unless accompanied by physical evidence that the last thorough examination has been carried out.

Revision

The Lifting Operations and Lifting Equipment Regulations 1998 require:

- lifting equipment to be suitable, strong, stable, marked with the safe working load
- load and lifting attachments to be strong, stable
- lifting operations to be supervised and safely conducted
- inspection, test, thorough examination – minimum, or devised by a competent person
- documents – record-keeping and evidence

10 The Manual Handling Operations Regulations 1992

Introduction

Significant obligations are placed on employers and the self-employed with regard to all manual handling tasks in the workplace. The Regulations are the result of the European Directive on minimum health and safety requirements for the manual handling of loads where there is a risk particularly of back injury to workers (90/269/EEC). The Regulations replace all former provisions on the subject, and apply to all those at work regardless of their particular workplace. Previous legislation proved to be inadequate, being only applicable to certain work industries and difficult to interpret and enforce. It employed vague terms related to single lifts of loads; some regulations covering particular industries applied load limits, again for single loads. The Regulations concentrate attention on ergonomic solutions, and take account of a range of relevant factors including repetitive lifting. Upper limb disorders are not addressed by the Regulations.

A clear step-by-step approach to removing hazards and minimising risks is provided. The most fundamental requirement is to avoid hazardous manual handling operations so far as is reasonably practicable. Automation, redesigning the task to avoid manual movement of the load and mechanisation have to be considered at this point, if there is a risk of injury and a manual handling operation (as defined) is involved. If risk remains, a suitable and sufficient assessment of the operation has to be made, and that risk has to be reduced so far as is reasonably practicable, preferably involving mechanical assistance.

Objectives of the Regulations

The objectives of the Regulations are to implement the Directive and apply an ergonomic approach to the prevention of injury while carrying out manual handling tasks.

Some important definitions

Injury – includes those resulting from the weight, shape, size, external state, rigidity (or lack of rigidity) of the load or from the movement or orientation of its contents. It does not include those resulting from the spillage or leakage of the contents (ie those areas already controlled by other regulations such as COSHH).

Load – includes any person and any animal. Generally, this would be a discrete, movable object and would include a patient receiving medical attention and an animal undergoing veterinary treatment. This definition also covers material supported on a shovel or fork. However, an implement, tool or machine such as a chainsaw is not included.

Manual handling operation – means any transporting or supporting of a load (including lifting, putting down, pushing, pulling, carrying or moving) by hand or bodily force (such as using the shoulder). The force applied must be human and not mechanical.

Requirements of the Regulations

The Regulations impose the same duties on the self-employed as employers (**Regulation 2(2)**) but do not apply to normal activities on seagoing ships under the direction of the master (**Regulation 3**).

Regulation 4 places duties on employers to make evaluations and then assessments of certain manual handling operations. They must, so far as is reasonably practicable, avoid the need for employees to carry out those operations which involve a risk of injury (**Regulation 4(1)(a)**). Where this cannot be done, **Regulation 4(1)(b)** requires them to:

- make and keep up to date a suitable and sufficient assessment of all such manual handling tasks (**4(1)(b)(i)**), considering the factors and questions specified in the **Schedule** to the Regulations. Assessments are to be reviewed and amended if appropriate – if they become invalid or there is a significant change in the manual handling operation to which they relate

- take appropriate steps to reduce the risk of injury to employees arising from any such operation to the lowest level reasonably practicable (**4(1)(b)(ii)**)
- take appropriate steps to provide employees who are carrying out manual handling operations with general indications and, where reasonably practicable to do so, precise information on the weight of each load and the heaviest side of any load whose centre of gravity is not centrally positioned (**4(1)(b)(iii)**).

Factors which must be taken into account when making assessments are specified in detail in Schedule 1, the use of which is compulsory.

Regulation 5 requires employees to make full and proper use of any system of work put in place by the employer to reduce the risk of injury during manual handling. This duty is in addition to the general duties of employees

under Regulations 7 and 8 of the Health and Safety at Work etc Act 1974.

Regulation 6 provides for the issue of exemption certificates by the Secretary of State for Defence for the armed forces in the interests of national security. **Regulation 7** extends the requirements to offshore work in territorial waters. **Regulation 8** revokes all the previous specific load-limiting regulations, and sections of the Factories Act 1961 and the Offices, Shops and Railway Premises Act 1963 referring to prohibitions on 'loads so heavy as to be likely to cause injury'.

NB: These requirements must be taken in conjunction with other duties imposed, particularly by the Management of Health and Safety at Work Regulations 1999, where the reader's attention is drawn to the need for capability assessment and training. Part 2 Section 4 gives more information on the practical implications of manual handling, and information on risk assessment is available in Part 1 Section 3.

Revision

5 requirements of the Manual Handling Operations Regulations 1992:
- avoid manual handling
- mechanise or automate process
- assessment
- risk reduction
- information

Self-assessment questions

1 Identify and list activities in your workplace where these Regulations apply.

2 Write notes on how these requirements affect a company supplying and installing double glazing.

11 The Health and Safety (Display Screen Equipment) Regulations 1992

Introduction

The Regulations implement a European Directive (90/270/EEC) on minimum health and safety requirements for work display screen equipment (DSE). The Regulations do not replace general obligations under the Health and Safety at Work etc Act 1974, the Management of Health and Safety at Work Regulations 1999 (MHSWR), the Workplace (Health, Safety and Welfare) Regulations 1992, and the Provision and Use of Work Equipment Regulations 1998.

There are some overlaps between general and specific legislation such as this. Where general duties are similar to specific ones, as in the case of general risk assessments and those required by these Regulations, the legal requirement is to comply with both the general and the specific duty. However, assessment of workstations required by Regulation 2 will also satisfy the general requirement under the MHSWR for risk assessment as far as those workstations are concerned. The employer still has to apply the general duties in other work areas, and other parts of work areas where display screen work is being carried out.

Objectives of the Regulations

The objectives of the Regulations are to improve working conditions at display screen equipment by providing ergonomic solutions, to enable certain regular users of the equipment to obtain eye and eyesight tests and information about hazards, risks and control measures associated with their workstations.

Some important definitions

Display screen equipment – any alphanumeric or graphic display screen, regardless of the display process involved. This includes cathode-ray tube display screens, liquid crystal and other new technologies. Screens showing mainly TV or film pictures are not covered. Microfiche screens are included in the definition. **Regulation 1(4)** states the following are not covered by these Regulations: window typewriters, calculators, cash registers and other equipment with small data displays, portable systems not in prolonged use, and DSE mainly intended for public use (cashpoints, etc), on a means of transport, and in drivers' or control cabs for vehicles or machinery.

Use – means in connection with work.

User – an employee who habitually uses DSE as a significant part of his or her normal work.

Operator – a self-employed person who habitually uses DSE as a significant part of his or her normal work.

Workstation – the immediate work environment around the DSE, including all accessories, desk, chair, keyboard, printer and other peripheral items.

More about users and operators

An individual will generally be classified as a user or operator if most or all of the following apply:

- depends on the use of DSE to do the job (no alternative means readily available to achieve the same results)
- has no discretion as to use or non-use of the DSE
- needs significant training and/or special skills in the use of DSE to do the job
- normally uses DSE for continuous spells of an hour or more at a time
- uses DSE as above more or less daily
- the job requires fast transfer of information between user and screen
- performance requirements of the system demand high levels of attention and concentration.

Some examples of those likely to be classified as users/operators are: word-processing pool workers, personal secretaries, data input operators, journalists, telesales and enquiry operators, librarians. If there is doubt, carrying out a risk assessment should assist in making the decision. The guidance material accompanying the Regulations contains a helpful discussion and examples of this classification.

Requirements of the Regulations

Regulation 2 requires every employer to perform a suitable and sufficient analysis of those workstations which are used by users or operators, to assess the health and safety risks to which they are exposed as a consequence. An assessment is to be reviewed by the employer if there has been a significant change in the matters to which it relates, or if the employer suspects it is no longer valid. The employer is then required to reduce the risks so identified to the lowest extent which is reasonably practicable. 'Significant change' includes a major change in the software used, the hardware, furniture, increase in time spent using the DSE, increase in task requirement such as speed and accuracy, relocation of the workstation and modification to the lighting.

Regulation 3 requires that all workstations used by users or operators must meet the requirements of the **Schedule** to the Regulations. This lists minimum requirements for the workstation; the matters dealt with are:

- display screen
- keyboard
- work desk/surface
- work chair
- space requirements
- lighting
- reflections and glare
- noise
- heat
- radiation (but no action is necessary)
- humidity
- computer–user/operator interface.

Regulation 4 requires the planning of activities of users by the employer so that daily work is periodically interrupted by breaks or activity changes. General guidance is given that these could be informal breaks away from the screen for a short period each hour.

Regulation 5 gives users or those about to become users the opportunity to have an appropriate eye and eyesight test as soon as practicable after request and at regular intervals. This entitlement also applies when a user experiences visual difficulties attributable to DSE work. Special corrective appliances are to be provided by the employer for users where normal ones cannot be used. ('Special' appliances will be those prescribed to meet vision defects at the viewing distance – anti-glare screens are not special corrective appliances.) The costs of tests and special corrective appliances are to be met by the employer of the user.

Regulation 6 requires the user's employer to provide adequate health and safety training in the use of any workstation he/she may be required to work on, and further training whenever the organisation of any such workstation is substantially modified. The employer is to provide operators and users with information about risk assessments and control measures concerning the health and safety aspects of their workstations (**Regulation 7**). Users are to be given information about breaks and activity changes, eye and eyesight tests, and training both initially and when the workstation is modified.

The Secretary of State for Defence can issue exemption certificates for the armed forces if in the national interest to do so (**Regulation 8**). The Regulations are extended outside Great Britain in parallel with the application of the Health and Safety at Work etc Act 1974 (**Regulation 9**).

Revision
5 requirements of the Health and Safety (Display Screen Equipment) Regulations 1992:
  • risk assessment
  • schedule compliance
  • work breaks/activity changes
  • eye and eyesight tests, corrective appliances
  • training and information

Self-assessment questions

1 List the equipment in your work area to which these Regulations apply, and identify any 'users'.

2 Explain the difference between a 'user' and an 'operator'.

12 The Personal Protective Equipment at Work Regulations 1992

Introduction
The Personal Protective Equipment at Work Regulations 1992 replaced much of the prescriptive requirements of legislation made prior to the Health and Safety at Work etc Act 1974, and implement the requirements of a European Directive. They place duties on both employers and the self-employed. There had long been piecemeal requirements in various statutes, but these have now been repealed or revoked and the present Regulations provide a framework for the provision of PPE in circumstances where assessment has shown a continuing need for personal protection. Generally, a breach of these Regulations can give rise to a civil claim against an employer.

PPE used for protection while travelling on a road, such as cycle helmets, crash helmets and motorbike leathers worn by employees, is not covered by the Regulations. However, the requirements do apply in circumstances where this type of equipment is worn elsewhere (ie other than the highway) at work. An example is the wearing of crash helmets by farm workers using all-terrain vehicles. Protective equipment used during the playing of competitive sports is not covered either. Self-defence equipment (personal sirens/alarms) or portable devices for detecting and signalling risks and nuisances (such as personal gas detection equipment or radiation dosimeters) are not within the scope of the Regulations. Offensive weapons are not items of PPE.

The requirements of the Regulations do not apply to most respiratory protective equipment, ear protectors and some other types of PPE because they are already covered by existing Regulations, such as COSHH, the Ionising Radiations Regulations 1999, the Control of Lead at Work Regulations 2002, the Control of Asbestos at Work Regulations 2002, and the Construction (Head Protection) Regulations 1989 (see later Sections in this Part). However, this legislation is amended in varying degrees by Schedule 1 of these Regulations, which introduces the concept of risk assessment where it was previously missing. The Regulations also amend earlier provisions by requiring equipment to comply with European conformity standards and inserting a requirement to return it to suitable storage after use.

Objectives of the Regulations
The objectives of the Regulations are to

implement the Directive and formalise the provision of PPE following the assessment of risks required in companion regulations.

PPE – the definition
Personal protective equipment means all equipment designed to be worn or held by a person at work to protect against one or more risks, and any addition or accessory designed to meet this objective.

Both protective clothing and equipment are within the scope of the definition, and therefore such items as diverse as safety footwear, waterproof clothing, safety helmets, gloves, high-visibility clothing, eye protection, respirators, underwater breathing apparatus and safety harnesses are covered by the Regulations. Ordinary working clothes or clothing provided which is not specifically designed to protect the health and safety of the wearer is not within the definition – such as clothes provided with the primary aim of presenting a corporate image and protective clothing provided in the food industry for hygiene purposes.

Requirements of the Regulations
The Regulations do not apply to the crews of ships while operational (**Regulation 3**). Shore-based contractors working on ships will be within the scope of the Regulations while inside territorial waters. The requirements apply to PPE used on ships operating on inland waterways and aircraft while on the ground and operating in UK airspace.

Regulation 4 requires every employer to provide 'suitable' PPE to each of his/her employees who may be exposed to any risk while at work, except where any such risk has been adequately controlled by other means which are equally or more effective. A similar provision applies to the self-employed.

PPE is to be used as a last resort. Steps should first be taken to prevent or control the risk at source by making machinery or processes safer and by using engineering controls and systems of work. Risk assessments made under the Management of Health and Safety at Work Regulations 1999 will help determine the most appropriate controls.

'Suitable' means:
- appropriate for the risks involved and the conditions
- takes account of ergonomic requirements and the state of health of the person wearing it
- it is capable of fitting the wearer correctly after adjustments
- it is effective to prevent or adequately control the risk without leading to any increased risk (particularly where several types of PPE are to be worn) so far as is practicable
- complies with national and European conformity standards in existence at the particular time. (This requirement does not apply to PPE obtained before the harmonising Regulations came into force, which can continue to be used without European Conformity marks, but the employer must ensure that the equipment he/she provides remains 'suitable' for the purpose to which it is put.)

Regulation 5 requires the employer and others to ensure compatibility of PPE in circumstances where more than one item of equipment is required to control the various risks.

Before choosing any PPE, the employer or the self-employed person must make an assessment to determine whether the proposed PPE is suitable (**Regulation 6**). This assessment of risk is necessary because no PPE provides 100 per cent protection for 100 per cent of the time (it

may fail in operation, it may not be worn correctly or at all, for example) and as with other assessments, judgement of types of hazard and degree of risk will have to be made to ensure correct equipment selection. The assessment must include:
- an assessment of any risks which have not been avoided by any other means
- the definition of the characteristics which PPE must have in order to be effective against the risks, including any risks which the PPE itself may create
- a comparison of the characteristics of the PPE available with the identified desired characteristics.

For PPE used in high-risk situations or for complicated pieces of PPE (such as some diving equipment), the assessment should be in writing and be kept available. Review of it will be required if it is suspected to be no longer valid, or if the work has changed significantly.

Regulation 7 requires all PPE to be maintained, replaced or cleaned as appropriate. Appropriate accommodation must be provided by the employer or self-employed when the PPE is not being used (**Regulation 8**). 'Accommodation' includes pegs for helmets and clothing, carrying cases for safety spectacles and containers for PPE carried in vehicles. In all cases, storage must protect from contamination, loss or damage (particularly from harmful substances, damp or sunlight).

Regulation 9 requires an employer who provides PPE to give adequate and appropriate information, instruction and training to enable those required to use it to know the risks the PPE will avoid or limit, and the purpose, manner of use and action required by the employee to ensure the PPE remains in a fit state, working order, good repair and hygienic condition. Information and instruction provided must be comprehensible.

By **Regulation 10**, every employer providing PPE must take all reasonable steps to ensure it is properly used. Furthermore, Regulation 10 requires every employee provided with PPE to use it in accordance with the training and instruction given by the employer. The self-employed must make full and proper use of any PPE. All reasonable steps must be taken to ensure the equipment is returned to the accommodation provided for it after use.

Employees provided with PPE are required under **Regulation 11** to report any loss or defect of that equipment to their employer forthwith. **Regulation 12** permits the granting of

exemption certificates by the Secretary of State for Defence in respect of the armed forces. **Regulation 13** extends the applicability of the Regulations to all work activities in territorial waters and elsewhere subject to the Health and Safety at Work etc Act 1974.

NB: By virtue of Section 9 of the Health and Safety at Work etc Act 1974, no charge can be made to the worker for the provision of PPE which is used only at work.

Revision

8 requirements of the Personal Protective Equipment at Work Regulations 1992:
- provision as 'the last resort'
- suitability assessment
- compatibility between items
- maintenance and replacement
- accommodation
- information, instruction and training
- proper use
- report of loss/defect

Self-assessment questions

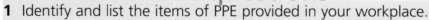

1 Identify and list the items of PPE provided in your workplace.

2 Write notes on the duties of employers under the Regulations.

13 The Fire Precautions Act 1971 and the Fire Precautions (Workplace) Regulations 1997

Introduction

Fire legislation, like other health and safety legislation, has evolved over a period of time, having its roots in the Industrial Revolution. It developed gradually, meeting individual and specific needs as and when it was necessary, embracing diverse premises. This approach to its development led to a fragmented body of legislation not just in relation to its requirements but also its enforcement. This lack of uniformity created pressure for rationalisation and consolidation.

Reform commenced with the report of the Holroyd Committee on the fire service in 1970. Members of this Committee recommended that the fire safety laws be consolidated into two distinct areas. The first dealt with fire precautions for new and altered premises (generally controlled under the Building Regulations) and the second dealt with premises already occupied. Control of this latter area was completed by the **Fire Precautions Act 1971**.

Readers should note that the Regulatory Reform (Fire Safety) Order 2005 makes substantial changes to these regulations, and comes into force on 1 April 2006. At the time of writing, guidance on the Regulations and their practical application is not available, but the HSE website (www.hse.gov.uk) will carry updates and should be checked regularly. Essentially, the Order reforms the law relating to fire safety in non-domestic premises. It replaces fire certification under the Fire Precautions Act 1971 with a general duty to ensure, so far as is reasonably practicable, the safety of employees. There is also a general duty, in relation to non-employees, to take such fire precautions as may reasonably be required in the circumstances to ensure that premises are safe, along with a duty to carry out a fire risk assessment. The Order imposes a number of specific duties in relation to the fire precautions to be taken. The Order also amends or repeals other primary legislation concerning fire safety. It follows that the material in this Section referring to fire certification will not apply after 1 April 2006.

The **Fire Precautions (Workplace) Regulations 1997** supplemented the Act by providing a uniform code applying to all workplaces not requiring a fire certificate under the exceptions laid down. In 1998 it was proposed to extend the application of the Fire Regulations, in response to concern expressed by the European Commission about the adequacy of implementation of Directive requirements.

One effect of the Fire Regulations was to require workplaces to be subject to specific fire risk assessment, but this did not apply to workplaces excepted from the application of the Regulations for various reasons – chiefly because they were required instead to hold a fire certificate. The European Commission's view was that regulations should be imposed on all employers without exception, fully reflecting the unconditional nature of the obligations contained within the Framework Directive.

The **Fire Precautions (Workplace) (Amendment) Regulations 1999** removed most of the exceptions, so that from 1 December 1999, almost all premises require fire risk assessments to be made and recorded, regardless of whether or not a fire certificate is held.

Objectives of the Act

The main objective of the Act is to provide a framework for the control of fire safety in all occupied premises designated by the Act, ensuring that adequate general fire precautions are taken, and appropriate means of escape and related precautions are present. (**NB:** The definition of 'occupied' extends to periods outside normal working hours, and includes 'occupation' by groups such as cleaners, security staff and maintenance contractors.)

How the objectives are met

In 1972, hotels and boarding houses were the first class of premises to be 'designated' under the Act, and as a result all but the smallest were required to have a fire certificate. By 1977, the Fire Precautions (Factories, Offices, Shops and Railway Premises) Order 1976 had required certain other premises to have a fire certificate as well, and for all premises in these categories which did not require a fire certificate because of their individual circumstances, the Fire Precautions (Non-Certified Factories, Offices, Shops and Railway Premises) Regulations 1976 specified basic safety requirements. The 1976 Order and Regulations were revoked in 1989.

The issue of fire certificates is the responsibility of fire authorities which enforce the Act and Regulations made under it in their local areas. For Crown premises, the responsibility falls to the Fire Service Inspectorates of the Home Office and the Scottish Home and Health Departments.

The Fire Safety and Safety of Places of Sport Act 1987 amended the 1971 Act, and allows certain premises to be exempted from the requirement to have a fire certificate by the fire authority. It also gives the fire authority power to issue improvement and prohibition notices. The Fire Precautions (Factories, Offices, Shops and Railway Premises) Order 1989 designated the use of premises of certain types under Section 1 of the 1971 Act, which thus require fire certificates. The uses designated are as factory, office, shop and railway premises where in each case people are employed to work. However, a fire certificate is not required for any of these premises where not more than 20 people work at any one time, and not more than 10 are at work on floors other than the ground floor, unless one or more of the following apply:

- the premises are in a building containing more than one set of premises used for the designated purposes and where more than 20 people in total are employed
- the premises are in a building containing more than one set of premises where the total number of employees working elsewhere than on the ground floor exceeds 10
- in the case of factory premises, explosive or highly flammable materials are stored or

used there (in these circumstances the fire authority may determine there is no serious additional risk and a certificate will not need to be issued).

Some premises, where hazardous materials, chemicals or processes are involved, are termed 'special premises' and are under the administrative control of the HSE. Detailed regulations have been implemented to certificate the use of such premises with regard to fire precautions. These are the Fire Certificates (Special Premises) Regulations 1976 – under which 15 types of premises are identified as requiring a fire certificate. The HSE is responsible for issuing the certificates under these Regulations, and the administrative control.

Hotels and boarding houses require fire certificates if six beds are provided (for staff, guests or both), or if any beds are provided below the ground floor or above the first floor.

Fire certificates

Application for a fire certificate must be made to the fire authority for the area in which the premises are situated, on form FP1(rev) 1989. While the application is pending, the owner/occupier is required by **Section 5(2A)** of the Act to ensure that the means of escape in case of fire at the premises can be used safely and effectively at all times when people are in the premises, that the provided firefighting facilities are maintained in efficient working order, and that those employed there receive instruction or training in what to do in case of fire.

A fire certificate will specify:
- the use of the premises
- the means of escape in case of fire
- the means for securing that the means of escape can be used safely at all relevant times
- the means for fighting fire, eg the number and type of fire extinguishers
- the means for giving warnings of fire, such as fire alarms
- particular requirements for any explosives or highly flammable materials stored or used on the premises (**Section 6**).

In addition, a fire certificate may specify:
- the required maintenance of the means of escape, including safe access to and along it
- means of other fire precautions
- training of employees and the keeping of training records
- limitation of the number of people allowed on the premises at any one time
- any other additional relevant fire precautions (**Section 6(2)**).

Where alterations are made to the structure of the building or to the working conditions carried on within that building, consultation is required (**Section 8**) with the fire authority before the work is carried out, and the fire authority may issue an amended certificate on completion of the work.

All occupied premises which are not required to have a fire certificate must be provided with an adequate means of escape in case of fire and a means of fighting fire (**Section 9A**).

The Fire Precautions (Workplace) Regulations 1997

Changes have been made to these Regulations by the Fire Precautions (Workplace) (Amendment) Regulations 1999. These amendments include alterations to enforcement practice and changes to the types of workplace that are exempt from the requirement to make a formal risk assessment. The most significant of the changes is that the assessment requirement now applies to premises with a fire certificate. The Amendment Regulations came into force from 1 December 1999.

For workplaces within the scope of the Regulations, risk assessments specifically addressing fire risks in the workplace must be made; this duty is in addition to the existing risk assessments required by the Management of Health and Safety at Work Regulations 1999 among others. Where five or more are employed, the findings must be formally recorded, and the findings must be passed to others affected, including employed staff.

Means of fire detection is required, and of giving warning, in each case as appropriate to the workplace and the circumstances. Means of escape in case of fire must be identified and properly signed, and firefighting equipment must be sufficient, maintained and tested, and of a suitable type appropriate to the nature of the risks. Training and induction must include information on what to do in the event of a fire emergency.

An HSE guidance document – *Fire safety: an employer's guide* – gives good basic advice on the requirements; the reader should also consult Part 2 Section 10 of this book for a treatment of some of the theoretical aspects of the subject.

Revision

A fire certificate is required when:
- more than 20 people are employed at work at any time
- more than 10 people are employed at work at any time elsewhere than on the ground floor
- flammable or explosive substances are stored or used in premises
- the premises are 'special premises'

Self-assessment questions

1 Does your own workplace require a fire certificate? Can you say why? Examine it carefully, and see how the certificate is worded and what (if any) special requirements have been added by the local fire authority.

2 List the contents of a fire certificate. Why are they crucial to planning for fire safety in the workplace?

14 The Control of Substances Hazardous to Health Regulations 2002 (as amended)

Introduction

People in the workplace are often exposed to substances which have the potential to damage their health. Many of these are present as a direct result of their use in a manufacturing process; some are formed by the process itself. Others are used in maintenance activities such as cleaning. Some hazardous substances occur naturally, such as microbiological agents, which can cause diseases like leptospirosis (leptospiral jaundice, also known as Weil's disease). The original move to control all such substances came as a result of EC Directive 80/1107/EEC on "the protection of workers from the risk related to exposure to chemical, physical and biological agents at work". These Regulations cover all people at work, from quarries to offices, and apply to virtually all substances hazardous to health. If the Dangerous Substances and Explosive Atmospheres Regulations 2002 (see Part 4 Section 27) also apply to a substance, the employer will have duties to comply with both sets of Regulations.

Originally introduced in 1988, the COSHH Regulations were revised and enlarged to incorporate provisions required by the Biological Agents Directive 90/679/EEC, and to prohibit importation into the UK of specified substances and articles from outside the European Economic Area. The latest complete revision of the Regulations – COSHH 2002 – came into force on 21 November 2002 and implemented the EC Chemical Agents Directive (98/24/EC).

As a result, there are legal definitions of basic terms, including 'hazard', 'risk' and 'risk assessment', as well as altered definitions of 'health surveillance', 'inhalable dust' and even 'substance hazardous to health'. The consequential revision of the ACoP and its guidance material requires detailed study to achieve an understanding of these comprehensive Regulations.

Amendments made in **2003** added a definition of 'mutagen' to Regulation 2(1) and extended some provisions to mutagens that previously referred only to carcinogens and biological agents. Seventeen dioxins were included in Schedule 1 as carcinogenic within the meaning of the Regulations.

In **2004**, substantial changes were made by the Amendment Regulations. They limited the supply and use of cement containing more than a very small amount of soluble chromium VI and imposed marking requirements, disapplied COSHH to all merchant ships in particular circumstances, and clarified the duty to maintain

exposure control methods by specifically extending the scope beyond plant and equipment to work methods and supervision. Arguably the most significant changes were to replace OESs and MELs with workplace exposure limits. See Part 3 Section 2 for a discussion of these terms. The amendments also included a new framework for the adequate control of exposure by specifying principles of good practice in Schedule 2A.

Objectives of the Regulations

The Regulations are intended to prevent workplace disease resulting from exposure to hazardous substances. Basic occupational hygiene principles are followed by the Regulations which introduce a control framework by requiring an adequate assessment of the risks to health arising from work activities associated with hazardous substances, the introduction of adequate control measures, maintenance of the measures and equipment associated with them, and monitoring the effectiveness of the measures and the health of employees.

What is a substance hazardous to health?

This term includes any material, mixture or compound used at work or arising from work activities, which is harmful to people's health in the form in which it occurs in the work activity. Categories specifically mentioned are:

- substances labelled as dangerous (toxic, very toxic, harmful, corrosive and irritant) under other statutory provisions

- substances assigned a WEL
- harmful micro-organisms
- substantial quantities of dust
- any substance creating a comparable hazard.

A 'substantial' concentration of dust is more than $10mg/m^3$ (8-hour time-weighted average (TWA)) of total inhalable dust, and more than $4mg/m^3$ similarly of respirable dust, where no lower value is given for the substance.

Part 3 contains more details on the terms above, and those used in the Regulations, which will not be given a further explanation here.

How the objectives are met by the Regulations

Duties are imposed on employers for the protection of employees who may be exposed to substances hazardous to health at work, and of other persons who may be affected by such work. Specific duties are also imposed on the self-employed and employees. An employer must not carry on any work which is liable to expose any employee to a substance hazardous to health unless a suitable and sufficient assessment has been made of the risks to health created by the substance at work and about the measures necessary to control exposure to it (**Regulation 6**).

This is an essential requirement; the assessment is a systematic review of the use of the substance present – its form and quantity, possible harmful effects, how it is stored, handled, used and transported as appropriate, the people who may be affected and for how long, and the control steps which are appropriate. The assessment should take all of these matters into account, taking care to describe actual conditions when assessing risk. **It is important to realise that collection of suppliers' material safety data sheets does not constitute making a risk assessment, but rather the gathering of data to assist in making it.** The assessment should be written down for ease of communication. All risk assessments should be reviewed every five years, in addition to the requirement to review them if they are suspected of being invalid.

The employer must ensure that the exposure of employees to substances hazardous to health is either prevented, or where this is not reasonably practicable, is adequately controlled (**Regulation 7**). This applies whether the substance is hazardous through inhalation, ingestion, absorption through the skin or contact with the skin. The primary control required is substitution, and where exposure prevention is not reasonably practicable, a prioritised list of control measures is specified in **Regulation 7(2)**.

Prevention of exposure must be attempted in the first place by means other than the provision and use of PPE. **Regulation 7(3)** only permits use of PPE where other control is not reasonably practicable, and then only in conjunction with other more effective controls. Any RPE supplied must comply with current requirements as to suitability (**Regulation 7(9)**).

Prevention of exposure can be achieved by removing the hazardous substance, replacing it with a less hazardous one or using it in a less hazardous form, enclosing the process, isolating the worker(s), using partial enclosure and extraction equipment, general ventilation and the use of safe systems of work.

Regulation 7(7) provides that control of exposure to a hazardous substance will only treated as adequate where the principles of good practice in Schedule 2A are applied; where any WEL for that substance is not exceeded; and, for substances carrying risk phrases 45, 46 or 49, or substances and processes listed in Schedule 1 or any asthma-inducing substance, exposure is reduced to as a low a level as is reasonably practicable.

Employers providing control measures must ensure they are properly used, and every employee must make full and proper use of what is provided, reporting any defects to the employer and doing their best to return equipment after use to any accommodation provided for it (**Regulation 8**).

Regulation 9 requires those control measures which are provided to be maintained in efficient working order and good repair. PPE must to be kept in a clean condition. If the controls are engineering controls, they must be examined and tested at prescribed intervals. For local exhaust ventilation equipment, this is defined as at least once in every 14 months generally. The Regulation also requires examination and testing of RPE at suitable intervals, where the equipment is provided under Regulation 7. Records of these tests and examinations have to be kept, and a summary of them has to be kept for at least five years.

Monitoring of exposure must be carried out (Regulation 10) where it is necessary to ensure that exposure is adequately controlled. For listed processes and substances, the frequency is specified. Records of this monitoring have to be kept for at least five years, or 40 years where employees can be personally identified. Monitoring would be required where:

- there could be serious risks to health if control measures should deteriorate
- it cannot be guaranteed without measurement that exposure limits are not being exceeded, or control measures are working properly.

Health surveillance is required where appropriate, which includes:

- where there is significant exposure to a substance listed in Schedule 5 to the Regulations and the employee is working at a process specified in the Schedule
- where there is an identifiable disease or adverse health effect which may be related to the exposure
- where there is a reasonable likelihood that this may occur under the particular work conditions
- where there are valid techniques for detecting indications of it (Regulation 11).

The surveillance can be done by doctors, nurses or trained supervisors making simple checks. Where the surveillance is carried out, records or copies of them must be kept for 40 years from the date of the last entry.

If any of his/her employees are or may be exposed to substances hazardous to health, the employer must provide them with suitable and sufficient information, instruction and training for them to know the health risks created by the exposure and the precautions which should be taken (Regulation 12). The information must include the results of any monitoring and the collective (non-personalised) results of health surveillance.

Regulation 13 gives supplementary requirements on emergency arrangements to deal with incidents to those contained in Regulation 8 of the Management of Health and Safety at Work Regulations 1999, broadly requiring mitigation of the situation where appropriate and limiting exposure to essential personnel.

Provision is made in Regulation 14 for the control of certain fumigation operations, generally requiring notices to be posted beforehand. The Ministry of Defence is exempt from certain requirements of the Regulations, and there are a number of exceptions to them which are mostly covered in Regulation 3. Notably, health surveillance requirements are not extended to non-employees, but the training, information and monitoring requirements of Regulations 10, 12(1), 12(2) and 13 are extended to them if they are on the employer's premises. Regulations 10 and 11 do not apply to the self-employed, but all the others do. The Regulations do not extend to normal shipboard activities of the crew of a seagoing ship or its master.

Regulations 6–13 do not have effect in some circumstances, covered in Regulation 5. These are:

- where the Control of Lead at Work Regulations 2002 apply
- where the Control of Asbestos at Work Regulations 2002 apply
- where the hazardous substance is hazardous solely because of its radioactivity, flammability, explosive properties, high or low temperature or high pressure

- where the health risk to a person arises because the substance is administered in the course of medical treatment.

Schedule 2A to the Regulations (which is best studied in the ACoP) contains eight principles of good practice for the control of exposure to substances hazardous to health. If correctly applied, the principles should ensure that any exposure is below the relevant WEL. They are:
- design and operate processes to minimise emission release and spread of substances hazardous to health
- take into account all relevant routes of exposure when developing control measures
- control exposure by measures proportionate to the risk to health
- select the most effective control options to minimise escape and spread of the substance
- provide PPE where adequate control cannot be achieved by other means
- check and review regularly all elements of control measures for continuing effectiveness
- inform and train all employees on the hazards and risks from these substances, and the controls in use
- ensure that the control measures introduced do not increase the overall risk to health and safety.

Revision

6 requirements of COSHH 2002:
- assessment of health risks and selection of suitable control measures
- practical control of the risks
- maintenance of control measures
- monitoring of exposure of employees, including health surveillance
- monitoring of effectiveness of controls
- information, instruction and training of the workforce

8 COSHH principles:
- process design and operation
- routes of possible exposure
- proportionate measures
- effective control options
- PPE last resort
- check and review
- inform and train
- no overall risk increase

Self-assessment questions

1 A new substance is to be introduced into your workplace. What basic steps should be taken to ensure compliance with the COSHH Regulations?

2 Select an area of your workplace and identify substances which are defined as hazardous to health. Attempt to find examples from the operational categories of process, raw materials, engineering, cleaning, service and by-product.

15 The Electricity at Work Regulations 1989

Introduction

The dangers of the use of electricity have been discussed previously in Part 2 Section 9. Legislative control of electrical matters in the past has been concerned with not just the fundamental principles of electrical safety but also specific and detailed requirements relating to particular plant and activities. For many years, the Electricity (Factories Act) Special Regulations 1908 and 1944 controlled the use of electricity at work. These Regulations were enforced under the Factories Act 1961, and continued under the Health and Safety at Work etc Act 1974. The limiting factor was that the Regulations were only applicable to 'factory' premises as defined, and thus applied to a limited number of people and premises. Also, some of the requirements of the Regulations had become outdated as technology advanced and working and engineering practices changed to accommodate it.

All the provisions of the Provision and Use of Work Equipment Regulations 1998 are relevant to these Regulations, which came into force on 1 April 1990. They removed the above drawbacks and provide a coherent and practical code which applies to all work areas and all workers.

Objectives of the Regulations

The Regulations introduce a control framework incorporating fundamental principles of electrical safety, applying to a wide range of plant, systems and work activities. They apply to all places of work, and electrical systems at all voltages.

Some important definitions

An **electrical system** is a system in which all electrical equipment is, or may be, electrically connected to a common source of electrical energy, and includes the source and the equipment.

Electrical equipment includes anything used, intended to be used or installed for use to generate, provide, transmit, transform, rectify/convert, conduct, distribute, control, store, measure or use electrical energy – a very comprehensive definition.

Danger means the risk of injury – where 'injury' in this context means death or personal injury from electric shock, burn, explosion or arcing, or from a fire or explosion initiated by electrical energy, where any such death or injury is associated with electrical equipment.

Requirements of the Regulations

The Regulations revoke the earlier Electricity Regulations, and generally consist of

requirements which have regard to principles of use and practice, rather than identifying particular circumstances and conditions. Action is required to prevent danger and injury from electricity in all its forms. Employers (including managers of mines and quarries), self-employed people and employees all have duties of compliance with the Regulations so far as they relate to matters within their control; these are all known as **duty holders**. Additionally, employees are required to co-operate with their employer so far is necessary for the employer to comply with the Regulations (**Regulation 3**).

All electrical systems must be constructed and maintained at all times to prevent danger, so far as is reasonably practicable (**Regulation 4(1) and 4(2)**).

Every work activity (including operation, use and maintenance of, and work near, electrical systems) shall be carried out so as not to give rise to danger, so far as is reasonably practicable (**Regulation 4(3)** – see also further requirements of **Regulations 12, 13, 14 and 16**).

Equipment provided for the purpose of protecting persons at work near electrical equipment must be suitable, properly maintained and properly used (**Regulation 4(4)**). The term 'protective equipment' as used here has a wide application, but typically it includes special tools, protective clothing or insulating screening equipment, for example, which may be necessary to work safely on live electrical equipment.

No electrical equipment shall be put into use where its strength and capability may be exceeded, giving rise to danger (**Regulation 5**). This requires that before equipment is energised the characteristics of the system to which it is connected must be taken into account, including those characteristics under normal, transient and fault conditions. The effects to be considered include voltage stress and the heating and electromagnetic effects of current.

Electrical equipment must be protected and constructed against adverse or hazardous environments (such as mechanical damage, weather, temperature or pressure, natural hazards, wet, dirty or corrosive conditions, and flammable or explosive atmospheres) (**Regulation 6**).

All conductors in a system which could give rise to danger must be insulated, protected or placed so as not to cause danger (**Regulation 7**). A **conductor** means a conductor of electrical energy. The danger to be protected against generally arises from differences in electrical potential (voltage) between circuit conductors and others in the system, such as conductors at earth potential. The conventional approach is to insulate them, or place them so that people cannot receive electric shocks or burns.

Precautions shall be taken, by earthing or by other suitable means, to prevent danger from a conductor (other than a circuit conductor) which may become charged, either as a result of the use of the system or of a fault in the system (**Regulation 8**). A **circuit conductor** means any conductor in a system which is intended to carry current in normal conditions, or to be energised in normal conditions. This definition does not include a conductor provided solely to perform a protective function as, for example, an earth connection. The scope

of Regulation 8 is, therefore, such that it includes as requiring to be earthed a substantial number of types of conductor, including combined neutral and earth conductors, and others which may become charged under fault conditions such as metal conduit and trunking, metal water pipes and building structures.

There are restrictions on the placing of anything which might give rise to danger (fuses, for example) in any circuit conductor connected to earth (**Regulation 9**).

Connections used in the joining of electrical systems must be both mechanically and electrically suitable (**Regulation 10**). This means that all connections in circuits and protective conductors, including connections to terminals, plugs and sockets and any other means of joining or connecting conductors, should be suitable for the purposes for which they are used. Applying equally to both temporary and permanent connections, this Regulation covers supplies to construction site conditions, for example.

Systems must be protected from any dangers arising from excess current (**Regulation 11**). The means of protection is likely to take the form of fuses or circuit breakers controlled by relays. Other means are also capable of achieving compliance.

There must be suitable means provided for cutting off energy supply to, and the isolation of, electrical equipment. These means must not be a source of electrical energy themselves (**Regulation 12(1)** and **(2)**). In circumstances where switching off and isolation is impracticable, as in large capacitors, for example, precautions must be taken to prevent danger so far as is reasonably practicable (**Regulation 12(3)**). Switching off can be achieved by direct manual operation or by indirect operation by 'stop' buttons. Effective isolation includes ensuring that the supply remains switched off and that inadvertent reconnection is prevented. This **essential** difference between 'switching off' and 'isolation' is crucial to understanding and complying with

Regulation 13. This requires that adequate precautions are taken to prevent electrical equipment made dead to be worked on from becoming electrically charged, for example, by adequate isolation of the equipment.

No person shall be engaged in any work near a live conductor (unless insulated so as to prevent danger) unless: (a) it is unreasonable in all circumstances for it to be dead; (b) it is reasonable in all circumstances for persons to be at work on or near it while it is live; and (c) suitable precautions are taken to prevent injury (**Regulation 14**).

Adequate working space, means of access and lighting must be provided at all electrical equipment on which or near which work is being done which could give rise to danger (**Regulation 15**).

Restrictions are placed on who can work where technical knowledge or experience is necessary to prevent danger or injury (**Regulation 16**). They must possess the necessary knowledge or experience or be under appropriate supervision having regard to the nature of the work.

Regulations 17–27 inclusive give detailed additional requirements relating to mines and quarries, specifically those premises covered by the Mines and Quarries Act 1954.

Regulation 29 provides a means of defence in legal proceedings (see Section 2), where a person charged can show that all reasonable steps were taken and all due diligence was exercised to avoid committing the offence.

The requirements of the Regulations do not extend to the master or crew of a seagoing ship, or to their employer in relation to normal crew shipboard activities under the direction of the master, and they do not extend to any person in relation to an aircraft or hovercraft moving under its own power (**Regulation 32**).

Revision

The Electricity at Work Regulations 1989 refer to:

- construction and maintenance of electrical equipment
- carrying out work activities near electrical systems
- provision of protective equipment
- putting electrical equipment into use
- protection of electrical equipment
- precautions required in relation to conductors
- suitability of connections
- protection from excess current
- switching off and effective isolation of current
- restriction of work on live conductors
- provision of adequate space, access and lighting
- restrictions on personnel to carry out electrical work

Self-assessment questions

1 What are the conditions under which live electrical working is permissible?

2 What are the qualities a person must have to carry out work with electrical apparatus?

16 The Confined Spaces Regulations 1997

Introduction

The Confined Spaces Regulations (SI 1997 No 1713) came into force on 28 January 1998, and apply in all premises and circumstances where the Health and Safety at Work etc Act 1974 applies. The Regulations do not apply to work in mines or to diving projects covered by the Diving at Work Regulations 1997. They do not apply to the normal shipboard activities of the crew of a seagoing ship, but where ship and shore workers are working side by side they will apply, and co-operation will be needed to ensure that the requirements of the Regulations are met. The HSE may grant exemptions for a person or space where it is satisfied that health and safety will not be affected by the issue of the exemption.

The Regulations repealed Section 30 of the Factories Act 1961 and parts of the Shipbuilding and Ship-repairing Regulations 1960. To comply with the Confined Spaces Regulations 1997, it will be necessary to consider other Regulations including the Control of Substances Hazardous to Health Regulations 2002, the Provision and Use of Work Equipment Regulations 1998, the Personal Protective Equipment at Work Regulations 1992 and the Management of Health and Safety at Work Regulations 1999.

The definition of 'confined space' is very wide and includes any places such as trenches, vats, silos, pits, chambers, sewers, wells or other similar spaces which because of their nature could give rise to a '**specified risk**'.

'Specified risks' are those defined in **Regulation 1**, and include injury from fire or explosion; loss of consciousness through a rise in body temperature or by asphyxiation; drowning; or asphyxiation of trapping caused by free-flowing solids. Confined space hazards arise because of the confined nature of the place of work and the presence of substances or conditions which, when taken together, increase risks to safety and health. It must be remembered that a hazard can be introduced into a space that would otherwise be safe.

A 'confined space' has two defining features. First, such a space is substantially (although not necessarily entirely) closed and, second, there is a foreseeable risk from hazardous substances or conditions within the space or nearby.

Although at first sight it may appear that a particular business has no contact with 'confined spaces', they can be found in many workplaces. Obvious examples include such places as manholes, shafts, inspection pits, cofferdams and brewing vats. Possibly less

obvious are ships' cargo holds and tanks, building voids, plant rooms, cellars, and the interiors of plant, machines or vehicles.

Likely hazards include:
- flammable substances, either from the contents of the space or a nearby area
- oxygen enrichment, eg from a leaking welding cylinder
- ignition of airborne contaminants
- fumes or sludge remaining from previous processes or contents; these may release toxic or flammable gases when disturbed
- oxygen deficiency, which can result from inert gas purging; from natural biological processes such as rusting, decomposition or fermentation; from processes such as burning and welding; from workers breathing within the space
- liquids entering the space from elsewhere, and solid materials which can flow into it
- heat exhaustion caused by working in the confined space or from nearby processes.

Objectives of the Regulations

The Regulations specify requirements and prohibitions to protect the health and safety of persons working in confined spaces and also those who may be affected by the work.

How the objectives are met by the Regulations

Regulation 3 imposes a general duty on employers to ensure that their employees comply with the Regulations. The self-employed must also comply with the Regulations, and both employers and the self-employed must ensure that those over whom they have control comply as far as their control permits. In many cases, the employer or the self-employed person will need to liaise with other parties to ensure that their duties under the Regulations are fulfilled.

Regulation 4 requires that no-one shall enter a confined space to carry out work for any purpose, unless it is not reasonably practicable to achieve the purpose without entering the space. Under the Health and Safety at Work etc Act 1974 and the Construction (Design and Management) Regulations 1994, there are duties on designers to ensure that articles and buildings are designed to minimise foreseeable risks to health and safety, and engineers, designers and others should aim to eliminate or minimise the need to enter confined spaces. Employers have a duty to prevent employees and others under their control from entering or working inside a confined space when it is reasonably practicable to do the work from outside.

Employers should consider modifying the space to avoid the need for entry, or allow work to be done from outside. Working practices may need to be changed to make atmosphere testing, cleaning or inspections possible from outside the space using suitable equipment, sight glasses or CCTV.

To comply with Regulation 4, risk assessment will be necessary. The priority in carrying out a confined spaces risk assessment is to identify the measures necessary to avoid work within the space. Where the assessment shows that it is not reasonably practicable to do the work without entering the space, it can be used to identify the precautions to be included in a safe system of work. A competent person must carry out the risk assessment, and in large or complex situations, more than one person may need to be involved.

The factors to be assessed will include:
- the general situation and the risks that may be present
- the previous contents of the space, including any residues and the effect of disturbing them
- the risk of contamination from adjacent spaces and nearby plant or machinery. Where confined spaces are below ground, there is a particular risk of contamination through soil strata or from outside machinery. The risk of ingress of substances from other areas should be considered
- the likelihood of oxygen deficiency or oxygen enrichment
- the physical dimensions and layout of the space, the possibility of gas accumulation at different levels, and constraints on access and rescue
- any hazards from the work process itself including chemicals and sources of ignition
- the requirements for rescue procedures.

Regulation 4(2) requires that no person shall enter a confined space unless a safe system of work is in place. Before deciding what precautions are needed for entry, priority should be given to eliminating any sources of danger. The factors to be considered in designing a safe system, and which may form the basis for a permit to work, will depend on the risk assessment and could include:
- necessary supervision levels, including the possible need to appoint a competent person to supervise the work
- competence and suitability of workers, including training and experience and physical and mental attributes
- provision of adequate communications arrangements to summon help in an emergency, and to allow those inside the space to communicate with each other and those outside
- how the atmosphere within the space will be tested, the choice of test equipment, the type of contaminants and level of oxygen
- whether the space will require purging with air or inert gas, and the amount and method of ventilation that will be required
- whether it is necessary to clean the space or remove residues and how this will be done

- the way in which gases, liquids and flowing materials, and mechanical and electrical equipment can be isolated
- selection and use of suitable equipment for lighting, work or rescue inside the space, including PPE and RPE; working time may need to be limited where RPE is being worn, or where heat and/or humidity are high
- whether petrol, diesel and gas-fuelled machinery can be excluded from the space
- control procedures for gas hoses and pipelines and measures to prevent static electricity build-up, including earthing and bonding
- arrangements for access, egress and emergency and rescue arrangements
- fire prevention, material storage and smoking control arrangements inside and around the space.

Regulation 5 prohibits entry into a space unless suitable rescue arrangements have been provided. The arrangements must reduce the risks to those involved in the rescue to the lowest reasonably practicable level, and should include the provision of resuscitation equipment where conditions require it. Regulation 5 also requires that where any emergency situation arises, the rescue arrangements should be put into operation immediately.

Regulation 6 deals with exemption certificates.

Regulation 7 provides a defence, whereby an employer can claim that rescue arrangements were not put into immediate effect through the fault of another person who was not in his/her employment, and that he/she took all reasonable precautions to prevent this happening. If the defence is successful, the other person may become guilty of the offence. The precise rescue arrangements will depend on the risks identified. When assessing rescue arrangements, general accidents such as incapacitation after falls should be considered as well as those resulting specifically from work in confined spaces.

Rescue equipment provided as part of the arrangements should be suitable for the risk, and could include 'self-rescue' equipment (where there will be time to react to an anticipated emergency situation), lifelines and lifting equipment, first aid equipment and breathing apparatus.

In the case of prolonged or complex work, and where the risks require it, it may be necessary to warn the emergency services and provide them with information before starting. Whether the emergency services are notified in advance or not, there should be a procedure in place to ensure that they can be alerted rapidly should an accident occur.

Workers and rescuers should be trained in the communication, emergency and rescue procedures, and the use of equipment. Training should include refresher training and rehearsals and drills, and cover the following areas:
- likely causes of an emergency
- use of equipment, including donning procedures, tests and function checks, malfunctions and defects, and maintenance
- emergency procedures and methods of raising the alarm, including contact and liaison with the emergency services
- methods of shutting down adjacent plant
- operation of firefighting equipment.

Revision

The Confined Spaces Regulations 1997 require:
- identification of a confined space
- the work to be done from outside where reasonably practicable
- formal risk assessment
- safe system of work
- rescue arrangements
- training

17 The Noise at Work Regulations 1989

Introduction

Although exposure to noise at work has been recognised as damaging to health for a number of years, until recently there has been little specific legal control. The Agriculture (Tractor Cabs) Regulations 1974, the Offshore Installations (Operational Safety, Health and Welfare) Regulations 1976 and the Woodworking Machines Regulations 1974 had specific controls included in them, and there has been a general use of the code of practice issued in 1972 by the Department of Employment (*Code of practice for reducing the exposure of employed persons to noise*) to determine compliance with the general duties imposed by the Health and Safety at Work etc Act 1974.

From 1 January 1990, the Noise at Work Regulations became effective, and introduced a framework for controlling exposure to workplace noise and ensuring compliance with the EC Directive on the protection of workers from risks relating to exposure to noise at work. The Regulations require the protection of all persons at work, with a broad general requirement for employers to reduce the risk to employees arising from noise exposure as far as is reasonably practicable (**Regulation 6**). The Regulations also affect the self-employed and trainees.

Objectives of the Regulations

The Regulations introduce a control framework by requiring assessment of the extent of the problem, by carrying out noise surveys to identify work areas and employees at risk; control by engineering measures, isolation or segregation of affected employees; supply and use of protective equipment and administrative means; and monitoring by reassessment at intervals to ensure that the controls used remain effective.

How the objectives are met by the Regulations

Regulation 6 imposes a general duty to reduce the risk of hearing damage to the lowest level that is reasonably practicable.

Three noise **action levels** are defined, which determine the course of action an employer has to take if his/her employees are exposed to noise at or above the levels. These are:

- First Action Level – daily personal noise exposure ($L_{EP,d}$) of 85dB(A)
- Second Action Level – daily personal noise exposure of 90dB(A)
- Peak Action Level – peak sound pressure level of 200 Pascals (Pa) or more. (**NB:** This level is important in circumstances where workers are subjected to small numbers of loud impulse noises during an otherwise relatively quiet day, eg during piling operations or when working with cartridge-operated fixing tools.)

Where the daily noise exposure exceeds the First Action Level, employers must:

- carry out noise assessments (**Regulation 4**) and keep records of these until new ones are made; competent persons only should carry out the assessments (**Regulation 5**)
- provide adequate information, instruction and training for the employees about the risks to hearing, the steps to be taken to minimise the risks, how employees can obtain hearing protectors if they are exposed to levels between 85 and 90dB(A) $L_{EP,d}$, and their obligations under the Regulations (**Regulation 11**)
- ensure that hearing protectors (complying with current requirements as to suitability) are provided to those employees who ask for them (**Regulation 8(1)**) and that the protectors are maintained or repaired as required (**Regulation 10(1)(b)**)
- ensure so far as is practicable that all equipment provided under the Regulations is used (**Regulation 10(a)**) – apart from the hearing protectors provided on request as noted above.

Where the noise exposure exceeds the Second Action Level or the Peak Action Level, employers must in addition to the requirements detailed above:

- take steps to reduce noise exposure so far as is reasonably practicable by means other than provision of hearing protection (**Regulation 7**); this is an important requirement, introducing good occupational health and hygiene practice to ensure hazards are designed out of a process as opposed to employee involvement only to reduce the risks
- establish hearing protection zones, marking them with notices so far as is reasonably practicable (**Regulation 9**)
- supply hearing protection (complying with current requirements as to suitability) to those exposed (**Regulation 8(2)**) and ensure they are worn by them (**Regulation 10(1)(a)**), so far as is practicable
- ensure that all those entering marked hearing protection zones use hearing protection, so far as is reasonably practicable.

Duties are imposed on workers by the Regulations to ensure their effectiveness. The requirements (**Regulation 10(2)**) are:

- on being exposed to noise at or above the First Action Level, they must use noise control equipment other than hearing protection which the employer may provide and report any defects discovered to the employer
- if exposed to the Second Action Level or Peak Action Level, they must in addition wear the hearing protection supplied by the employer.

The Regulations also extend duties placed on designers, importers, suppliers and manufacturers under Section 6 of the Health and Safety at Work etc Act 1974 to provide information on the noise likely to be generated, should an article (which can mean a machine or other noise-producing device) produce noise levels reaching any of the three Action Levels (**Regulation 12**). In practice, this means supplying data from noise tests conducted on the machine, in addition to other information regarding safe use, installation, etc, already required by Section 6.

Revision

The Noise at Work Regulations 1989 require:

- adequate assessment
- assessment records
- reduction of the risk of hearing damage
- reduction of noise exposure, starting with engineering controls
- provision and maintenance of hearing protection
- provision of information and training for employees
- manufacturers and others to provide noise data

Self-assessment questions

1 The noise exposure of some of your workers has been measured and found to fall between the First and Second Action Levels. What steps should be taken to ensure compliance with the Noise at Work Regulations?

2 What are the main differences between the action required when the Second or Peak Action Levels are exceeded, and that required when the First Action Level is exceeded?

18 The Control of Vibration at Work Regulations 2005

Introduction

These Regulations came into force on 6 July 2005, implementing for Great Britain the Directive 2002/44/EC on exposure of workers to the risks arising from physical agents (vibration). They apply both to hand–arm and whole-body vibration. The former occurs as a result of using hand-held power tools and is the cause of significant ill health, producing disorders of blood vessels, nerves, joints and muscles of the hands and arms that are known collectively as hand–arm vibration syndrome (HAVS). Whole-body vibration affects those who ride in or on vehicles, especially over rough terrain, and is a factor in back pain.

The Directive's main requirements are for employers to reduce exposure to a minimum, provide information and training, assess exposure levels, carry out a programme of reduction and introduce health surveillance when exposure reaches an action value, and to keep exposure below the limit value.

Some important definitions

Daily exposure means the quantity of mechanical vibration to which a worker is exposed during a working day, normalised to an 8-hour reference period, which takes account of the magnitude and duration of the vibration.

Exposure action value means the level of daily exposure set out in Regulation 4 which, if reached or exceeded, requires specified action to be taken to reduce the risk.

Exposure limit value means the level of daily exposure set out in Regulation 4 for any worker which must not be exceeded, except in specified circumstances.

Working day means a daily working period, regardless of when it starts or finishes.

Objective of the Regulations

The Regulations impose duties on employers to protect employees who may be exposed to risk from exposure to vibration at work, and other persons who might be affected by the work (whether they are at work or not).

How the objective is met by the Regulations

Regulations 1 and **2** deal with their citation, commencement and interpretation. They apply throughout Great Britain, and beyond as extended by the Health and Safety at Work etc Act 1974 (Application outside Great Britain) Order 2001 – this means that they apply offshore.

Regulation 3 includes transitional provisions as well as the application of the Regulations. One of their key requirements, in Regulation 6(4), covers the duty to ensure an exposure limit value is not exceeded and to reduce exposure if it is. This does not take effect until 6 July 2010 where the work equipment was provided to employees before 6 July 2007 and does not permit compliance. However, the employer still has to comply with Regulation 6(2) and cannot ignore the issue as a result. For the agricultural and forestry sectors, Regulation 6(4) does not apply to whole-body vibration until 6 July 2014 for work equipment first provided to employees before 6 July 2007 and does not permit compliance – the above proviso applies here as well.

Where the employer has a duty in respect of his or her employees, that duty is extended to any other person, at work or not, who may be affected by the work carried out, except that this does not extend to health surveillance under Regulation 7 on information, training and instruction, unless those persons are on the premises where the work is carried out.

Except for Regulation 7, these Regulations apply to the self-employed as if they were an employer and an employee.

The customary exemption is applied to the master and crew of a ship in relation to normal shipboard activities under the direction of the master.

Regulation 4 describes exposure limit values and action values, which are:
- hand–arm vibration
  - daily exposure limit value of 5ms^{-2} A(8)
  - daily exposure action value of 2.5ms^{-2} A(8)
- whole body vibration
  - 1.15ms^{-2} A(8)
  - daily exposure action value of 0.5ms^{-2} A(8).

Daily exposure is calculated on the basis set out in Schedules 1 and 2 respectively.

(**Note:** The root mean square or A8 method uses units of ms^{-2} normalised to 8 hours, producing a cumulative exposure using an average acceleration adjusted to represent an 8-hour working day. This is readily calculated from work pattern data and manufacturers' vibration data. Vibration calculators can be found on the HSE web pages.)

Regulation 5 requires employers who carry out work liable to expose employees to risk from vibration to make a suitable and sufficient risk assessment, which must identify the measures needed to comply with the Regulations. In so doing, the employer must assess daily exposure to vibration by means of:
- observation of specific working practices
- reference to relevant information on the probable magnitude of the vibration related to the particular equipment and working conditions
- measurement where necessary.

The risk assessment has to include consideration of:
- magnitude, type and duration of exposure
- exposure to intermittent vibration or shocks
- effects of exposure on those whose health is at particular risk from the exposure
- any effects of vibration on the workplace and work equipment
- any information provided by manufacturers of work equipment
- availability of replacement equipment designed to reduce such exposure
- extension of whole-body exposure beyond normal working hours

- specific working conditions such as low temperatures
- appropriate information from health surveillance.

The risk assessment must be reviewed regularly in the same manner as those required by the Management of Health and Safety at Work Regulations 1999 (MHSWR), and the significant findings must be recorded together with the control measures that will be put in place.

Regulation 6 requires elimination at source of the risk from exposure to vibration as far as is reasonably practicable, or reduced to as low a level as is reasonably practicable. Where risk at source is not reducible as required, and an exposure action value is likely to be reached or exceeded, the employer must run a programme of organisational and technical steps appropriate to the activity. In both cases the employer is to have regard to the Schedule to MHSWR giving the principles of prevention, including:
- other working methods eliminating or reducing the exposure
- choice of appropriate ergonomic work equipment to produce the least possible vibration
- providing auxiliary equipment which reduces the risk
- appropriate maintenance programmes
- design and layout of workstations, workplaces and rest facilities
- information and training for employees
- limitation of the duration and magnitude of exposure
- appropriate work schedules with adequate rest periods
- providing clothing to protect from cold and damp.

Regulation 6(4) states that the employer shall ensure employees are not exposed to vibration above an exposure limit value, and if they are, then he or she must identify the reason for that limit being exceeded and modify the control measures to prevent it happening again. This does not apply where employee exposure is usually below the exposure action value but varies markedly from time to time and may occasionally exceed the value, provided that:

- exposure averaged over one week is less than the limit value
- there is evidence to show that the actual risk from the exposure pattern is less than the corresponding risk from constant exposure at the limit value
- risk is as low as is reasonably practicable under the circumstances
- the employees concerned are subject to increased health surveillance where this is appropriate.

Regulation 7 covers health surveillance. This is required if there is an assessed risk to the health of exposed or potentially exposed employees, or if they are likely to be exposed to vibration at or above an exposure action value. The health surveillance is designed to diagnose any health effect linked with exposure to vibration, and as with the COSHH Regulations it is 'appropriate' where exposure is such that a link can be established between vibration exposure and an identifiable disease of adverse health effect, where it is probable that the disease or health effect may occur under the work conditions, and where there are valid techniques for detecting the disease of effect.

A health record must be made for those undergoing surveillance and kept available in a suitable form. Employees are to be allowed access to it, giving reasonable notice, and it must be produced on demand by the enforcing authority. Where the health surveillance shows a positive link as above, the employer must ensure that the employee is advised of this by a qualified person and is given information and advice. The employer must also make the employee aware of the significant findings of the health surveillance, review the risk assessment and measures taken to comply with Regulation 6, consider assigning the employee

to alternative work without the vibration exposure risk, and provide for a health review of any other employee similarly exposed. Employees are under a duty to present themselves for necessary surveillance procedures triggered by this Regulation.

Regulation 8 addresses information, instruction and training. There is a general requirement for this where the initial risk assessment shows that there is a health risk to exposed employees. In particular this must include the following:

- all the measures taken to comply with Regulation 6
- exposure limit values and action values contained in Regulation 4
- significant findings of the risk assessment, including any measurements, with an explanation of the findings
- why and how to detect and report signs of injury
- entitlement to health surveillance under Regulation 7 and what its purpose is
- safe working practices to minimise vibration exposure
- collective and anonymised results of any health surveillance carried out by the employer under Regulation 7.

The information must be kept up to date to take account of significant changes in the work or the working methods used.

Regulations 9–11 provide for exemptions to be made by the HSE for individuals or classes of people in certain circumstances which include taking steps to reduce the risks.

The Regulations are accompanied by two technical Schedules, to which the reader is referred. They contain formulae for calculating exposure; discussion of them is beyond the scope of this book.

19 The Health and Safety (First Aid) Regulations 1981

Introduction

Giving first aid to injured persons once an accident has occurred is of vital importance. It can mean the prevention of further injury, or even death. First aid has two functions. First, it provides treatment for the purpose of preserving life and minimising the consequences of injury or illness until medical help (from a doctor or nurse) can be obtained. Second, it provides treatment of minor injuries which would otherwise receive no treatment, or which do not need the help of a medical practitioner or nurse. This definition of first aid is included in **Regulation 2**. To ensure the appropriate steps have been taken to provide adequate facilities (trained personnel as well as equipment) in all places of work, the Health and Safety (First Aid) Regulations were implemented on 1 July 1982. They replaced several pieces of legislation which imposed specific requirements for specific places of work, and set out rules covering all places of work.

Objectives of the Regulations

The Regulations are supported by an ACoP, which was revised in March 1997. They provide a framework for first aid arrangements which incorporates flexibility by setting objective standards to be achieved. Different types of premises, processes and industries can thus be covered by the same set of Regulations, requiring them to develop effective first aid arrangements after making an assessment of the risks involved and the likely use of the facilities.

How the objectives are met by the Regulations

The Regulations meet their objectives by requiring that every employer must provide equipment and facilities which are adequate and appropriate in the circumstances for administering first aid to his/her employees (**Regulation 3(1)**).

To ensure compliance with this requirement, an employer must make an assessment to determine the needs. The assessment is aided by the inclusion of a useful Appendix to the ACoP. Consideration of the following is required:

- **different work activities** – these need different provisions. Some, such as offices, have relatively few hazards and low levels of risk; others have one or more specific hazards (construction or chemical sites). First aid requirements will be dependent on the type of work being carried out
- **difficult access to treatment** – the provision of an equipped first aid room may be required if ambulance access is difficult or likely to be delayed. The ambulance service should be informed in any case if the work is hazardous
- **employees working away from employers' premises** – the nature of the work and its risk will need to be considered, and whether there is a work group or single employees
- **employees of more than one employer working together** – agreement can be made to share adequate facilities, with one employer responsible for their provision. Such agreement should be in writing, with steps taken by each employer to inform his/her employees of the arrangements
- **provision for non-employees** – the Regulations do not require an employer to make first aid provision for any person other than his/her employees, but liability issues and interpretation placed on the Health and Safety at Work etc Act 1974 may alter the situation, as, for example, in the case of a shop or other place where the public enter.

According to the supplied checklist in Appendix 1 of the ACoP, the minimum first aid provision for each workplace is:

- a suitably stocked first aid container
- a person appointed to take charge of first aid arrangements
- information for employees on those arrangements.

The checklist will assist employers in deciding whether they need to make any additional provision. The ACoP outlines minimum standards for the contents of first aid containers (no longer called 'boxes') – at least one will always be required. Additional facilities such as a stretcher or first aid room may also be appropriate.

There is no mandatory list of contents for the container but these will depend on the information obtained from the employer's assessment of needs. The latest version of the ACoP suggests the inclusion in the container of a pair of disposable gloves. Equipment such as disposable aprons, scissors and adhesive tape can be stored separately (ie not actually in the container) as long as they are available for use if required.

The employer must ensure that adequate numbers of 'suitable persons' are provided to administer first aid. 'Suitable persons' are those who have received training and acquired qualifications approved by the HSE, and any additional training which might be appropriate under the circumstances, such as in relation to any special hazards (**Regulation 3(2)**). All relevant factors have to be taken into account when deciding how many 'suitable persons' will be needed. These include:

- **situations where access to treatment is difficult** – first aiders would be required where work activities are a long distance from accident and emergency facilities
- **sharing first aiders** – arrangements can be made to share the expertise of personnel. Usually, as on a multi-contractor site, one contractor supplies the personnel
- **employees regularly working away from the employer's premises**
- **the numbers of the employees**, including fluctuations caused by shift patterns. The more employees there are, the higher the probability of injury
- **absence of first aiders** through illness or annual leave.

In circumstances where the first aider is absent – 'in temporary and exceptional circumstances' – such as through sudden illness (but not through planned annual leave), an employer can appoint a person to take charge in an emergency and take charge of the equipment and facilities provided (**Regulation 3(3)**). Also, in appropriate circumstances, an employer can provide an 'appointed person' instead of a first aider. He/she must first consider the nature of the work, the number of employees and the location of the workplace (**Regulation 3(4)**). The 'appointed person' is someone appointed by the employer to take charge of the situation (eg to call an ambulance) if a serious injury occurs in the absence of a first aider. It is recommended that the 'appointed person' be able to administer emergency first aid and be responsible for the equipment provided. In some situations, by virtue of the location, nature of work and the (small) number of employees, provision of an 'appointed person' only will be adequate. However, **as a minimum** an employer must provide an 'appointed person' at all times when employees are at work.

The ACoP emphasises that there are no fixed rules about how many first aiders will be needed, but it suggests that where more than 50 people are employed at least one should be provided unless the assessment justifies otherwise. Table 1 of the ACoP gives suggestions as to the number of first aiders in particular circumstances; compliance with its guidance would normally satisfy the Regulations.

An employer must inform his/her employees about the first aid arrangements, including the location of equipment, facilities and identification of trained personnel (**Regulation 4**). It is necessary to do this during induction training for new employees and when employees start work in a new area. This is normally done by describing the arrangements in the safety policy statement, and the displaying of at least one notice giving details of the location of the facilities and trained personnel.

Training of first aiders is discussed in the ACoP. Employers are 'strongly recommended' to consider the need for emergency first aid training for appointed persons, although this is not mandatory and HSE approval for the training is not required. Training courses provide a basic curriculum in a range of competencies which are listed in Appendix 2 of the ACoP.

Self-employed people must ensure that adequate and suitable provision is made for administering first aid while at work (**Regulation 5**). Again, an assessment has to be made of the likely hazards and risks to determine the extent and nature of what needs to be provided. It is also possible for the self-employed to make agreements with employers to share facilities.

These Regulations do not apply where the Diving Operations at Work Regulations 1981, the Merchant Shipping (Fishing Vessels) Regulations 1974 or the Merchant Shipping (Medical Scales) Regulations 1974 apply. The Regulations do not apply to vessels registered outside the UK, to mines (of coal, stratified ironstone, shale or fire clay), or to the armed forces (**Regulation 7**). **Regulations 8** and **9** make provision for mines not excluded by Regulation 7 and for work offshore respectively. **Regulation 10** deals with repeals, revocations and modifications, the more notable of which include the repeal of Section 61 of the Factories Act 1961 and Section 24 of the Offices, Shops and Railway Premises Act 1963.

Revision

The Health and Safety (First Aid) Regulations 1981 require:
- the provision of adequate first aid equipment
- the provision of adequately trained personnel
- giving of information on first aid provision to employees

Self-assessment questions

1 What considerations would have to be made to determine first aid requirements for a construction site?

2 Explain the difference between a first aider and an appointed person.

20 The Reporting of Injuries, Diseases and Dangerous Occurrences Regulations 1995

Introduction

Information about the types of accidents which happen is a very useful tool to work with in the prevention of future events of a similar kind. The information gained can be used to indicate how and where problems occur, and demonstrates trends of time.

It is important to distinguish between accidents, incidents and injuries – they are not the same. Injury can occur as a result of an incident; the injury and the incident together amount to an accident (the common term).

Since 1985, when the first version of these Regulations replaced the Notification of Accidents and Dangerous Occurrences Regulations (NADO), the law has properly recognised in the title of the Regulations that it requires the collection of specified information about incidents which result in specified types of injury; in some cases, it also requires information about specified incidents with the potential to cause serious physical injury, whether or not they produced such injury.

The enforcing authorities are interested in assembling such information because it gives them knowledge of trends and performance (failure) statistics. It also highlights areas for research, enforcement or future legislation.

Following consultation on the functioning of the previous version, these Regulations were implemented on 1 April 1996.

Objectives of the Regulations

The main purpose of the Regulations is to provide enforcing authorities with information on specific injuries, diseases and dangerous occurrences arising from work activities covered by the Health and Safety at Work etc Act 1974. The authorities are able to investigate only a proportion of the total, so the Regulations aim to bring the most serious injuries to their attention quickly.

How the objectives are met by the Regulations

The Regulations cover employees, self-employed people and those who receive training for employment (as defined by the Health and Safety at Work etc Act), and also members of the public, pupils and students, hotel residents and other people who die or suffer injuries or conditions specified, as a result of work activity.

The Regulations meet their objectives by requiring the following:

Where any person dies or suffers any of the injuries or conditions defined in Schedule 1, or where there is a 'dangerous occurrence' as defined in Schedule 2, as a result of work activities, the 'responsible person' must notify the relevant enforcing authority. This must be done by the quickest practicable means (usually the telephone) and a written report must be sent to the authority within 10 days (**Regulation 3(1)**). If the personal injury results in an absence of more than three calendar days, but does not fall into the categories specified as 'major', the written report alone is required (**Regulation 3(2)**). The day of the accident is not counted when calculating absence, but any days which would not have been working days are counted.

Where the injured person is a member of the public, and where the person is either killed or the injury is sufficient to warrant the injured person being taken directly from the scene to hospital, the injury is reportable by the occupier of the premises to the relevant enforcing authority (see below).

Regulation 2 contains interpretation of all significant terms used within the Regulations. These include:

Accident – the term is still relevant as a generic descriptor of the event causing the injury. The definition now includes 'a non-consensual act of physical violence done to a person at work', and 'an act of suicide which occurs on or in the course of operation of a relevant transport system' (railway, tramway, trolley vehicle system or guided transport system, all of which are further defined in Section 67 of the Transport and Works Act 1992).

Enforcing authority – the body responsible for the enforcement of health and safety legislation relating to the premises where the injury or disease occurred. Usually, this will be the HSE or the local authority's environmental health department. In case of doubt, reports should be made to the HSE, which will forward the information to the correct enforcing authority if this is not the HSE.

Major injury – the following injuries are classified as major:
- any fracture, except to fingers, thumbs or toes
- any amputation
- dislocation of shoulder, hip, knee or spine
- loss of sight (temporary or permanent)
- eye injury from chemical or hot metal burn, and any penetrating eye injury
- injury from electric shock or electric burn leading to unconsciousness, or requiring resuscitation or admittance to hospital for more than 24 hours
- loss of consciousness caused by asphyxia or by exposure to a harmful substance or biological agent
- acute illness requiring medical treatment, or loss of consciousness, arising from absorption of any substance by inhalation, ingestion or through the skin
- acute illness requiring medical treatment where there is reason to believe it has resulted from exposure to a biological agent or its toxins, or infected materials.

Responsible person – this may be the employer of the person injured, a self-employed person, someone in control of premises where work is being carried out, or someone who provides training for employment. The 'responsible person' for reporting any particular injury or dangerous occurrence is determined by the circumstances, and the employment or other relationship of the person who is killed or suffers the injury or condition.

Where death results within one year of a notifiable work accident or condition, the person's employer must notify the relevant enforcing authority in writing (**Regulation 4**). There is no prescribed form for this purpose.

When reporting injuries and dangerous occurrences, the approved form must be used (F2508). The reporting of diseases (on form F2508A) which are specified in **Schedule 3** is required only when the employer receives a written statement or other confirmation from a registered medical practitioner that the affected person is not only suffering from a listed disease but also that it has arisen in the manner specified in the Schedule (**Regulation 5**).

Incidents involving death or major injury arising from the supply of flammable gas must be notified to the HSE forthwith and a written report must be sent on the approved form (F2508G) within 14 days (**Regulation 6**). Gas fittings used by consumers and defined by the Gas Safety (Installation and Use) Regulations 1994 which are found on examination to be dangerous must also be reported.

Reports can now be made by telephone to the HSE Accident Hotline – doing so satisfies the notification requirements in full for all notifiable injuries, diseases and dangerous occurrences. Those using the facility are sent a copy of the report generated for record purposes.

Records must be kept by employers and others of those injuries, diseases and dangerous occurrences which require reporting (**Regulation 7**). Records can be kept in the form of entries made in the accident book (form BI 510) with reportable injuries and occurrences clearly highlighted, by keeping photocopies of reports sent to the enforcing authorities, or on computer provided that they can be retrieved and printed out. Keeping of computer records of this type will require registration under the Data Protection Act if individuals can be identified.

Records should be kept at the place of work or business, for at least three years from the date they were made. The enforcing authorities may request copies of such records which must then be provided (**Regulation 7(4)**). Further information may be requested by the enforcing authorities about any injury, disease or dangerous occurrence – in some situations, analysis of the initial reports might highlight patterns or categories indicating a need for further legislation or action, and a deeper study requiring more detailed information may then be carried out by the HSE with the approval of the HSC.

There are additional provisions in **Regulation 8** for mines and quarries. **Regulation 11** contains what amounts to a 'due diligence' defence provision, whereby a person can escape conviction for an offence under the Regulations if he/she can prove that he/she was not aware of the event which should have been reported and that he/she had taken all reasonable steps to have such events brought to his/her notice. (The presence of a suitable reporting procedure detailed as an arrangement within an organisation's health and safety policy would, of course, constitute such steps.)

The Regulations do not extend to cover:
- patients who die or are injured while undergoing treatment in hospital, dental or medical surgeries
- members of the armed services killed or injured while on duty
- people killed or injured on the road (except where the injury or condition results out of exposure to a substance conveyed by road, unloading or loading vehicles, or by maintenance and construction activities on public roads)
- people killed or injured during train travel (**Regulation 10**).

They do, however, cover dangerous occurrences on public and private roads.

Eight Schedules are attached to the Regulations. An extended and extensive list of dangerous occurrences can be found in Schedule 2, which also contains special parts relating specifically to mines and quarries, offshore workplaces and relevant transport systems. Schedule 3 contains the list of occupational diseases and their corresponding work activities which make them reportable. Schedules 5 and 6 hold additional requirements for mines and quarries and offshore workplaces respectively.

Revision

The Reporting of Injuries, Diseases and Dangerous Occurrences Regulations 1995 require:
- reporting of certain injuries, diseases and dangerous occurrences, as defined
- reporting of gas-related incidents and dangerous gas fittings
- record-keeping of reports/notifications sent to the enforcing authority
- further assistance to be given to the enforcing authority which they may require

Self-assessment questions

1 For your place of work, find out the enforcing authority, and the system followed to ensure that reportable injuries are properly notified.

2 Are the following reportable under the Regulations?
a A resident in a nursing home assaults a nurse, causing absence for a week.
b Two cars are involved in an accident on a motorway. One of the drivers, who is driving on business during normal working hours, is killed.
c An employee is unloading materials from a lorry parked in a road outside a building site. He is hit by a passing car, admitted to hospital for observation and released two days later.
d A child falls in a caravan park while playing football and is taken to hospital by his parents. Examination reveals no damage other than a twisted knee and the child is discharged.

21 Introduction to the Construction Regulations

Introduction

The requirements of the Factories Act 1961 applied to 'building operations' and 'works of engineering construction' (ie construction work), but few of the Act's requirements actually dealt with the safety of workers in construction processes. Specific regulations were made with the intention of protecting employees against the particular dangers arising in the industry.

Most recently, a major revision of the regulations was introduced, resulting in revocation of those dating back to the mid-1960s and earlier.

Currently applicable Regulations

The major sets of regulations currently applicable to construction working conditions are:

- Confined Spaces Regulations 1997
- Construction (Health, Safety and Welfare) Regulations 1996
- Construction (Head Protection) Regulations 1989
- Construction (Design and Management) Regulations 1994
- Provision and Use of Work Equipment Regulations 1998
- Lifting Operations and Lifting Equipment Regulations 1998
- Work at Height Regulations 2005.

Other 'mainstream' regulations apply to construction activities and processes, including:

- Dangerous Substances and Explosive Atmospheres Regulations 2002
- Health and Safety (First Aid) Regulations 1981
- Control of Lead at Work Regulations 2002
- Control of Asbestos at Work Regulations 2002
- Control of Substances Hazardous to Health Regulations 2002 (as amended)
- Electricity at Work Regulations 1989
- Noise at Work Regulations 1989
- Management of Health and Safety at Work Regulations 1999
- Manual Handling Operations Regulations 1992
- Health and Safety (Display Screen Equipment) Regulations 1992
- Personal Protective Equipment at Work Regulations 1992.

The requirements of these Regulations and those of the Health and Safety at Work etc Act 1974 are discussed in other Sections of this Part. The Workplace (Health, Safety and Welfare) Regulations 1992 do not apply to construction work or to site offices.

Obligations under the Regulations

The application of the Construction (Health, Safety and Welfare) Regulations 1996 (CHSWR) is to construction work 'carried out by a person at work', normally at a construction site. A comprehensive definition of 'construction work' follows that first found within the Construction (Design and Management) Regulations 1994 (CDM). General duties under the CHSWR are laid on the employer of construction workers and the self-employed, and also on any other person who controls the way in which construction work is done. This is obviously a major change, with potential consequences for clients, managing contractors and others who were previously able to argue about the application of regulations to them. Obligations regarding provision of welfare facilities and the making of necessary inspections are precisely targeted in the Regulations.

They also describe the general duties owed by employees working in the industry, which are to comply with those regulations which require something to be done or not done, and to report any defect in plant or equipment without unreasonable delay.

The law and control of contractors

Decisions of the courts have emphasised the role of the client in organising and sharing responsibility for the safety of work carried out by contractors. Although criminal prosecutions have been brought mostly under Section 3 of the Health and Safety at Work etc Act 1974 rather than the various construction regulations, mention is made of them here because employers in other industries are often unaware of their potential liability when employing construction contractors.

In 1995, the prosecution of Rhône-Poulenc Rorer Ltd concerned a breach of the Construction (Working Places) Regulations 1966, involving a fall through fragile materials. There was also an alleged breach of Section 3(1) of the Act. It was held by the Court of Appeal that the Section can be complied with by warning an independent contractor of the dangers and co-ordinating the work, but that where a contractor provides labour only to a factory occupier the duty is much greater. In this case, it was decided that because the occupier had been in breach of Regulations which applied to its own employees, it was also in breach of the Section as far as the contractor's employees working alongside him were concerned.

The case of Associated Octel (1996) went to the House of Lords via the Court of Appeal. It is a landmark case as it decided definitively the position of the employer of contractors, who is not able to delegate by contract or otherwise the duty placed on him by Section 3(1) of the Act. The duty extends to persons not in his employment, requiring him to conduct his undertaking so as to ensure, so far as is reasonably practicable, that such persons who may be affected thereby are not exposed to risks to their health or safety. This duty extends to incidental aspects of the employer's business which may be thought of by the employer as being entirely unconnected with it. The duty certainly extends to such 'construction work' as repair, maintenance, renovation and refurbishment carried out by a contractor, and to the risks which that work produces.

Octel was prosecuted following a serious burn injury to a contractor's employee. RGP, an independent contractor, was hired to repair a tank lining at Octel's premises at Ellesmere Port. Octel had authorised the use of a flammable solvent for cleaning, the vapour of which was then ignited by an electric light nearby. Although other risks had been identified and PPE had been provided by Octel, the courts found that the risks associated with the use of the solvent were neither spotted nor controlled by Octel.

Octel's argument was that the injury did not arise out of the conduct of its undertaking, but of the contractor's, and it had no case to answer. The Crown court disagreed and Octel was fined £25,000 plus £60,000 costs. The appeal, which went as high as the House of Lords, turned entirely on the question of the extent of the Section 3(1) duty. Octel said that as it had hired a competent contractor to do the work it could not be conducting its undertaking and could not be liable. It claimed that custom and practice in the industry is that the employer has no right to control how such a contractor works, and there is normally no common law liability for the actions of an independent contractor. The conduct of an undertaking must be limited to doing things over which it has control.

In judgement, it was said that what constitutes part of an undertaking is a matter of fact to be decided in each case. 'The place where the activity takes place will usually be very important', if not decisive. The duty is subject to reasonable practicability, which will be a matter of fact and degree in each case and also subject to the question of control.

In particular cases, therefore, the employer may have no control of a contractor. That could be where the employer has no expertise and hires a specialist. But it is suggested that if the employer attempts any control other than establishing the competence of a proposed contractor, then he/she will be likely to be viewed as having the potential for control. Then it will be a matter of fact as to whether the activity generating the risk is part of the employer's undertaking. If it is, the burden of proof rests on the employer to show that all that was reasonably practicable had been done to prevent injury (Section 40 of the Act).

Objectives of the Regulations

The Regulations were designed to require protection against the specific dangers arising from construction work. The types of dangers that arise are many and varied, but accident records show that construction workers are mostly killed or injured as a result of falls from height or during high-risk activities such as demolition, roof work and excavation. It is in these areas that the regulations set standards to be achieved. Little information is available on the more general occupational health risks of construction work. In particular areas, much research has been carried out and used to formulate control measures contained in legislation affecting industry as a whole in relation to specific hazards, such as asbestos.

Exemption may be granted from any or all of the requirements to any plant or equipment, or to any work.

Research and experience has shown that an underlying cause of accidents in the industry has been lack of organisational control and clear lines of responsibility touching all parties from the client to the employee. These matters are now addressed by the CHSWR, and requirements for the management aspects of construction work, to comply with the EC Temporary and Mobile Construction Sites Directive, are contained in the CDM Regulations, discussed in Section 25.

22 The Construction (Health, Safety and Welfare) Regulations 1996

Introduction

These Regulations (CHSWR) took effect from 2 September 1996. They replaced the Construction (General Provisions) Regulations 1961, the Construction (Health and Welfare) Regulations 1966 and the Construction (Working Places) Regulations 1966, which were revoked. They give effect to those parts of Annex 4 of the Construction Directive 92/57/EEC which deals with minimum health and safety requirements at construction sites. The Construction (Design and Management) Regulations 1994 (CDM) also came from the Directive. CHSWR provides the 'construction equivalent' of the Workplace (Health, Safety and Welfare) Regulations 1992 – which do not apply to construction sites. For that reason, CHSWR is written much less prescriptively than its predecessors, frequently using terms such as 'suitable', 'sufficient' and 'adequate' in relation to requirements.

Several concepts new to regulations are developed, which in general have less to do with physical features than the hazards and risks arising from them. For example, there is no longer a regulation entitled 'Scaffolding' but there is one entitled 'Prevention of drowning'. Detailed specifications are generally absent from the Regulations; if they are provided, they are normally to be found in one of the nine Schedules to the Regulations rather than within the 35 Regulations. There is no ACoP to interpret the Regulations.

With the introduction of the Work at Height Regulations 2005, the CHSWR have been amended to remove references to this type of work and details of the new work at height requirements are given in Section 23. For ease of comprehension, in the summary which follows, the Schedules are placed in context.

Some important definitions

Construction site – a place where the principal work activity is construction work.

Construction work – a significant term which is repeated from the CDM Regulations and is covered in detail in Section 25.

Excavation – includes any earthwork, trench, well, shaft, tunnel or underground working.

Loading bay – a new term to the construction industry, which is simply 'any facility for loading/unloading equipment or materials'.

Vehicle – includes mobile plant and locomotives, and any vehicle towed by another vehicle.

Objectives of the Regulations

The Regulations promote the health and safety of employees, the self-employed and others who may be affected by construction activities by setting out the physical issues which must be addressed in order to reduce risks.

How the objectives are met by the Regulations

Regulation 3 restricts the application of the Regulations to persons at work carrying out construction work. CHSWR does not extend to workplaces on a construction site set aside for non-construction purposes, such as storage areas and site offices. The Regulations covering traffic routes, emergency routes, exits and procedures, fire, cleanliness and site boundary markings only apply to construction work carried out on site.

General duties to comply with the Regulations are laid on employers and the self-employed as far as they affect him/her or persons under his/her control, or as far as they relate to matters under his/her control (**Regulation 4**). CHSWR assigns 'controllers' of construction work general duties to comply with the Regulations, by **Regulation 4(2)**, regardless of their status as an employer of workers.

Employees must also comply with the Regulations, and all at work are to co-operate with duty holders. When working under the control of another person, they must report any defect which may endanger anyone. The terms 'contractor' and 'employer of workmen' are no longer used.

Under **Regulation 5**, the requirement for a safe place of work extends to 'every place of work', and 'every other place provided for the use of any person while at work'. It includes the duty to prevent risks to health and for there to be sufficient working space suitably arranged for any person working there. Where the safe place cannot be provided, access to a place which is not safe must be barred. All of this Regulation's requirements are to be observed so far as is reasonably practicable, except for the fourth part which disapplies them to persons working on making the place of work safe, provided that all practicable steps are taken to ensure their safety.

Regulations 6–8 and Schedules 1–5 have been revoked by the Work at Height Regulations 2005.

The stability of structures is governed by **Regulation 9**. Practicable steps are to be taken where necessary to prevent the accidental collapse of a structure which might result in danger to any person, not only to construction workers. The Regulation prohibits loading any part of a structure so as to render it unsafe to any person, and any erection or removal of supportwork is restricted to being done only under supervision of a competent person.

Suitable and sufficient steps are required to ensure that any demolition or dismantling of structures is done so as to prevent danger, as far as practicable (**Regulation 10**), under the supervision of a competent person. The same 'suitable and sufficient' steps have to be taken to ensure no person is exposed to risk of injury from the use of explosives, including material ejected by the explosion (**Regulation 11**).

Excavations are covered at length in **Regulation 12**. There is no Schedule to this Regulation. The maximum depth before support (1.2m) quoted previously in the General

Provisions Regulations has been removed; the requirement has been effectively upgraded, as 'all practicable' steps are now required to prevent danger to any person, ensuring that accidental collapses do not occur.

The taking of suitable and sufficient steps to prevent burying or entrapment by material falling, so far as is reasonably practicable, is required by **Regulation 12(2)** and the duty extends to 'any person'. Support of excavations is to take place as early as practicable in the work, where it is necessary, using adequate support material. Support for excavations is to be installed or altered only under the supervision of a competent person. Suitable and sufficient steps are to be taken to prevent falls by anything or anyone into excavations. Nothing is to be placed or moved near an excavation where it could cause a collapse.

A provision concerns the identification and prevention of risk from underground services – the identification steps are to be suitable and sufficient and they are to prevent so far as is reasonably practicable any risk of injury from an underground cable or service.

Cofferdams and caissons must be of sufficient capacity and suitable design, as well as of suitable strong and sound material and properly maintained (**Regulation 13**). Supervision by a competent person is required for all structural work on them.

The prevention of drowning is addressed by **Regulation 14**, which requires the taking of reasonably practicable steps to prevent falls into, and to minimise the risk of drowning in, water or any liquid where drowning could occur. Suitable rescue equipment is required. Transport by water to or from a place of work must be done safely, and any vessel used must be suitably built and maintained, not overcrowded or overloaded, and in the control of a competent person.

Specific provisions on the safety of traffic routes are contained in **Regulation 15**. They are largely the equivalent of Workplace Regulation

17 (see Section 7 of this Part). There are some differences; in the first paragraph 'reasonably practicable' is added to modify the burden of compliance. The second paragraph on suitability is identical to the Workplace Regulation. The main burden of the Regulation is to detail what constitutes a 'suitable' traffic route, which is one where:

- pedestrians and/or vehicles can use it without causing danger to persons near it
- doors or gates leading onto a traffic route are sufficiently separated to allow users to see approaching traffic while still in a place of safety
- there is sufficient separation between vehicles and pedestrians, or where this cannot be achieved reasonably then there are other means of protection for pedestrians and effective warnings of approaching vehicles
- any loading bay has at least one exit point for the exclusive use of pedestrians
- any gate intended mainly for vehicle use has at least one pedestrian gate close by, which is clearly marked and free from obstruction.

Other details of Regulation 15 include a prohibition on driving vehicles on an obstructed route, and on one with insufficient clearance, so far as is reasonably practicable. Where it is not, steps must be taken to warn the driver and any passengers of the obstructions or lack of clearance. This can be done by signs, which are also required to indicate the traffic route where necessary for reasons of health or safety.

Doors, gates and hatches receive attention in **Regulation 16**, which does not apply to those on mobile plant and equipment. The basic requirement is for the fitting of suitable safety devices where necessary to prevent risk of injury to anyone. Whether the door, gate or hatch complies with the suitability of the devices is covered in the second part of the Regulation, which lists four features which must be present. An example is a device to stop a sliding gate coming off its track during use.

Regulation 17 contains requirements on vehicles and their use. Suitable and sufficient steps must be taken to prevent unintended movement, and warning is to be given by the effective controller of the vehicle to anyone liable to be at risk from the movement. Safe operation and loading is specified. Riders on vehicles must be in a safe place provided for the purpose, and a safe place must be provided if anyone remains on a vehicle during loading and unloading of any loose material. Where vehicles are used for excavating or handling/tipping materials, they must be prevented from falling into an excavation or pit, or into water, or overrunning the edge of an embankment or earthwork. Derailed vehicles (the definition of 'vehicle' includes locomotives) must only be moved or replaced on a track by use of suitable plant or equipment.

The contents of **Regulation 18** deal with prevention of risk from fire, 'etc' – the latter extends the reasonably practicable suitable and sufficient steps to be taken to cover the risk of injury from any substance liable to cause asphyxiation, as well as risks from fire, explosion or flooding. Carbon monoxide is an example of an asphyxiant, so in order to meet the requirement of this Regulation the use of internal combustion-powered equipment in confined spaces will have to be examined closely at the planning stages of projects. Emergency routes and exits are to be provided where necessary in sufficient numbers to enable any person to reach a place of safety quickly in the event of danger (**Regulation 19**). These are to lead as directly as possible to an identified safe area, be kept clear and provided with emergency lighting where necessary. A list of factors to be taken into account when determining suitability of these exits and routes is provided in paragraph 4.

Regulation 20 covers emergency procedures, which are arrangements to deal with any foreseeable emergency. It is aimed at, but not restricted to, sites where CDM health and safety plans are in force, and amounts to a requirement for them to take these matters into account when drawing up the plan and otherwise controlling the work, and also informing all those to whom the arrangements are likely to extend of the details, and testing the procedures at suitable intervals.

Fire detection and firefighting equipment (which works, is maintained and examined at appropriate intervals, and is suitably located) is to be provided on sites where necessary (**Regulation 21**). The list of factors to be taken into account is the same as given in Regulation 19(4). Paragraphs 5 and 6 require instruction in the use of firefighting equipment for all persons on site so far as is reasonably practicable. This can be done during induction training. Special instruction will be required for welders, roofers, felters and others carrying out hot work. Suitable signs are required to indicate the location of equipment.

Regulation 22 sets out the requirements for the provision of welfare facilities. The duty to comply with the Regulation rests on the person in control of the site who must comply so far as is reasonably practicable. This is a new requirement, rendering the person in control vulnerable to prosecution more directly than was possible before, and taking the duty away from individual employers. No numbers of toilets and washbasins are prescribed – 'suitable and sufficient' will be the numbers determined by the risk assessment made by the person in control.

Schedule 6

The principles to be observed are listed in Schedule 6, to which all the paragraphs of the Regulation refer. The reader is referred to the Schedule for detailed study. The main points which it contains are:

- **toilets** – no numbers are specified in relation to the facilities provided, but they are required to have adequate ventilation and lighting, and to be in a clean state. Separate male and female toilets are not required if they are each in a separate room with a door which can be secured from the inside
- **washing facilities** – these are to include showers if required by the nature of the work or for health reasons. They are to be in 'readily accessible places' and to be provided in the immediate vicinity of every toilet and

every changing room, and include a supply of clean hot and cold, or warm, water (which is to be running water so far as is reasonably practicable), a supply of soap or other suitable cleanser, towels or drying facility, and be sufficiently ventilated and lit, and in a clean and orderly condition. The Schedule provides that washing facilities can be unisex where only one person at a time can use them and they are in a room where the door can be secured from the inside. Otherwise, facilities for males and females are to be separate, unless they are only used for washing hands, forearms and faces

- **drinking water** – a supply of wholesome drinking water is to be readily accessible at suitable places, and so far as is reasonably practicable, marked with a conspicuous sign, with a supply of cups or other drinking vessels where the drinking water is not from a jet
- **accommodation for clothing** – this has to be made available for normal clothing not worn at work, and for special clothing which is not taken home. This accommodation has to include a drying facility so far as is reasonably practicable, and provide changing facilities where special clothing is required for the work and for reasons of health or propriety persons cannot be expected to change elsewhere
- **rest facilities** – must be suitable and sufficient, readily accessible and in compliance with Schedule 6(14) so far as is reasonably practicable. This states that there must be one or more rest rooms or areas, including those with suitable arrangements to protect non-smokers from discomfort caused by tobacco smoke and, where necessary, include suitable facilities for workers who are pregnant or nursing mothers to allow them to rest. The facilities must include suitable arrangements for preparation and eating of meals, including the means for boiling water.

There is no specific requirement for the means of heating food. Nothing in the Regulation or the Schedule appears to prevent contractors using facilities reasonably adjacent to a site, whether public or private.

Under **Regulation 23**, sufficient fresh or purified air is required to ensure, so far as is reasonably practicable, that every workplace or approach thereto is safe and without risks to health. Any plant used to achieve this must be fitted with a device to give an audible or visual warning of any failure.

Regulation 24 requires that a reasonable temperature is to be ensured indoors, having regard to the purpose of the workplace. Outdoors, the place of work should provide protection from the weather, taken in relation to reasonable practicability and any PPE provided, as well as the purpose of use of the workplace.

Lighting generally on site is covered by **Regulation 25**. Every place of work (and approach to it) and traffic routes must be lit, and the lighting shall be by natural light so far as is reasonably practicable. Regard is to be had for the ability of artificial light to change perception of signs or signals, so the colour of the lighting is to be controlled with this in mind. Secondary lighting shall be provided to any place where there would be a health or safety risk to any person in the event of failure of primary artificial lighting (paragraph 3).

'Housekeeping' is known as 'good order' in **Regulation 26**; good order and a reasonable standard of cleanliness is required, so far as is reasonably practicable (paragraph 1). Where necessary, the identification of the site perimeter by signs, and the extent of the site must be readily identifiable. Timber or other material with dangerous projecting nails shall not be used or be allowed to remain in a place where it may be dangerous (paragraph 3).

Regulation 27 was directed at plant and equipment, and has been revoked by PUWER 98.

The training requirement in **Regulation 28** introduces no age limits and incorporates all of a number of previous specific training requirements. Wherever training, knowledge or experience is necessary, it shall be possessed or the person must be under an appropriate degree of supervision by someone who already has it.

Inspection requirements are contained in **Regulation 29**. The main Regulation refers the seeker after inspection details to Schedule 7. The Regulation itself requires that a place of work specified in the Schedule cannot be used to carry out construction work unless it has been inspected by a competent person as required by the Schedule and that person is satisfied that the work can be done safely (paragraph 1). In particular (paragraph 2), where the place of work is a part of an excavation, cofferdam or caisson, anyone controlling the way the work is done must ensure it is stable, of sound construction and that the safeguards required by the Regulations are in place before his/her employees or persons under his/her control first use the workplace.

Please see the Work at Height Regulations 2005 (Part 4 Section 23) for the inspection requirements for scaffolding.

Paragraph 3 requires that, where the competent person is not satisfied that the work can be done safely from the workplace and the inspection was carried out on behalf of another person, he/she shall inform that person of anything he/she is not satisfied about and the place of work is not to be used until the defects are remedied. Inspections of a place of work must include plant, equipment and any materials which affect the safety of the place of work (paragraph 4).

Schedule 7
As above, the Schedule lists the places of work requiring inspection, and the times for the inspections. These are:
- any working platform or part thereof (not just scaffolding), and any personal suspension equipment, which are to be inspected:
  - before being taken into use for the first time
  - after substantial addition, dismantling or other alteration
  - after any event likely to have affected its strength or stability
  - at intervals not exceeding seven days since the last inspection

This applies to platforms from which falls of more than 2m can occur

- any excavation which is supported as required by paragraphs 1, 2 or 3 of Regulation 12:
  - before work at the start of every shift
  - after any event likely to have affected their strength or stability
  - after any accidental fall of rock, earth or other material
- cofferdams and caissons, which are to be inspected:
  - before work at the start of every shift
  - after any event likely to have affected their strength or stability.

Inspections, though, are not required to be the subject of reports in every case. **Regulation 30** sets out those cases in which reports are required and the manner of the reporting. The required contents of the report of inspection are detailed in Schedule 8, which supplements this Regulation. Note that the old inspection register used in the construction industry (Form 91 Part 1) does not allow for the recording of all the particulars now required, and there is no statutory form listed to replace it.

Schedule 8

The prescribed contents to be included in reports are:

- name and address of the person on whose behalf the inspection was carried out
- location of the work inspected
- description of the place of work inspected
- date and time of the inspection
- details of matters found that could give rise to a risk to the health or safety of any person
- details of any action taken as a result of finding those matters
- details of any further action considered necessary
- name and position of person making the report.

The basics of the requirements on reports in Regulation 30 are:

Paragraph 1 – where an inspection is required, the report, conforming to Schedule 8, must be prepared by the person who carried out the inspection before the end of the working period in which the inspection was made.

Paragraph 2 – the person preparing the report must provide a copy of it to the person on whose behalf the inspection was carried out within 24 hours of completing the inspection.

Paragraph 3 – the report or a copy of it must be kept at the place of work where the inspection was carried out, and after the work is completed it must be kept at the office of the person for whom it was prepared for three months after the date of completion.

Paragraph 4 – reports are to be open to inspection by inspectors, and can be required to be sent to an inspector.

Paragraph 5 – no report is required following inspection of a working platform or alternative means of support where falls of less than 2m could occur.

Paragraph 6 – no report is required for:
- mobile towers, unless they remain erected in the same place for seven days or more. Note that if a tower has to be dismantled and then re-erected because it has been moved, then it is probably no longer 'in the same place'
- for additions, dismantling or other alteration, not more than one report every 24 hours is required.

For excavations, cofferdams and caissons, an inspection is required at the start of every shift, but a report is needed only every seven days.

A main consequence of this Regulation is that inspections and reports are required on scaffolding over 2m high *before first use by anyone*.

The remaining **Regulations 31–35** deal with issue of exemption certificates, application of the Regulations outside Great Britain, enforcement of the Regulations on fire (which is done by the fire authority only where a construction site forms part of or is within premises occupied by people other than those doing the construction work), and modifications and revocations.

Revision

The Construction (Health, Safety and Welfare) Regulations 1996 cover:

- safe places of work
- stability of structures
- demolition
- explosives
- excavations
- cofferdams
- drowning
- traffic routes
- doors and gates
- vehicles
- fire
- emergency routes
- emergency procedures
- fire risks
- welfare facilities
- fresh air
- temperature
- lighting
- good order
- training
- inspection
- reports

23 The Work at Height Regulations 2005

Introduction

Falls from height remains the single most frequent cause of fatalities and major injuries in the workplace. Until now, legal controls have remained piecemeal, and contained in law of broader scope, notably the Construction (Health, Safety and Welfare) Regulations 1996 (CHSWR). The new Regulations (SI 2005 No 735) came into force on 6 April 2005. They gave effect to EC Directive 2001/45/EC and amended regulations transposed from other directives dealing with the use of work equipment, minimum requirements for workplaces and temporary or mobile construction sites. In the UK they were implemented late, and were preceded by rumours that they would ban or severely limit the use of ladders and stepladders. Neither is true and there was nothing throughout the extensive consultation process to suggest that either would be. Some contractors have chosen to ban stepladders from their sites entirely, on the grounds of best practice, but there is nothing in these Regulations to support such action.

There are 19 Regulations and eight Schedules that make up this legislation; the approach is based on risk assessment. Many of the requirements are qualified by the phrase 'so far as is reasonably practicable'.

Some important definitions

Access and **egress** include ascent and descent.

Fragile surface means a surface which would be liable to fail if any reasonably foreseeable load were to be applied to it.

Work at height applies to any place, including at or below ground level, and obtaining access to or egress from it while at work, except by a staircase in a permanent workplace.

Objectives of the Regulations

The Regulations impose health and safety requirements with respect to work at height for all workplaces, including organising and planning the work, competency issues, avoidance of risk from work at height and from fragile surfaces, falling objects and danger areas. Six sets of earlier regulations are revoked in part; these mostly involve the removal of the '2 metre rule' and changes in some definitions within the CHSWR.

How the objectives are met by the Regulations

Regulations 1–3 deal with their citation, commencement, interpretation and application. They apply throughout Great Britain, and outside as extended by the Health and Safety at Work etc Act 1974 (Application outside Great Britain) Order 2001 – this means that they apply offshore. They also apply to work in most industries and sectors. However, Regulations 4–16 do not apply to:

- the master and crew of a ship in respect of some ship-board activities
- dock operations as specified in Regulation 7(6) of the Docks Regulations 1988
- any place specified in Regulation 5(3) of the Loading and Unloading of Fishing Vessels Regulations 1988
- the provision of instruction or leadership in caving or climbing.

Regulation 4 requires that all employers ensure that work at height is properly planned, appropriately supervised, and carried out in a manner which so far as is reasonably practicable is safe. The planning of the work must include the selection of work equipment as specified in **Regulation 7**, and planning for emergencies and rescue. The new requirement to ensure work at height is only carried out when weather conditions do not pose a risk to health and safety does not apply to the emergency services acting in an emergency.

Regulation 5 requires that employers ensure that any person who participates in any aspect of work at height, including its organisation, planning and supervision, is competent to do so. People undergoing training must have competent supervision.

Regulation 6 reinforces the general principle established by Regulation 3 of the Management of Health and Safety at Work Regulations 1999 that a risk assessment must be carried out before any work at height is attempted, and that this must be suitable and sufficient. Work must not be carried out with persons working at height if it is reasonably practicable to do it some other way. If work at height cannot be avoided then employers must ensure that they take 'suitable and sufficient' measures to prevent, so far as is reasonably practicable, any person falling a distance liable to cause personal injury.

If work can be carried out from an existing place of work and/or via an existing means of access and egress this must be used, as long as it complies with **Schedule 1**. If this cannot be complied with then the employer must provide sufficient work equipment for preventing a fall from occurring. If it is not reasonably practicable to prevent a fall from occurring then the distance and consequences of a fall must be minimised. If the distance of a fall cannot be minimised then its consequences must be. Training and instruction must also be given so as to prevent anyone falling a distance liable to cause personal injury.

Regulation 7 covers selection of work equipment for work at height, and requires that collective prevention measures such as guardrails or nets be given priority over personal protection measures such as harnesses. The Regulation includes a list of matters to be taken into account when selecting appropriate measures. These include:

- the working conditions and the risks to safety at the place where the equipment is to be used
- the distance involved if the equipment is to be used for access and egress
- the distance and potential consequences of a fall
- duration and frequency of use of the work equipment

- the need for easy and timely emergency evacuation and rescue requirements
- any additional risks posed by the installation, use or removal of the equipment.

This is not an exhaustive list: the Regulation also requires that in all other respects the equipment selected is the most suitable.

Regulation 8 contains requirements for particular equipment and refers to the various Schedules (see below) for detailed requirements, for example the distance between guardrails.

Regulation 9 deals with passage on or near, and work on or near, fragile surfaces. The starting point is to avoid it if that is reasonably practicable. If not, then suitable and sufficient platforms, coverings, guardrails or similar means of effective support or protection must be provided and used. If there is still a risk of falling then suitable and sufficient steps must be taken to minimise the distances and consequences of a fall.

Notices warning of the fragile surface are required to be posted at the approach to the surface, or if this is not reasonably practicable then other means of warning must be found and used. This requirement is not applicable to the police, fire, ambulance and other emergency services acting in an emergency.

Regulation 10 requires that steps be taken to prevent, as far as is reasonably practicable, the fall of any material or object. It also forbids the throwing or tipping of objects from height where this might cause injury, and deals with the storage of material or objects so as to prevent risk arising from any unintended movement of them.

Regulation 11 introduces the requirement to post warning signs and physically restrict unauthorised entry to places where there is a risk of falling or being struck by a falling object.

Regulation 12 sets out the inspection requirements of equipment to which Regulation 8 and Schedules 2–6 apply. This must be inspected before first use in the position in which it will be used and at suitable intervals depending on the circumstances, which include adverse weather conditions and any exceptional circumstances.

If a working platform is used for construction work and is at a height from which a person could fall 2m or more, it must only be used if it has been inspected within the previous seven days. In addition a report must be compiled before the end of the working period containing the particulars set out in **Schedule 7** to the Regulations. If the inspection was carried out on behalf of someone else they must be provided with a copy of the report within 24 hours of the completion of the inspection. A copy of the report must be held at the site where the inspection was carried out until the construction work is completed and for three months after that at the office of the person it was carried out for. 'Inspection' in this case means a visual or more rigorous inspection by a competent person as appropriate for safety purposes.

Regulation 13 introduces an inspection requirement for surfaces and parapets, permanent rails or other fall protection measures of every place of work at height on each occasion before the place is used, so far as is reasonably practicable.

Regulation 14 sets out the duty of anyone at work to report anything connected with work at height that he or she knows is likely to endanger his or her safety or that of another person. It also puts a duty on workers to use work equipment and safety devices that they have been provided with for working at height in accordance with the training and instructions they have been given.

Regulations 15–19 cover exemptions, amendments, repeals and revocations.

The Regulations are accompanied by extensive Schedules, to which the reader is referred for a complete understanding. **Schedule 1** contains requirements for existing places of work and means of access or egress at height. These must:

- be stable and of sufficient strength and rigidity
- rest on a stable, strong surface
- be of sufficient dimensions to permit the safe passage of persons and the safe use of plant or materials and to provide a safe working area
- possess suitable and sufficient means of preventing a fall
- possess a surface which has no gap:
  - through which a person could fall
  - through which materials or objects could fall
  - which could give rise to other risks of injury to a person
- be constructed, used and maintained so as to prevent the risk of slipping or tripping or a person being caught between it and any adjacent structure
- be prevented from moving inadvertently during work at height (where there are moving parts).

Schedule 2 contains requirements for guardrails, toeboards, barriers and similar collective means of protection. These must be:
- of sufficient dimensions, strength and rigidity for the purpose
- placed, secured and used so as to ensure that they do not become displaced
- placed so as to prevent the fall of any person, material or object.

In the case of construction work, the top guardrail must be at least 950mm above the edge from which a person might fall (at least 910mm for any means of protection already fixed when the Regulations came into force). Toeboards must be suitable and sufficient for the purpose, and an intermediate guardrail must have a gap that does not exceed 470mm between it and other means of protection (such as the toeboard or top guardrail).

Openings are only allowed where they are for a ladder or stairway. Means of protection are only to be removed when necessary to gain access or to perform a particular task and at such times effective compensatory safety measures must be in place.

Schedule 3 covers requirements for working platforms, in two parts. The first part addresses all working platforms, the second part applies to scaffolding.

The supporting structure must rest on a surface which is stable, of sufficient strength and of suitable composition to support the structure, the working platform and anything that might be placed on it. It must be suitable and of sufficient strength and rigidity, and if wheeled be prevented from moving inadvertently during work. Where it is not wheeled it must be securely attached or have an anti-slip device to stop it slipping. It must be stable while being erected, used and dismantled, and not overloaded. It must remain stable when altered or modified.

The working platform is subject to effectively the same requirements regarding stability and strength. Its dismantling must take place in such a way as to prevent accidental displacement. As far as safety of the platform is concerned:
- it must be large enough to permit the safe passage of persons and allow the safe use of plant and materials
- it must not have a gap through which a person or materials could fall, or give rise to any other risk of injury to a person
- it must be in such a condition so as to prevent the risk of slipping or tripping, or of any person being caught between the platform and any structure
- it must not be overloaded.

The second part provides additional requirements for scaffolding:
- strength and stability calculations must be carried out unless they are already available or the scaffolding is assembled in conformity with a generally recognised standard configuration
- an assembly, use and dismantling plan must be drawn up by a competent person. The complexity of the plan will depend on the complexity of the scaffolding selected for the job. This could be in a standard form as long as it is supplemented by site-specific details. The plan must be available for the use of anyone who needs it. The dimensions and layout of the scaffolding decks must be appropriate for the tasks to be performed and suitable for the loads to be supported. Scaffolding that is incomplete, including because it is being assembled, dismantled or altered, must have warning signs displayed that comply with the Health and Safety (Safety Signs and Signals) Regulations 1996 and access to the danger zone must be prevented
- scaffolding must only be assembled, dismantled or altered under the supervision of a competent person and by appropriately trained people who know the risks inherent in such operations and in particular have an understanding of the work plan, permissible loadings and the safe system of work for the operations they are carrying out.

Schedule 4 gives requirements for collective safeguards for arresting falls (including air bags). These must only be used where a risk assessment has shown that the work activity can be performed safely while using it, that the use of other, safer equipment is not reasonably practicable, and enough people have been adequately trained in its use and in rescue procedures.

Any safeguard must be suitable and of sufficient strength to arrest safely the fall of any person who might fall. If the safeguard is designed to be attached it must be attached securely by all the anchor points and these must be sufficient to support any foreseen load. Airbags, beanbags and the like must be stable. There must be sufficient clearance under methods that distort, such as nets. The safeguard itself, so far as is practicable, must not cause injury itself.

Schedule 5 deals with detailed requirements for personal fall protection systems, in five parts. Again, these systems can only be used if a risk assessment has shown that it is safe to do so and that the use of other, safer equipment is not reasonably practicable. The user and a sufficient number of available persons must have received adequate training in the operations involved, including rescue procedures.

Any personal fall protection system must be suitable and of sufficient strength for its intended purpose. It must also be a good fit for whoever is wearing it and must be designed to minimise injury in usage. Anchorage points must be secure and able to support any foreseeable loading.

Part 2 covers additional requirements for work positioning systems – these can only be used if the system includes a suitable backup system. Where this is a line, the user must be connected to it. If there is no reasonably practicable backup system, all practicable measures must be taken to ensure the work positioning system does not fail.

Part 3 covers additional requirements for rope access and positioning techniques. These need a safety line as well as a working line; the user must be connected to both by a suitable harness; the working line must have a self-locking system to prevent falls; and it must have a mobile fall protection system which travels with the user. Provision has to be made for a seat matching ergonomic requirements of the work. However, the system can be a single rope, where risk assessment shows that a second line poses a higher risk and where appropriate safety measures are taken.

Part 4 covers additional requirements for fall arrest systems. There must be a suitable means of limiting forces applied to the user's body by absorbing energy, and there must be no risk of a line being cut. These systems require the establishment of a safety zone beneath, allowing for any pendulum effect.

Part 5 covers additional requirements for work restraint systems, which must be designed, if used correctly, to prevent the user from getting into a position where a fall can occur.

Schedule 6 gives requirements for ladders. Ladders can only be used for work at height after a risk assessment has been carried out which shows that the use of more suitable equipment is not justified because the work is of low risk and of short duration, or because of existing features

on site which cannot be altered. The guidance on the Regulations for the construction industry states: "Where ladders and stepladders are used they should only be used as a workplace for light work of short duration."

The surface on which any ladder rests must be stable, firm, sufficiently strong and of suitable composition to support the ladder so that its rungs or steps remain in a horizontal position.

Displacement and swinging of a suspended ladder must be prevented unless it is designed to be flexible. Portable ladders must be prevented from slipping by securing the stiles at their upper or lower ends, the use of an effective stability device, or "any other arrangement of equivalent effectiveness".

A ladder used for access must protrude sufficiently above the place of landing unless there is some other firm handhold. If a ladder or ladder run rises 9m or more above its base, sufficient safe landing areas or rest platforms must be provided at suitable intervals, where reasonably practicable.

A ladder must be capable of being used in a way that affords a secure handhold, including when carrying a load. There is an exception to this for a stepladder, if the maintenance of a handhold is not practicable when a load is being carried and a risk assessment has demonstrated that the use of a stepladder is justified because of the low risk and the short duration of use.

Schedule 7 details the particulars to be included in a report of an inspection. These are:
- name and address of the person for whom the inspection was carried out
- location of the equipment inspected
- description of the equipment inspected
- date and time of the inspection
- details of anything identified that could give rise to a risk to health and safety
- details of any action taken
- details of any further action that is required
- name and position of person making the report.

Schedule 8 deals with revocations.

Further information can be obtained from the following two publications:

- *Question and answer brief for the construction industry on the Work at Height Regulations 2005* – www.hse.gov.uk/construction/pdf/fallsqa.pdf
- *The Work at Height Regulations 2005: a brief guide* – www.hse.gov.uk/pubns/indg401.pdf or from HSE Books

24 The Construction (Head Protection) Regulations 1989

Introduction

Accident statistics for the construction industry indicate that a small but significant number of injuries are caused by failure to prevent objects falling from heights, and by operations with lifting appliances. A number of activities are also regarded as high risk, notably demolition and excavation. The industry has had a poor experience of voluntary use of safety helmets, so that legislative control finally became necessary. This was also welcomed by the industry because of its standardising effect, which would increase the acceptability of controls by the workforce.

In February 1995, the HSE announced that it estimated around 20 deaths each year have been avoided since 1990 – around 90 deaths in total with an estimated annual net benefit of about £17 million. This was one of the key findings of an evaluation into the effectiveness of the Regulations, based on a survey in 1992. In the survey, between 80 per cent and 100 per cent of workers wore suitable head protection on 69 per cent of the sites visited, which contrasted with the results of a similar survey in the early 1980s. This had found that, despite the voluntary agreement, safety helmets were not worn widely on construction sites. The decline in head injury rates in the industry is attributed to the success of the Regulations in encouraging the use of safety helmets.

The Regulations came into force on 30 March 1990.

Objectives of the Regulations

The objectives of the Regulations are to ensure the provision, maintenance and use of adequate head protection during construction work.

How the objectives are met by the Regulations

The basic principle of the Regulations is to require everyone (except turban-wearing Sikhs) working in the industry to wear suitable head protection whenever there is a risk of injury to the head from falling objects or hitting the head against something. The head protection is not required to protect against head injury caused by a fall of a person.

These Regulations apply to the same activities as the the Construction (Health, Safety and Welfare) Regulations 1996 (**Regulation 2**), but not to diving operations.

Regulation 1 defines what is meant by 'suitable' head protection. It is that which:

- is designed to provide protection so far as is reasonably practicable against the risk of head injury – this means conformance to current standards, presently BS EN 397:1995; the risk must be foreseeable given the environment in which the work takes place
- fits the wearer, after suitable adjustment
- is suitable for the activity in which the wearer may be engaged.

Sections 11 and 12 of the Employment Act 1989 have been used to exempt members of the Sikh religion who are wearing turbans from any requirement to wear head protection on a construction site. A Sikh not wearing a turban is required to comply with these Regulations in all respects. Sikhs choosing to wear turbans deny themselves use of adequate head protection – turbans are said to be remarkably ineffective in protecting against impacts – and provision is made to limit the employer's liability in the event of a claim because of this.

Each employer must provide suitable head protection for each employee, and must ensure that it is properly maintained and replaced whenever necessary (**Regulation 3**). Self-employed persons must provide their own, and equally ensure its proper maintenance and replacement when necessary. Any head protection provided must comply with any legislation implementing provisions on design or manufacture with respect to health or safety in any relevant directive applicable to head protection. Before choosing head protection, an employer or self-employed person must make an assessment to determine whether it is suitable (**Regulation 3(4)**), the process being identical to that required for PPE generally by the Personal Protective Equipment at Work Regulations 1992.

The assessment must involve:
* the definition of the characteristics which head protection must have in order to be suitable
* comparison of the protection available with the above characteristics.

The assessment must be reviewed if there is reason to suspect it is no longer valid, or if there has been a significant change in the work to which it relates (**Regulation 3(6)**). Every employer and self-employed person must ensure that appropriate accommodation is available for head protection when it is not being used (**Regulation 3(7)**).

Every employer is to ensure, so far as is reasonably practicable, that each of his/her employees wears suitable head protection, unless there is no foreseeable risk of injury to the head other than by falling (**Regulation 4**). Every person – employer, employee or self-employed – who has control over someone else is also responsible for seeing that head protection is worn where there is a foreseeable risk of injury (except from falling).

A person in charge of a site may make rules applying to the site, or give directions to ensure his/her employees comply with Regulation 4. This is a requirement of **Regulation 5**, which also requires these rules to be in writing, and to be brought to the attention of all those who may be affected by them. Once this is done, **Regulation 6** gives the rules the force of law, because it requires every employee to wear head protection when required to do so by rules or instructions. Similar duties are placed on the self-employed to comply with the instructions and rules of those in charge of the site. The process of rule-making is now familiar, as the drawing up of the health and safety plan is required by the Construction (Design and Management) Regulations 1994 (CDM), described in the next Section. It should be noted, though, that these Regulations apply to all construction work, not only that within the scope of CDM.

Full and proper use must be made of the head protection, so it must be worn properly, ie as the manufacturer intended – not backwards. Employees must also take all reasonable steps to return it to the accommodation provided for it after use (**Regulation 6(4)**).

Employees given head protection for their use must take reasonable care of it, and report any loss or damage to their employer without delay (**Regulation 7**).

Exemptions may be granted to persons or activities by the HSE, but not if granting an exemption might have an adverse effect on the health and safety of those likely to be affected by the exemption (**Regulation 9**).

Revision

The Construction (Head Protection) Regulations 1989 require:
* provision and replacement by the employer of suitable head protection
* suitability assessment
* the employer and those in charge to ensure it is worn where risks of head injury exist
* compliance by employees with any rules the employer may make on wearing head protection in areas
* accommodation to be provided
* the employee to make full and proper use of the head protection
* the employee to return it to its accommodation after use, and report loss or defects

25 The Construction (Design and Management) Regulations 1994

Introduction
The Construction (Design and Management) Regulations 1994 (CDM) implement the design and management content of the Temporary or Mobile Construction Sites Directive, which was adopted on 24 June 1992. The Directive contains two parts: the first applies the Framework Directive provisions to construction sites; the second refers the Workplace Directive to sites. The Construction (Health, Safety and Welfare) Regulations 1996 (CHSWR) implement the second part of the Directive.

The Temporary or Mobile Construction Sites Directive had to be implemented in EU member states by local regulations by 1 January 1994. After substantial consultation, the UK Regulations implementing the first part of the Directive were made on 19 December 1994, and laid before Parliament on 11 January 1995. They took effect on 31 March 1995.

Recognising that the construction 'system' in the UK has characteristics not anticipated by the Directive, the CDM Regulations simplify it by requiring the client to appoint central figures for the planning and the carrying out of construction work falling within the scope of the Regulations.

These are the planning supervisor and the principal contractor. The link between them is the health and safety plan, initiated by the planning supervisor and adjusted by the principal contractor to provide health and safety details relevant to the work – risks, timings, information – arrangements which together make up a comprehensive 'how to do it safely' document for the work to be done. There is also a requirement for a health and safety file to be kept and given to the eventual owner of the construction work – a 'how we built it' collection of relevant information which will be of use if future modification is required. The Regulations give detailed job descriptions for identified parties to the contract, in addition to duties placed on them under the Management of Health and Safety at Work Regulations 1999 (MHSWR).

Like the Construction (Head Protection) Regulations 1989, the CDM Regulations apply to work in territorial waters, but not further offshore, nor to the drilling and extractive industries, except indirectly.

Objectives of the Regulations
The Regulations introduce a control framework covering design, commissioning of work, its planning and execution, which applies to all construction work likely to pose significant risks to workers and other parties.

Application
Generally, the Regulations apply to construction work which:
- lasts for more than 30 days
- will involve more than 500 person-days of work
- includes any demolition work (regardless of duration or size)
- involves five or more workers being on site at any one time.

CDM always applies to any design work for the construction process. The Regulations do not apply where the work to be done is minor in nature. 'Minor work' is done by people who normally work on the premises. It is either not notifiable, or entirely internal, or carried out in an area which is not physically segregated, where normal work is carrying on and the contractor has no authority to exclude people while it is being done. Maintaining or removing insulation on pipes or boilers, or other parts of heating and hot water systems, is not classed as 'minor work'.

Premises normally inspected by the local authority include those where goods are stored for wholesale or retail distribution, or sold. These premises have included car windscreen, tyre and exhaust system replacement operations, exhibition displays, offices, catering services, caravan and temporary residential accommodation, animal care facilities except vets, farms and stables, consumer services, launderettes and sports premises. The boundaries of enforcement jurisdiction have changed, and will no doubt change again over time, so these examples should be regarded as illustrative of what has been seen as 'low risk' work and workplaces, rather than definitive.

Some important definitions

Agent means any person who acts as agent for a client in connection with the carrying on by the person of a trade, business or other undertaking, whether for profit or not.

Client means any person for whom a project is carried out, whether the project is carried out in-house or by another person. One of a number of clients, or the agent of a client, can volunteer to accept the duties of Regulations 6 and 8–12. This acceptance has to be made by way of a declaration to the HSE when the project is notified.

Cleaning work means the cleaning of any window or transparent/translucent wall, ceiling or roof in or on a structure where the cleaning involves a risk of falling more than 2m.

Construction work means the carrying out of any building, civil engineering or engineering construction work, and includes any of the following:
- construction, alteration, conversion, fitting out, commissioning, renovation, repair, upkeep, redecoration or other maintenance (including cleaning which involves the use of water or an abrasive at high pressure or the use of corrosive or toxic substances), decommissioning, demolition or dismantling of a structure
- preparation for an intended structure including site clearance, exploration, investigation (but not site survey) and excavation, and laying or installing the foundations of the structure
- assembly or disassembly of prefabricated elements of a structure
- removal of a structure or part of a structure,

or of any product or waste resulting from demolition or dismantling of a structure or disassembly of prefabricated elements of a structure
- installation, commissioning, maintenance, repair or removal of mechanical, electrical, gas, compressed air, hydraulic, telecommunications, computer or similar services which are normally fixed within or to a structure.

However, it does not include mineral resource exploration or extraction activities.

Contractor means any person who carries on a trade or business or other undertaking, whether for profit or not, in connection with which he/she undertakes to or does manage construction work, or arranges for any person at work under his/her control (including any employee, where he/she is an employer) to carry out or manage construction work. This definition can therefore be applied to the self-employed.

Design in relation to any structure includes drawing, design details, specification and bill of quantities (including specification of articles or substances) in relation to the structure.

Designer means any person who carries on a trade, business or other undertaking in connection with which he/she prepares a design or arranges for any person under his/her control to prepare a design relating to a structure or part of a structure.

Health and safety file means a file or other record in permanent form containing the information – required by Regulation 14(d) – about the design, methods and materials used in the construction of a structure which may be necessary for appropriate third parties to know about for their health or safety.

Project means a project which includes or is intended to include construction work.

Structure means any building, steel or reinforced concrete structure (not being a building), railway line or siding, tramway line, dock, harbour, inland navigation, tunnel, shaft, bridge, viaduct, waterworks, reservoir, pipe or pipeline (regardless of intended or actual contents), cable, aqueduct, sewer, sewage works, gasholder, road, airfield, sea defence works, river works, drainage works, earthworks, lagoon, dam, wall, caisson, mast, tower, pylon, underground tank, earth retaining structure or structure designed to preserve or alter any natural feature, and any similar structure to these, and any formwork, falsework, scaffold or other structure designed or used to provide support or means of access during construction work, and any fixed plant in respect of work which is installation, commissioning, decommissioning or dismantling and where that work involves a risk of falling more than 2m.

The requirements of the Regulations

Regulation 4 covers circumstances where a number of clients (or their agents – persons acting with the client's authority) are about to be involved in a project. One of the clients, or the agent of a client, can take on the responsibilities of the effective sole client. Clients appointing agents must be reasonably satisfied about their competence to carry out the duties given to clients by the Regulations. A declaration is sent to the HSE, which is required to acknowledge and date the receipt of it.

Regulation 5 applies CDM to the developer of land transferred to a domestic client and which will include premises intended to be occupied as a residence. Where this happens, the developer is subject to Regulations 6 and 8–12 as if he/she were the client.

Regulation 6 requires every client to appoint a planning supervisor and principal contractor in respect of each project. These appointments must be changed or renewed as necessary so that they remained filled at all times until the end of the construction phase. The planning supervisor has to be appointed as soon as is practicable after the client has enough information about the project and the construction work to enable him to comply with Regulations 8(1) and 9(1). The principal contractor, who must be a contractor, has to be appointed as soon as practicable after the client has enough information to enable

him/her to comply with Regulations 8(3) and 9(3). The same person can be appointed as both planning supervisor and principal contractor provided he/she is competent to fulfil both roles, and the client can appoint himself/herself to either or both positions provided he/she is similarly competent.

Regulation 7 requires the planning supervisor to give written notice of notifiable projects to the HSE, as soon as practicable after his/her appointment as planning supervisor (Parts I and II) and as soon as practicable after the appointment of the principal contractor (Parts I and III), and in any event before the start of the construction work. Where work which is notifiable is done for a domestic client and a developer is not involved, then the contractor(s) doing the work have the responsibility of notifying the HSE, and one of these can notify on behalf of any others. Again, notification must be made before any work starts.

Regulation 8 covers the requirements for competence of the planning supervisor, designers and contractors. The client must be reasonably satisfied that a potential planning supervisor is competent to perform the functions required by these Regulations, before any person is appointed to the role (Regulation 8(1)). No designer can be hired by any person to prepare a design unless that person is reasonably satisfied the designer is competent to do so, and similarly no contractor can be employed by anyone to carry out or manage construction work unless the person employing the contractor is reasonably satisfied as to the contractor's competence (Regulation 8(3)). Those under a duty to satisfy themselves about competence will only discharge that duty when they have taken 'such steps as it is reasonable for a person in their position to take' – which include making reasonable enquiries or seeking advice where necessary – to satisfy themselves as to the competence.

'Competence' in this sense refers only to competence to carry out any requirement, and not to contravene any prohibition placed on the person by any relevant regulation or provision.

'Provision' for health and safety in a wide sense is covered by **Regulation 9**. Clients must not appoint any person as planning supervisor unless they are reasonably satisfied that the potential appointee has allocated or will allocate adequate resources to enable the functions of planning supervisor to be carried out. Similarly, anyone arranging for a designer to prepare a design must be reasonably satisfied the potential designer has allocated or will allocate adequate resources to comply with Regulation 13. No-one is allowed to appoint a contractor to carry out or manage construction work unless reasonably satisfied that the contractor has allocated or will allocate adequate resources to enable compliance with statutory requirements.

The client is required by **Regulation 10** to ensure so far as is reasonably practicable that a health and safety plan which complies with Regulation 15(4) has been prepared in respect of the project before the construction phase starts. There is no duty on either the client or the planning supervisor to ensure that the plan continues to be in compliance with Regulation 15 once work has begun; this duty to keep it up to date is laid on the principal contractor by Regulation 15(4).

Regulation 11 obliges the client to ensure that the planning supervisor is provided with information relevant to his/her functions about the state or condition of any premises where relevant construction work will be carried out. This will be information which the client has, or which the client could obtain by making reasonable enquiries, and it has to be provided as soon as reasonably practicable but certainly before work starts to which the information relates.

The client is required by **Regulation 12** to take reasonable steps to ensure that the information in any health and safety file which is given to him/her is kept available for inspection by anyone who may need the information in the file to comply with any law. A client who sells the entire interest in the property of the structure can satisfy this requirement by handing over the file to the purchaser and ensuring the purchaser is aware of the significance and contents of the file.

Designers' duties are given in **Regulation 13**. Except where the design is prepared in-house, employers cannot allow employees to prepare a design, and no self-employed person can prepare a design, unless the employer has taken reasonable steps to make the client for the project aware of the duties of the client within these Regulations and any practical HSC guidance on how to comply.

The designer is firstly to ensure that any design he/she prepares and which he/she knows will be used for construction work includes adequate regard to three needs:
* to avoid foreseeable risks to health and safety of those constructing or cleaning the structure at any time and anyone who may be affected by that work
* to combat at source any risks to constructors or cleaners of the structure at any time, or to anyone who may be affected by that work
* to give priority to measures which will protect all such persons over measures which only protect each person at work.

Secondly, the designer is to ensure the design includes adequate information about any aspect of the project, structure or materials to be used which might affect the health or safety of constructors, cleaners or anyone who may be affected by their work.

Thirdly, a (stronger) duty to co-operate with the planning supervisor and any other designer is placed on a designer, so far as is necessary to enable each of them to comply with health and safety laws.

The duties are then qualified by Regulation 13(3), which allows the first two of them to include the required matters only to the extent that it is reasonable to expect the designer to deal with them at the time the design is prepared, and as far as is reasonably practicable.

Regulation 14 details the main duties of the planning supervisor. This duty holder must:

- ensure as far as is reasonably practicable the design of any structure in the project complies with the needs specified in Regulation 13 and includes adequate information
- take reasonable steps to ensure co-operation between designers to enable each to comply with Regulation 13
- be in a position to advise any client and any contractor to enable them to comply with Regulations 8(2) and 9(2) (competence of designer) and to advise any client on compliance with Regulations 8(3), 9(3) and 10 (competence of contractor and readiness of the health and safety plan)
- ensure that a health and safety file is prepared in respect of each structure in the project, reviewing and amending it over time, and finally delivering it to the client on completion of construction work on each structure.

Regulation 15 sets out the specific requirements for the health and safety plan. This is to be prepared by the appointed planning supervisor so as to contain the required information and be available for provision to any contractor before arrangements are made for the contractor to carry out or manage construction work on the project.

The required information to go into the health and safety plan is:

- a general description of the construction work
- details of the intended timescale for the project and any intermediate stages
- details of any risks to the health and safety of constructors known to the planning supervisor or which are reasonably foreseeable
- any other information which the planning supervisor has or could reasonably get which a contractor would need to show he/she has the necessary competence or has or will get the adequate resources required by Regulation 9
- information which the principal contractor and other contractors could reasonably need

to satisfy their own duties under the Regulations.

The principal contractor must take reasonable steps to ensure the health and safety plan contains, until the end of the construction phase, the required features, which are:

- arrangements for the project which will ensure so far as is reasonably practicable the health and safety of all constructors and those who may be affected by the work, taking account of the risks involved in the work and any activity in the premises which could put any people at risk
- sufficient information about welfare arrangements to enable any contractor to understand how he/she can comply with any duties placed on him/her in respect of welfare
- arrangements which will include where necessary the method of managing the construction work and monitoring of compliance with health and safety regulations.

Regulation 16 specifies the duties and powers of the principal contractor. These are firstly to take reasonable steps to ensure co-operation between contractors so far as is necessary to enable each to comply with requirements imposed. This includes, but is not limited to, those sharing the construction site for the purposes of Regulation 11 of the MHSWR.

Secondly, the principal contractor must ensure so far is reasonably practicable that every contractor and every employee complies with any rules in the health and safety plan. The principal contractor can make any reasonable written rules and include them in the plan, and give reasonable directions to any contractor.

Thirdly, the principal contractor must take reasonable steps to ensure that only authorised persons are allowed where construction work is being carried out.

Fourthly, he/she must ensure that required particulars are displayed in any notice covered by Regulation 7, and are displayed in a readable condition where they can be read by any person at work on the project.

Finally under this Regulation, he/she must provide the planning supervisor promptly with any information he/she possesses or could reasonably find out from a contractor which the planning supervisor does not already possess and which could reasonably be believed necessary to include in the health and safety file.

Duties on the giving of information and training requirements are set out in **Regulation 17**. The principal contractor must (as far as is reasonably practicable) ensure that every contractor is provided with comprehensible information on the risks to himself/herself and any employees or persons under his/her control which are present as a result of the work. In the same terms, the principal contractor must ensure that every contractor who employs people on the work provides his/her employees with the information and training required by Regulations 10 and 13(2)(b) of the MHSWR.

Provision is made by **Regulation 18** for receipt of advice from employees and the self-employed by the principal contractor, who must ensure that there is a suitable mechanism in place for discussing and conveying their advice on health and safety matters affecting their work. Arrangements for the co-ordination of these views of employees or their representatives are to be made with reference to the nature of the work and the size of the premises concerned.

Contractors' duties are covered by **Regulation 19(1)**, where in relation to a project they must co-operate with the principal contractor as necessary, and provide the principal contractor with any relevant information which might affect anyone's health or safety while on the project or who could be affected by it. This information includes relevant risk assessments, and information which might prompt a review of the health and safety plan for the project. Contractors must also comply with any directions of the principal contractor, and any applicable rules in the plan.

The Regulation requires contractors to provide the principal contractor with any information which is notifiable by the contractor to the enforcing authority under RIDDOR – details of injuries, diseases and dangerous occurrences as defined which are related to the project. Other information to be supplied by the contractor to the principal contractor includes anything which he/she knows or could reasonably find out which the principal contractor does not know and would reasonably be expected to pass to the planning supervisor if he/she did know, in order to comply with Regulation 16(e) – to amend the health and safety file.

Regulation 19 contains in parts 2, 3 and 4 general requirements to be observed by employers with employees working on construction work, and the self-employed. No employer can allow his/her employees to work on construction work, and no self-employed person can work, unless the employer has been given specific pieces of information:

- the names of the planning supervisor and principal contractor
- the health and safety plan or relevant parts of it.

A defence is provided against prosecution in part 5, which allows that this duty can be satisfied by showing that all reasonable enquiries had been made and the employer or self-employed person reasonably believed either that the Regulations did not apply to the particular work being done or that he/she had in fact been given the information required.

The Regulations have the same coverage as the Health and Safety at Work etc Act 1974 by **Regulation 20**, and, except for two of them, do not confer a right of civil action (**Regulation 21**). This means that breach of the Regulations can only result in criminal prosecution, and, except for two of them, cannot be used in civil claims by injured people. This provision is the same as that contained in the MHSWR. The enforcing authority for the Regulations is exclusively the HSE (**Regulation 22**).

Revision

6 **duty holders:**
- client
- developer
- planning supervisor
- principal contractor
- designer
- contractor

2 **pieces of documentation:**
- health and safety plan
- health and safety file

Duty holders must be judged competent by the client and others

5 **duties of designers:**
- avoid foreseeable risks
- combat risks at source
- protect all people rather than individuals
- include information
- co-operate with the planning supervisor

5 **duties of the client:**
- appoint the planning supervisor and the principal contractor
- be satisfied about their competence
- give information about the structure
- establish the health and safety plan is in place before work starts
- keep the health and safety file available for inspection

26 The Gas Safety (Installation and Use) Regulations 1998

Introduction
The most recent issue of these Regulations updates, consolidates and amends the original 1994 version of the same title. The Gas Safety (Installation and Use) Regulations 1998 (SI 1998 No 2451) came into force on 31 October 1998. Much of the present version was already in place in the former regulations. They address the safe installation, maintenance and use of gas systems, which includes gas fittings, appliances and flues mainly in domestic and commercial premises. They include both natural gas and LPG, subject to some exceptions, and place responsibilities on those working with appliances and fittings as well as both suppliers and users of gas.

Objectives of the Regulations
The Regulations cover broad issues of competence, suitability of fittings, dealing with danger from gas release, specification of restrictions and protective measures during design and installation, provision of instructions for use and the giving of information following safety checks. Because of the wide-ranging scope of the Regulations, they interface with many of the other Regulations discussed in this Part. As usual, an authoritative interpretation of the Regulations can be made only by consulting the ACoP, and this step is commended to the reader.

Some significant definitions
Appropriate fitting means a fitting which will give a gastight seal when fitted, and which is reasonably tamper-proof.

Emergency control means a valve for shutting off the gas supply. The valve must be available for use by the consumer.

Flue pipe is a pipe which forms part of a flue. This definition does not include chimney liners and gas appliance ventilation ducts.

Gas appliance means an appliance designed for use by a consumer, eg for heating, lighting or cooking, but does not include portable appliances which are supplied with gas from a cylinder.

Gas fitting covers all pipework, valves, meters and appliances.

Operating pressure is the gas pressure at which an appliance is designed to operate.

Room-sealed appliance means an appliance where the combustion system is sealed from the room in which it is located.

Work on a gas fitting includes installation, removal, refitting, maintaining, disconnecting, repairing and altering the fitting. 'Work' also includes changing the position of the fitting.

How the objectives are met by the Regulations
Regulation 3 imposes duties on those who carry out the work and their employers. Any person who carries out work on gas fittings and equipment must be competent and approved by the HSE. In practical terms, as CORGI is the only organisation currently recognised by the HSE, workers must be CORGI-registered. Gas fitters doing work on their own behalf need to be registered in their own right. Work on LPG installations requires separate CORGI training.

Anyone who works on gas fittings must be competent, and the level of competence must be suitable for the type of work being carried out. DIY installers and those doing 'favours' must still have the appropriate level of competence, it being an offence for any person to carry out any work in relation to a gas fitting or gas storage vessel unless competent to do so.

Those who employ gas-fitting operatives have a duty to ensure that they are competent. As well as ensuring adequate levels of training and experience, employers must carry out ongoing performance checks to ensure adequate levels of competence. The duty to ensure competence extends to other employers with control over the workplace concerned, eg certain contractors and those having gas fitting work done in their own workplaces. The national accreditation certification scheme introduced in January 1998 requires reassessment of competence at five-year intervals. It is an offence under the Regulations, and thus a criminal offence, to falsely pretend to be CORGI-registered.

Regulation 4 requires employers and the self-employed in control of premises where the work is to take place, and those who have some control over the work, eg contractors, to take reasonable steps to ensure that workers are CORGI-registered directly or through their employer.

Regulation 5 deals with materials and workmanship. Gas fittings must be appropriate for the apparatus, and of the correct size, strength, construction and material to ensure safety. No person may carry out work on a gas fitting except in such a way as to prevent danger to any person and in accordance with the appropriate standards. Gas installers should be acquainted with the required standards and ensure that the fittings they use are appropriate. Non-metallic connectors used with movable gas appliances must conform to the appropriate standards. Lead piping must not be installed for supplying gas, but appliances or meters may be connected to existing lead pipes provided that the pipe is in a safe condition. Non-metallic pipes used in buildings must be sheathed in metal to minimise the escape of gas. Polyethylene pipes buried underground do not have to be sheathed, but those under floorboards do. Free-standing cookers are not classed as readily movable, and their flexible hoses should have metal sheathing.

General safety precautions are prescribed in **Regulation 6**. Gas must not be released while working on a gas appliance unless steps are taken to prevent danger to any person. No gas fitting work shall be left unattended unless incomplete gas fittings have been sealed or it is otherwise safe. Where a gas appliance is left unattended during fitting, it should be sealed with an appropriate fitting which will ensure that the gas supply cannot be readily restored until it is safe for this to happen – closing an emergency

control (supply valve) is not sufficient. The meaning of 'unattended' depends on the circumstances, but broadly means being left a significant time so that the supply could be reconnected by someone else. Where any appliance is removed, the end of the pipe must be sealed with an appropriate fitting. No-one working on a gas fitting may smoke or use sources of ignition, such as blowlamps or hot air guns, which could lead to a risk of explosion.

Regulation 6(5) specifies that ignition sources must not be used for leak testing. This applies to householders and the public as well as to gas fitters. Leaks should be located by smell, detection instruments approved for use in flammable atmospheres, leak detection fluids or pressure testing equipment only.

Where work on a gas fitting may affect gastightness, the gastightness should be tested immediately afterwards at least as far as the nearest valves.

LPG storage vessels must be located so that they can be used, filled or refilled without causing a danger to anyone.

Regulation 7 deals with protection of the fitting or appliance against damage. Gas fittings should be installed in accordance with current standards, and must be properly supported and placed or protected so as to avoid undue risk of damage to the fitting. Where there is a risk from dust or foreign matter, filters or other protection will be necessary. Protective devices should be suitable for the appliance. Where there is a risk of corrosion, the fittings should be constructed of a suitable resistant material. Hazardous environments include those where water, salt spray, damp, corrosive chemicals or soot are likely to be present.

No person may make an alteration to premises where a gas fitting or a storage vessel is fitted, if such an alteration would affect the safety of the gas equipment (**Regulation 8**). This imposes a wide duty in relation to property where a gas appliance is fitted, and concerns work on the property itself rather than work on the gas fittings. The Regulation applies to everyone, including householders and general builders. Alterations which may affect the safety of the gas installation include changes to windows, air bricks, extractor fans and chimneys.

Modifications to gas fittings must be carried out by a competent person. Modifications to premises where there may be an effect on safety must be checked by a competent person before the gas fitting is taken into use. It is recommended that a responsible person such as a foreman or site manager is appointed to ensure that work is carried out in a safe and controlled manner, that potential risks to gas safety are identified and that a competent person is consulted where necessary.

Regulation 8(2) prevents anyone from doing anything which may affect a gas fitting, flue, or means of ventilation connected with the fitting, where such an action may cause danger. This part of the Regulation does not apply to changes to premises, but includes actions such as blocking an appliance air vent or flue.

No person may enable gas to be supplied to premises for the first time unless an emergency control valve is fitted (**Regulation 9**). This valve must be fitted for each individual premises and must be readily accessible to consumers (ie not located in a basement, cellar or locked cupboard). The person who connects the gas must ensure that users are aware of the emergency valve and the action to be taken in the event of a gas emergency. In rented property, the landlord or the managing agent should ensure that tenants are given this information. Where there are several buildings on a common supply, only one valve is necessary provided that one person or organisation is in control. Where the emergency control is not fitted by the meter, a sign must be fitted at the control identifying it and giving emergency instructions.

Regulation 10 requires temporary electrical bonding to be fitted where it is necessary to prevent danger while working on a gas fitting.

In practice, this means that temporary bonding must be fitted to maintain electrical continuity and must remain in place until permanent electrical continuity has been restored.

Regulations 11–17 give technical requirements for the fitting and siting of meters and regulators. It should be noted that meters must not be installed where they may affect the means of escape in the event of fire. Every gas supply to a meter must have a regulator fitted to prevent over-pressurising of appliances.

Regulation 18 places duties on those who install gas pipework. When installing supply pipes, the installer must take account of the position of other services, eg pipes, cables and sewers, and the risk to safety that may result. Installers also have duties with regard to electrical bonding.

Equipotential bonding must be provided where required. It is most commonly required when fitting pipes into new premises or when fitting a new pipe into existing premises. The person installing the pipe which connects to a meter, whether or not the meter is installed, has a duty to inform the person responsible for the building (the occupier, owner or builder) if equipotential bonding is required. This advice should be in writing.

Enclosed pipes are dealt with specifically in **Regulation 19**. Where pipes are installed through walls or floors, they must take the shortest practicable route and must be protected to prevent any leakage passing into cavities and voids. Pipes must not be installed in cavities, and are only allowed to enter them when passing from one side of the wall to the other. Hidden pipes should be fitted with sleeved protection to allow leaked gas to be vented to a position outside the space. No pipe may be installed in a wall or floor unless it is adequately protected against structural movement.

Regulation 20 covers protection of buildings. Where structural strength or fire resistance may be affected by the installation of pipes, installation in such a way is prohibited.

Regulation 21 prevents the installation of pipework in which a deposit of liquid or solid may form, unless the pipework is fitted with a suitable device to collect the deposit. Natural gas and LPG will not normally require this fitting.

Regulation 22 requires that gas pipes are tested and purged after work. Where a person carries out work which may affect the gastightness of the pipe, a gastightness test is required immediately afterwards. Joints must be tested before being painted or coated against corrosion. After testing, the pipework must be purged to ensure that all air and gas contained in it is removed in a safe manner. Where the pipe will not be put into immediate use, the end should be sealed with a gastight fitting. Where purging is carried out through a loosened connection, the connection must be tested for gastightness immediately.

Regulation 23 applies only to non-domestic premises, and requires that where pipes are readily accessible, they must be clearly marked by the installer as containing gas. The pipe must remain marked for as long as it is used for conveying gas.

Regulation 24 gives additional technical requirements for pipes and valves where gas is supplied to large buildings.

Regulation 25 defines 'flue pipe' and 'operating pressure'. In **Regulation 26**, a list of safety precautions is given, some of which have been covered elsewhere in the Regulations. They include:

- no person may install a gas appliance unless it can be installed without danger to anyone
- appliances and flues should be checked to ensure that they are safe to use
- used gas appliances must not be installed unless they are in a safe condition for use; this applies to all appliances which are moved, even when they are moved within the same room
- appliances and flues must be installed in accordance with the manufacturer's instructions; this places an important duty

on the installer to ensure that the appliance itself, the way in which it is installed, and fittings, flues and other factors (eg ventilation) will not cause danger when the appliance is in use
- no person may connect a flued domestic appliance to the gas supply except by means of fixed rigid pipe
- no person may install a gas appliance unless it complies with relevant legislation; new appliances must conform to the Gas Appliances (Safety) Regulations 1995
- gas appliances must not be left connected to the gas supply unless they are safe to use, or the gas supply has been sealed off
- where appliances are fitted with signs or stamps to indicate that they comply with legislation, nothing shall be done to make the appliance cease to comply, and the sign or stamp must not be removed or defaced
- when work has been done on a gas appliance, the person doing the work must examine the flue, supply of combustion air, operating pressure and safe functioning immediately the work is complete; any defects must be brought to the attention of the owner or person responsible for the premises.

The requirements for flues are listed in **Regulation 27**. No person shall connect an appliance to a flue unless the flue is suitable and safe for use with the appliance. Requirements of building regulations must be met where applicable. Where the flue is fitted in a chimney, the seal between the flue and the chimney must be sited so that it can be inspected. Enclosures around flues must be sealed so that products of combustion cannot pass into a room or space. Power-operated flue systems may not be fitted unless they prevent the operation of the appliance if the draught fails. Flues must be installed in a safe position.

Regulation 28 stipulates that gas appliances must be installed in a position where they will be readily accessible for operation, inspection and maintenance.

The installer of a gas appliance is required to leave a copy of the manufacturer's instructions with the owner or occupier of the premises (**Regulation 29**).

Regulation 30 covers room-sealed appliances and prohibits the fitting of non-room-sealed appliances in bath or shower rooms and accommodation used for sleeping purposes. This prohibition extends to cupboards which vent into the rooms.

Regulation 31 prohibits the installation of suspended gas appliances unless the appliance is designed to be suspended and the pipework is strong enough.

Regulation 32 lays down the requirements for flue dampers. Dampers must be interlocked with the gas supply so that the burner will not operate if the damper fails. Dampers must be inspected immediately after installation. Manually-operated dampers must not be fitted to domestic appliances unless permanently fixed in the open position.

The appliance installer is required by **Regulation 33** to carry out testing. Tests must include pipes, fittings, appliances and flues. Their purpose is to ensure that:

* the appliance has been fitted in accordance with the Regulations
* the operating pressure is as specified by the manufacturer
* the appliance has been installed in accordance with the manufacturer's instructions
* all gas safety controls are in proper working order.

Testing procedures must be in accordance with the appropriate standards. Where tests show that adjustments are necessary, the installer must either carry out the adjustments or disconnect the appliance.

Regulation 34 prohibits the person responsible for the premises (ie the occupier or owner) from using or allowing the use of any appliance where he/she knows or has reason to suspect that it cannot be used safely. Also, any person engaged in work on a gas main, service pipe or gas fitting who knows or suspects that a gas

appliance cannot be used without danger must immediately take all reasonably practicable steps to notify the person responsible for the premises. Examples of potentially unsafe situations are given for guidance only in Appendix 2 of the ACoP. The ACoP also contains appendices which give requirements for appliances and flues, summarise all relevant legislation, list information sources and provide diagrams of typical installations.

Regulation 35 places a duty on employers and self-employed persons to ensure that any gas appliance, pipework or flue installed in a workplace under his/her control is maintained in a safe condition.

Regulation 36 imposes a number of duties on **landlords**. Landlords have a duty to ensure that appliances and flues are checked within 12 months of installation and at maximum intervals of 12 months thereafter. Landlords must ensure that gas appliances are safe before reletting premises on the expiry of a lease. Records of checks must be maintained for two years and the record, or a copy, must be made available to occupants on request. A copy must be issued to existing tenants within 28 days of the inspection, and to new tenants before they commence occupation. Where tenants occupy premises for less than 28 days, a copy of the inspection report must be posted in a prominent position. The contents of the record are specified.

Where accommodation is sublet, the landlord retains the duty if the 'subletter' is also an occupant, eg someone who rents a room or has a lodger. Where a 'proper' sublet occurs, the two landlords must make a clear allocation of responsibilities by contractual arrangement.

Landlords cannot delegate duties to the tenant, eg to arrange for inspections or work to be carried out. The responsibility rests with the landlord at all times, regardless of the kind of tenancy agreement in place. Even where there is a full repairing and insuring lease, it cannot be used to transfer responsibility to the tenant. The landlord can only transfer duties to the tenant by contractual arrangement where the gas appliance is fitted in part of non-residential premises.

Landlords must take all reasonable steps to ensure that access is available for maintenance and safety checks. Landlords should keep records of all steps taken to discharge their duties, and tenants should comply with their landlord's requests. Forced entry is not required by the Regulations. The landlord's duty also extends to ensuring safety checks are carried out on appliances which serve parts of a property where there is common occupancy by tenants and other people, such as common bathrooms, kitchens and living rooms.

In summary, the two main duties imposed on landlords are annual safety checks on gas appliances and flues, and provision of ongoing maintenance and remedial work. Safety checks and all other gas work must be done only by CORGI-registered gas installers.

Regulation 37 deals with gas escapes from LPG and other fuel gases, but does not cover the escape of natural gas from any network. This means that almost all escapes of natural gas are covered not under these Regulations but under the Gas Safety (Management) Regulations 1996.

Regulation 38 places a duty on users who have gas plant which may cause pressure fluctuations in the gas supply sufficient to cause danger to other consumers. Such users are required to notify the gas supplier at least 14 days before connection and to comply with any instructions given by the supplier.

There are some exceptions to liability under some of the Regulations, which are covered in **Regulation 39** and can be used by those who can show that they took all reasonable steps to prevent a breach of the relevant Regulation(s). Whether any particular use of an exception amounts to a valid defence is a matter for a court to decide.

Revision

7 main topics to manage:
- competence
- suitability of fittings
- dealing with danger from gas release
- specification of restrictions
- specification of protective measures during design and installation
- provision of instructions for use
- giving information following safety checks

2 duties for landlords:
- annual safety checks
- carrying out maintenance work

27 The Dangerous Substances and Explosive Atmospheres Regulations 2002

Introduction
These Regulations came into effect fully on 30 June 2003, putting in place requirements for eliminating or reducing risks to safety from fire, explosion or other events triggered by the properties of any 'dangerous substance' in connection with work activities. They give effect in the UK to the EC Directives on Chemical Agents (98/24/EC) and Explosive Atmospheres (99/92/EC). Repeals and revocations carried out by the Regulations include some requirements that have been in place for more than 80 years (dealing with celluloid and cinema film work) and other more recent and possibly more familiar regulations, including those on dry cleaning. Perhaps most significantly, the Highly Flammable Liquids and Liquefied Petroleum Gases Regulations 1972 have disappeared. All these topics and others are, of course, covered within the wide scope of the DSEAR. As usual, the reader is advised to consult the full text of the Regulations (SI 2002 No 2776), together with the ACoP, for full details of classifications, criteria and Schedules, as only a summary can be attempted here.

Objectives of the Regulations
The Regulations cover broad requirements for control of risk in relation primarily to chemical agents and explosive atmospheres, importantly by introducing a specific (and additional) requirement for risk assessment where dangerous substances are or may be present at the workplace. Zoning requirements are specified where explosive atmospheres may be present, and employers are required to make arrangements to deal with any incidents.

Some significant definitions
Dangerous substance means a substance or preparation meeting the classification criteria for explosive, oxidising, extremely flammable, highly flammable or flammable substances (whether or not so classified under the CHIP Regulations); and any other substance or preparation with physical or chemical properties which create a risk because of the way the substance is used or present in the workplace; and any other dust which can form an explosive mixture with air or an explosive atmosphere.

Explosive atmosphere means a mixture under atmospheric conditions of air and one or more dangerous substances in the form of gases, vapours, mists or dusts in which, following ignition, combustion spreads to the entire unburnt mixture.

Preparation means a mixture or solution of two or more substances.

Application
For workplaces where explosive atmospheres can occur, in use before July 2003, all the requirements must be met by July 2006. Any modification to the workplace after July 2003 must be compliant. Workplaces coming into use for the first time after July 2003 must be fully compliant.

The Regulations do not apply to the master and crew of a ship, nor to ship-repair except when carried out in dry dock. The zoning and co-ordination of response requirements of Regulations 7 and 11 do not apply to:
- medical treatment areas for patients
- non-industrial gas appliances used for cooking, heating, lighting, washing, refrigeration and hot water production where the normal water temperature does not exceed 105 degrees Celsius
- other domestic gas fittings
- the manufacture, handling, use, storage and transport of explosives or chemically unstable substances
- any work activity at a mine, quarry, borehole or offshore installation
- all common means of transport.

How the objectives are met by the Regulations
Generally, the DSEAR clarify existing requirements within the Management of Health and Safety at Work Regulations 1999 and expand them in some areas.

Regulation 4 extends the employer's duty towards his/her employees under the Management Regulations towards all non-employees who might be affected by the employer's work, so far as is reasonably practicable. This extension of duty does not apply to the provision of work clothing and PPE to non-employees, nor does it extend to dealing with accidents (Regulation 8) involving them and the provision of information, instruction and training (Regulation 9), except that in the latter case the Regulation must be applied according to the nature and degree of risk. Self-employed people are classed as both employee and employer for the purposes of the Regulations.

Where a dangerous substance is or may be present at a workplace, Regulation 5 requires the employer's risk assessment to take account specifically of the following 10 points:
- the hazardous properties of the substance
- supplier's safety information
- circumstances of the work – processes, substances used and possible interactions, the amount involved, risks from any combinations of substances, and arrangements for safe handling, storage and transport of dangerous substances and waste containing them
- activities such as maintenance for which there is a high risk potential
- the effect of measures taken to comply with these Regulations
- the likelihood that an explosive atmosphere will occur and its persistence
- the likelihood that ignition sources will be present and become active and effective
- the scale of the anticipated effects of a fire or explosion
- any places that can be connected via openings to places where explosive atmospheres may occur
- any necessary additional information needed to complete the risk assessment.

Note that the assessment is required regardless of the quantity of dangerous substance present, and also that non-routine activities such as maintenance and overhauls must also be included in the assessment, as these can have a greater potential for fire and explosion.

The significant findings of the assessment are to be recorded (where five or more are employed) as soon as practicable following it, and the record must also include specified information to show compliance with PUWER, and where it relates to a potential explosive atmosphere it must give details of zones classified in compliance with Regulation 7(1) and verifications required by Regulation 7(4), together with the aims and details of any co-ordinating measures required by Regulation 11.

Elimination or reduction of risks from dangerous substances is covered in Regulation 6. This requires reasonably practicable elimination or reduction of those risks, by substitution as a preference. Where risks cannot be eliminated in these ways, appropriate control measures consistent with the risk assessment must be applied, which also mitigate the detrimental effects of fire, explosion and other harmful physical effects. The risk control measures are specified in order of priority (Regulation 6(4)):
1 reduction of the quantity of dangerous substances to a minimum
2 avoidance or minimising of a release
3 control of the release at source
4 prevention of the formation of an explosive atmosphere (including ventilation)
5 ensuring that any release is suitably collected, contained, removed to a safe place or otherwise rendered safe
6 avoidance of ignition sources, including electrostatic discharges, and adverse conditions
7 segregation of incompatible dangerous substances.

The mitigation measures are also specified in order of priority (Regulation 6(5)):
1 reduction to a minimum of the number of employees exposed
2 avoidance of propagation of fires and explosions
3 provision of explosion relief arrangements
4 provision of explosion suppression equipment

5   provision of plant able to withstand likely explosion pressures
6   provision of suitable PPE.

The employer is to ensure that any conditions required by the Regulations are maintained, and also to arrange for the safe handling, storage and transport of dangerous substances and waste containing them. **Schedule 1** contains further specified measures covering workplaces and work processes, and organisational measures which must be complied with, the latter being written instructions and the use of permits to work.

Regulation 7 deals exclusively with places where explosive atmospheres may occur. The employer is to classify all such places as either hazardous or non-hazardous in accordance with **Schedule 2**. Hazardous places are to be zoned according to the classification provided, and where necessary marked with signs at entry points in accordance with **Schedule 4**. Note that any place in which an explosive atmosphere may occur in such quantities as to require special precautions to protect the health and safety of the workers concerned is deemed to be hazardous – and conversely, where special precautions are not needed, the place will be deemed non-hazardous within the meaning of the Regulations. **Schedule 3** applies to equipment and protective systems in places classed as hazardous. Overall, explosion safety measures must be certified by a competent person before first use. In hazardous places, antistatic work clothing must be provided.

The arrangements for dealing with accidents, incidents and emergencies are specified in detail within **Regulation 8**. In summary, procedures and safety drills are required to be in place and tested at regular intervals, information on emergency arrangements must be available, suitable warning systems are to be in place to enable an appropriate response to be made

immediately such an event occurs, and escape facilities identified by the risk assessment are to be provided and maintained. The information is to be displayed unless the risk assessment results indicate this is not necessary, and it must be made available to internal and external emergency services.

Regulation 8(3) deals with the post-incident situation, where the employer must ensure that immediate steps are taken to mitigate the effects, restore the situation to normal and inform affected employees. During the mitigation, only essential personnel are permitted in the affected area, with the necessary equipment including PPE.

Where the risk assessment shows the quantity of each dangerous substance present is such that only a 'slight risk' (not defined) to employees exists, and the employer has complied with **Regulation 6(1)** sufficiently to control that risk, then the first three paragraphs of Regulation 8 will not apply.

Appropriate and specific information, instruction and training is required by **Regulation 9**. This includes the details of any dangerous substance and the risk it presents, access to any relevant data sheet, and details of legislative provisions concerning the hazardous properties of the substance, together with the significant findings of the risk assessment.

Identification of hazardous contents of containers and pipes is covered by **Regulation 10**, which requires clear identification of contents and hazards where the container or pipe is not marked as required by the legislation listed in **Schedule 5**. The employer responsible for the workplace where an explosive atmosphere may occur is responsible for co-ordinating all the measures where two or more employers share it, whether permanently or temporarily (**Regulation 11**).

Revision

The DSEAR contain provisions for:

- dangerous substances and explosive atmospheres
- specific risk assessment
- prioritised control measures
- prioritised mitigation measures
- identification and marking of zones in explosive-hazardous areas
- specific information, instruction and training

Where exposure to a dangerous substance could occur, the Regulations also cover:

- warning and communications systems
- escape facilities if necessary
- emergency procedures
- equipment and clothing for essential personnel
- practice drills

28 Consultation with employees

Introduction

The desirability of a co-operative approach to health and safety in the workplace has been recognised for many years. A main recommendation of the Robens Committee in 1972 was that an internal policing system should be developed, whereby workforce representatives would play an active part in drawing hazards to the attention of workers and management, and play a positive role in explaining health and safety requirements to employees. Provision for these Regulations was made in the Health and Safety at Work etc Act 1974, which originally contained two subsections dealing with the subject of employee rights to consultation by the appointment of statutory representatives with whom the employer would be required to enter into dialogue.

One of the subsections allowed for the election of such representatives from among the workforce; the other gave the right to nominate safety representatives to trade unions. The provision for the general election of such representatives was repealed by the then Labour Government's Employment Protection Act of 1975, and the right to appoint safety representatives is now restricted to recognised independent trade unions. The Safety Representatives and Safety Committees Regulations 1977 (SRSCR) provided the detail of the general entitlement contained in Section 2 of the Act, and came into effect on 1 October 1978. They were further amended by the Management of Health and Safety at Work Regulations 1999 (MHSWR).

The first exception to the restriction of the right to election of safety representatives is contained in the equivalent regulations made for offshore installations in 1989, which allow for free election.

In 1996 the Health and Safety (Consultation with Employees) Regulations (HSCER) came into force. The European Court of Justice had ruled that consultation with the employer on health and safety issues cannot be limited to consultation with trade unions and their appointed safety representatives. Compliance with the Framework Directive requires more general consultation, and so these Regulations now require employers to consult (where there are employees who are not represented by safety representatives appointed by recognised trade unions under the SRSCR provisions), either employees directly or representatives elected by them.

In general, the SRSCR entitlements and functions given to safety representatives are more extensive than those given under the HSCER. In this Section, both sets of Regulations are summarised.

Objectives of the Regulations

The SRSCR and their accompanying ACoP provide a set of entitlements to consultation to nominees of recognised independent trade unions. They are given the right to make a number of kinds of inspection, to consult with the employer, and to receive information on health and safety matters. The Regulations also provide for training and time off with pay to carry out the functions of safety representation. Non-unionised employees also have consultation rights provided within the HSCER, which give elected employees less extensive functions.

The Safety Representatives and Safety Committees Regulations 1977

The right to appoint safety representatives is restricted to trade unions recognised by the employer for collective bargaining or by the Arbitration and Conciliation Advisory Service (ACAS) (**Regulation 3**). The presence of only one employee belonging to such a union is sufficient to require the employer to recognise that person (on application by his/her union) as a safety representative. The Regulation places no limit on the number of such representatives in any workplace, although the associated ACoP and Guidance Notes observe that the criteria to consider in making this decision include total number of employees, variety of occupations, type of work activity, and the degree and character of the inherent dangers. These matters should be negotiated with the appropriate unions and the arrangements made for representation should be recorded.

Unions wishing to make appointments of safety representatives must make written notification to the employer of the names of those appointed, who must be employees of the employer except in extremely limited circumstances (**Regulation 8**), which include membership of the Musicians' Union and Equity. On appointment in this way, safety representatives acquire statutory functions and rights, which are set out in **Regulation 4**. The employer cannot terminate an appointment; the union concerned must notify the employer that an appointment has been terminated, or the safety representative may resign, or employment may cease at a workplace whose employees he or she represents (unless still employed at one of a number of workplaces where appointed to represent employees).

Safety representatives are not required to have qualifications, except that whoever is appointed should have been employed for the preceding two years by the employer, or have two years' experience in 'similar employment'. The right to time off with pay during working hours for safety representatives in order to carry out their functions and to undergo 'reasonable' training is given in **Regulation 4(2)**. What is 'reasonable' in the last resort can be decided by an employment tribunal, to which the representative may make complaint (see below).

The MHSWR have inserted a further Regulation. **Regulation 4A** requires employers to consult safety representatives in good time, in respect of those employees they represent, concerning:
- introduction of any measure at the workplace which may substantially affect health and safety
- the employer's arrangements for appointing or nominating 'competent persons' as required by MHSWR Regulations 7(1) and 8(1)(b)
- any health and safety information the employer is required to provide to employees

- planning and organisation of any health and safety training the employer is required to provide
- health and safety consequences of the introduction of new technologies into the workplace.

Regulation 4A(2) requires every employer to provide such facilities and assistance as safety representatives may reasonably require to carry out their functions, which are given in Section 2(4) of the main Act and in the body of the SRSCR.

Safety representatives have the functions of representation and consultation with the employer as provided by Section 2(4) of the Health and Safety at Work etc Act 1974, and the following:
- investigation of potential hazards, dangerous occurrences and causes of accidents at the workplace
- investigation of complaints by employees represented on health, safety or welfare matters
- making representations to the employer on matters arising from the above
- making representations to the employer on general matters of health, safety or welfare
- carrying out inspections of the workplace regularly and following notifiable accidents, dangerous occurrences or diseases
- inspecting documents
- representing employees in workplace consultations with inspectors of the appropriate enforcing authority
- receiving information from those inspectors in accordance with Section 28(8) of the Act
- attending safety committee meetings in the capacity of safety representative in connection with any function above.

To assist in carrying out the functions, employers must provide 'such facilities and assistance as the safety representatives may reasonably require', and may be present during inspections. The facilities required include independent investigation and private discussion with employees.

Inspections

Regular routine inspections of the workplace can be carried out by entitlement every three months, having given reasonable previous notice in writing to the employer (**Regulation 5**), or more frequently with the agreement of the employer. Where there has been a 'substantial change' in the conditions of work, or new information has been published by the HSE relative to hazards of the workplace, further inspection may take place regardless of the time interval since the previous one. Defects noted are to be notified in writing to the employer; there is a suggested form for the purpose.

Safety representatives have a conditional right to inspect and copy certain documents by **Regulation 7**, having given the employer reasonable notice. There are restrictions on the kinds of documents which can be seen, which include commercial confidentiality, information relating to an individual unless this is consented to, information for use in legal proceedings, that which the employer cannot disclose without breaking a law, and anything where disclosure would be against national security interests.

The employer has to make additional information available so that statutory functions can be performed, and this is limited to information within the employer's knowledge. The ACoP contains many examples of the kind of information which should be provided, and the information which need not be disclosed.

'Duties'

Although the Regulations give wide powers to the safety representative, they specifically impose no additional duty. Representatives are given immunity from prosecution for anything done in breach of safety law while acting as a safety representative. It has been suggested that circumstances where this immunity might apply would include agreement during consultation on, for example, a system of work proposed by the employer which later turned out to be inadequate and became the subject of prosecutions of individuals involved in the decision to use the system.

Tribunals

Safety representatives may present claims to an employment tribunal if it is believed that the employer has failed to allow performance of the functions laid down by the Regulations, or to allow time off work with pay to which there was an entitlement (**Regulation 11**). If the tribunal agrees, it must make a declaration, and can award compensation to the employee payable by the employer. There is no right of access to the tribunal for the employer who feels aggrieved or who wishes to test in advance the arrangements he/she proposes to make.

Safety committees

Despite the title of the Regulations, their only reference to safety committees is in **Regulation 9**, which requires that a safety committee must be established by the employer if at least two safety representatives request this in writing. The employer must post a notice giving the composition of the committee and the areas to be covered by it in a place where it can be read easily by employees. The safety committee must be established within three months of the request for it.

The ACoP and the Guidance Note contain information and advice concerning the structure, role and functions of safety committees, which should be taken into account by non-statutory safety committees as well.

The Health and Safety (Consultation with Employees) Regulations 1996

Regulation 3 requires that, where there is no representation by safety representatives under the SRSCR, the employer must consult employees in good time on matters relating to their health and safety at work. In particular, employers are directed to do so about:

- the introduction of any measure at the workplace which may substantially affect the health and safety of those employees
- his/her arrangements for nominating 'competent persons' in accordance with the MHSWR to assist the employer on health

and safety matters, and to take charge of measures to combat identified serious and imminent danger at the workplace
* any statutory health and safety information which he/she has to provide
* the planning and organisation of any health and safety training he/she has to provide
* the health and safety consequences for those employees of the introduction of new technologies into the workplace.

Regulation 4 requires the consultation to be either with the employees directly, or with representatives elected by any group of employees. Those elected are referred to as 'representatives of employee safety' (ROES). It is important to appreciate that the choice of which form of consultation to adopt is left to the employer. If the employer decides to consult ROES, he/she must inform their constituents of their names and the group of employees represented. The employer must not consult an individual as a ROES in four circumstances, which are where:
* the person has notified the employer that the person does not intend to represent the group
* the person is no longer employed in the group the person represents
* the period for which the person was elected has expired without the person being re-elected

* the person has become incapacitated from carrying the functions under the Regulations.

If the consultation is discontinued for any of these reasons, the employer must inform the employees in the group concerned.

Paragraph 4 of Regulation 4 requires the employer to inform employees and ROES if he/she decides to change the consultation method to one of direct consultation.

Regulation 5 obliges the employer to make necessary information available to employees consulted directly to enable them to participate effectively, and where ROES are consulted as well as this, the employer is required to make available records he/she must keep in compliance with RIDDOR.

As with the SRSCR, the employer does not have to disclose information which would be against national security, which could contravene any (statutory) prohibition on the employer, which relates to any individual without their consent, which would cause substantial injury to an undertaking, or where the information was obtained by the employer in connection with legal proceedings against him/her. Inspection rights are not available over documents not related to health or safety.

Revision

The Safety Representatives and Safety Committees Regulations 1977:
* allow recognised trade unions to appoint safety representatives without limit on numbers
* permit them to make inspections regularly, after incidents and changes in work
* permit the inspection and copying of certain documents
* require the employer to provide reasonable facilities for inspections
* provide for time off with pay for safety representatives to carry out their functions
* impose no duty on safety representatives

* provide a detailed consultation mechanism between employers and employees
* require a safety committee to be set up when two or more safety representatives request this

The Health and Safety (Consultation with Employees) Regulations 1996:
* require the employer to consult non-union employees on health and safety matters
* allow the consultation to be either direct or with elected representatives
* require the employer to inform employees if the consultation method is to change

29 The Health and Safety (Safety Signs and Signals) Regulations 1996

Introduction
A safety sign is one which gives a message about health and safety by means of a combination of geometric form, safety colour and symbol or text (words, letters, numbers) or both. The Regulations now cover a variety of methods of communication of health and safety information, in addition to the traditional safety signs which have been in use in a common design since 1986. For example, they now include codes of signalling and the use of audible warning devices. The Regulations implement EU Directive 92/58/EEC, which gives minimum requirements for provision of safety signs at work and standardises signs throughout member states of the European Union. They apply to all places of work and activities where people are employed, but do not apply to transport operations or the marking and supply of dangerous goods.

The previous Safety Signs Regulations 1980 introduced four kinds of safety sign, which are now in general use, and each of which is distinguished by colour and shape:

- **prohibition** – certain behaviour is prohibited, evacuate, stop
- **warning** – gives warning of a hazard
- **mandatory** – indicates a specific course of action is to be taken
- **safe condition** – gives information about safe conditions, doors, exits, escape routes.

The 1996 Regulations build on this system, which is both continued and strengthened. The Regulations are consistent with BS 5378 ('Safety signs and colours') and BS 5499 ('Fire safety signs, notices and graphic symbols'). Where fire safety signs contain symbols which comply with the latter, no changes are necessary. Advice can be obtained from the relevant enforcing authority for fire safety, which may be the local authority's environmental health department, building control officer, the fire officer, or, in some cases, the HSE. Risk assessments will be significant in deciding whether the Regulations apply in particular cases. When the control measures identified in the risk assessment have been applied, there may be one or more 'residual risks' remaining which employees need to be warned of. If one or more of these risks is significant, then a safety sign will be needed, otherwise not.

It should be noted that some fire signs may be specified as, for example, requirements of a fire certificate, and in these cases compliance is mandatory regardless of the employer's views on risk.

There is no duty within these Regulations on employers or occupiers to provide signs to warn non-employees of risks to their health or safety. Duties to do so are present in the Management of Health and Safety at Work Regulations 1999 and the Health and Safety at Work etc Act 1974, and safety signs are likely to be seen as a convenient way of meeting general obligations towards third parties. The Regulations came into force on 1 April 1996.

Objectives of the Regulations
The Regulations provide the legal means to require employers to provide safety signs in a variety of situations where there is a significant risk to health or safety which has not been avoided or controlled satisfactorily by other methods. They also cover hand, verbal and acoustic signals intended to communicate health and safety information. The principle is that health and safety information should be presented to employees in a uniform and standardised way, keeping the use of words to a minimum. The reasons for doing so are that international trade and travel has increased, resulting in a need for international uniformity especially because of the development of international workforces which do not share a common language.

How the objectives are met by the Regulations
Regulation 4 deals with the provision and maintenance of safety signs following risk assessment, so that appropriate signs to warn or instruct, or both, are to be placed where employers cannot avoid or adequately reduce

risks to employees without them. Generally, 'adequate' reduction of risks is where there is no longer a significant risk of harm. The safety signs placed must comply with Schedule 1 to the Regulations. Appropriate hand signals or verbal communication must be ensured so far as is reasonably practicable, as described in Schedule 1 or Schedule 2. Traffic signs prescribed for a particular situation under the Road Traffic Regulation Act 1984 are to be used in a place of work where there is a risk to employees from traffic movement or presence, even though that Act may not apply to that place of work. Suitable and sufficient instruction and training in the meaning of safety signs (the term includes signalling systems) and what to do about them is to be given to each employee (**Regulation 5**), together with comprehensible and relevant information about the signs.

Schedule 1 to the Regulations is in nine parts, and is both detailed and comprehensive. It is contained within the guidance to the Regulations, and should be consulted in full. Among the (minimum) specific requirements are the following selected points of interest:
- there is a required specification for marking pipes and containers; colour coding of pipes is not covered in the Regulations

- traffic routes are to be marked permanently with a safety colour (eg yellow means 'take care')
- placing of too many signs together should be avoided, and two acoustic signals are not to be used at the same time
- small differences from the pictograms shown in the Schedule are acceptable as long as they do not affect or confuse the message conveyed
- text may be used to supplement a safety sign
- a sign must be removed when the situation to which it refers ceases to exist
- pipework signs are to be placed on the visible side(s), in unpliable, self-adhesive or painted form, and in the vicinity of the most dangerous points such as valves and joints, and at reasonable intervals
- firefighting equipment must be identified by being coloured red and either by placing a location signboard or by painting the place where it is kept (or an access point) red; the mandatory signboards must be used
- acoustic evacuation signals must be continuous
- signalmen giving hand signals must wear distinctive identifying clothing items which are brightly coloured and for their exclusive use.

Revision

The Health and Safety (Safety Signs and Signals) Regulations 1996 require:
- signs and signals where significant risk cannot otherwise be reduced

4 types of safety sign:
- prohibition and fire
- mandatory
- caution
- safe condition

9 parts to Schedule 1, giving minimum requirements for:
- safety signs and signals at work
- signboards
- signs on containers and pipes
- identification and location of firefighting equipment
- signs used for obstacles and dangerous locations, and marking traffic routes
- illuminated signs
- acoustic signals
- verbal communication
- hand signals

Relevant standards

The following British Standards give guidance on safety signs in addition to BS 5378:

BS 5499	Parts 1 and 3 – Fire safety signs, notices and graphic symbols (contains (statutory) requirements for signs on fire precautions and means of escape)
BS 6736:1986	Code of practice for hand signalling for use in agricultural operations
BS 7121	Code of practice for safe use of cranes (Part 1:1989: General)
BS 1710:1984	Specification for identification of pipelines and services
BS 7443:1991	Specification for sound systems for emergency purposes

Table 1: Requirements for safety signs

Safety colour	Warning purpose	Examples of use	Contrasting colour	Symbol colour	Description of sign
Red	Stop Prohibition	Stop signs Identification and colour of emergency shut-down devices	White	Black	Circular red band and cross bar Red to be at least 35 per cent of sign area
	Firefighting	Showing fire hose, ladder, emergency telephone, fire extinguisher	White	White	Red rectangle or oblong, red to be at least 50 per cent of sign area. Supplementary 'this way' signs for equipment are single white direction arrows on a red sign
Yellow	Caution, risk of danger	Identification of hazards (fire, explosion, chemical, radiation, etc) Warning signs Identification of thresholds, dangerous passages, obstacles Risk of collision	Black	Black	Triangle with black band Yellow to be at least 50 per cent of sign area
Blue	Mandatory action	Obligation to wear PPE Mandatory signs	White	White	Circular blue disc Blue to be at least 50 per cent of sign area
Green	Safe condition	Identification of safety showers, first aid points, emergency exit signs	White	White	Green square or oblong Green to be at least 50 per cent of sign area

Supplemental text may be added if required, in the contrasting colour on the safety colour, or as black text on a white background

30 The Ionising Radiations Regulations 1999

Introduction

These Regulations came into force fully on 13 May 2000. They replaced the 1985 Regulations of the same name, and implement in part the provisions of EC Directives 96/29 Euratom (basic safety standards for workers), 90/641 Euratom (operational protection of outside workers) and 97/43 Euratom (dangers of ionising radiation in relation to medical exposure), which prescribe the basic safety standards for the health protection of the general public and workers against the dangers of ionising radiation.

The Regulations are only concerned with ionising radiation arising from a work activity, and not from the natural background, such as cosmic radiation, external radiation of terrestrial origin and internal radiation from natural radionuclides in the body. (See Part 3 Section 8 for further information.)

Radiation protection is based on three general principles which are incorporated into the Regulations:

- every practice resulting in an exposure to ionising radiation shall be justified by the advantages it produces
- all exposures shall be kept as low as reasonably achievable
- the sum of doses received shall not exceed certain limits.

The earlier (1985) Regulations were among the first to require assessment of risks to be made, well in advance of the general duty to do this first set out in the Management of Health and Safety at Work Regulations 1992. (For those interested, other early assessment requirements can be found in the Regulations controlling lead (1980) and asbestos (1987) at work.)

These Regulations are arranged in seven parts and nine Schedules. The parts are:

Part 1 Interpretation and general
Part 2 General principles and procedures
Part 3 Arrangements for the management of radiation protection
Part 4 Designated areas
Part 5 Classification and monitoring of persons
Part 6 Arrangements for the control of radioactive substances, articles and equipment
Part 7 Duties of employees and miscellaneous.

Objectives of the Regulations

The primary aim of the Regulations is to introduce conditions whereby doses of ionising radiation can be maintained at an acceptable level. Details of acceptable methods of meeting the requirements of the Regulations are contained in an ACoP. The limiting of doses is achieved by setting limits on the number and size of doses received in any calendar year by various categories of people liable to be exposed.

How the objectives are met by the Regulations

In Part 1, **Regulations 1–4** deal with the interpretation of terms and some general requirements. They define the terms used in the text (**Regulation 2**), such as 'ionising radiation' (gamma rays, X-rays or corpuscular radiations which are capable of producing ions either directly or indirectly) and 'radioactive substance', and explain the scope of the provisions (**Regulation 3**).

Part 2 deals with the general requirements. **Regulation 5** requires 'radiation employers' to hold a licence (authorisation) from the HSE to use electrical equipment to produce X-rays for industrial radiography, product processing, research and medical treatment. Except for work specified in Schedule 1, employers must notify the HSE of their intention to work with ionising radiation (**Regulation 6**). Specific risk assessment of radiation accident potential is always required, sufficient to show that all hazards have been identified and the type and level of the risks have been evaluated (**Regulation 7**). The employer's duty is then to take reasonably practicable steps to prevent any such accident, limit the consequences should one occur, and provide employees with the information and equipment to restrict their exposure to ionising radiation.

Regulation 8 contains requirements on dose limitation. Every employer is required to take all necessary steps to restrict so far as is reasonably practicable the extent to which employees and other persons are exposed to ionising radiation. Limits are also imposed here (and in **Regulation 11**) on the doses of ionising radiations which employees and other persons may receive in any calendar year. **Regulation 9** specifies the use of approved PPE. Every employer must ensure that appropriate accommodation is provided for PPE when it is not being used, and employees must take all reasonable steps to return any PPE issued to the accommodation after use.

Regulation 10 specifies 'proper and suitable' maintenance and keeping of records for that equipment, and all other controls and features provided. In the event of a radiation accident being reasonably foreseeable, contingency plans are required by **Regulation 12**, to be rehearsed as appropriate.

Part 3 details arrangements for the management of radiation protection, principally by the appointment of one or more radiation protection advisers and supervisors. These people have the duty of making local rules for the safe conduct of the work, and ensuring the work is properly supervised. The advisers and supervisors may be from external organisations, working under contract to the employer. Employers must ensure that adequate information, instruction and training is given to employees and other persons potentially affected by the work. **Regulation 13** controls the appointment of radiation protection advisers, **Regulation 14** covers general training and information, and **Regulation 15** requires co-operation between employers.

Part 4 covers designation of controlled and supervised areas, and their monitoring. **Regulations 16–19** specify that areas in which

persons are likely to receive more than the specified doses of ionising radiation are to be designated as 'controlled areas' or 'supervised areas', and entry into these is restricted to specified people in specified circumstances. Supervised areas are those which might turn into controlled areas by reason of the doses that could be received by persons working there. Radiation protection supervisors enforce local rules made in respect of both kinds of area.

In Part 5, those employees who are likely to receive more than the specified doses must be designated as 'classified persons' (**Regulation 20**). **Regulations 21–26** deal with dosimetry and medical surveillance. Exposures to ionising radiation received by classified and certain other specified persons are to be assessed by one or more dosimetry services approved by the HSE. Records of the doses received are to be made and kept for each such person.

These Regulations also require classified persons and certain other employees to be subject to medical surveillance. Overexposure must be notified and investigated by the employer, who must keep the record for at least 50 years from the date on which it was made (**Regulation 25**). For overexposed employees, modified dose limits are provided.

Part 6 contains provisions requiring any radioactive substance used as a source of ionising radiation to be in the form of a sealed source as far as is reasonably practicable (**Regulation 27**). Any articles embodying or containing radioactive substances must be suitably designed, constructed, maintained and tested.

Also required is the accounting for (**Regulation 28**), keeping and moving (**Regulation 29**) of radioactive substances. Notification of escape or release, loss or theft of radioactive substances must be reported (**Regulation 30**). **Regulation 31** imposes duties on manufacturers, installers and others to ensure that articles for use in work with ionising radiations are designed, constructed and installed so as to restrict so far as is reasonably practicable exposure to ionising radiation. Similar duties rest with employers in relation to equipment used for medical exposures.

Employers are required to investigate any defect in medical equipment which may have resulted in the overexposure of those undergoing the treatment, record the detailed investigation results and keep them for at least 50 years (**Regulation 32**). Interference with sources of ionising radiations is prohibited (**Regulation 33**.)

Part 7 deals with employees' duties (**Regulation 34**) and miscellaneous provisions, including approval of dosimetry services (**Regulation 35**), defences available to prosecutions under the Regulations, transitional and other incidental provisions applying to offshore installations, and modifications relating to the Ministry of Defence and others.

Revision

The Ionising Radiations Regulations 1999 contain requirements on:

- dose limitation
- regulation of work with ionising radiation
- dosimetry and medical surveillance
- control methods for radioactive substances
- monitoring of ionising radiation
- risk assessments and notifications
- record-keeping
- safety of articles and equipment

31 The Control of Pesticides Regulations 1986

Introduction

Pesticides are chemical substances and certain micro-organisms such as bacteria, fungi or viruses which are prepared and used to destroy pests of all types, including animals, plants and other organisms. The term 'pesticide' extends to cover herbicides and fungicides, and for the purposes of these Regulations the term also encompasses a wide variety of pest control products including wood preservatives, plant growth hormones, soil sterilisers, bird or animal repellents and masonry biocides, as well as anti-foul boat paints.

Objectives of the Regulations

The Regulations provide a framework for the legal control of the use of pesticides in the United Kingdom. They are designed to protect people and the environment, and enable informed official approval of pesticides as well as control of their marketing and use. The control strategy is detailed within the body of the Regulations, with a series of detailed requirements contained in attached Schedules. They were first introduced on 6 October 1986 and were fully implemented on 1 January 1989.

How the objectives are met by the Regulations

The Regulations apply to all pesticides used in agriculture, forestry, food storage, horticulture, animal husbandry, wood preservation, domestic gardens, kitchens and larders. They also apply to those pesticides used in public hygiene and pest control (for parks, sports grounds, waterways, road and rail embankments, etc), masonry treatment and vertebrate chemosterilant products. They extend to cover anti-fouling paints and surface coatings used on boats, on structures below water, and on nets or floats/apparatus used in fish cultivation.

Exceptions within the Regulations exclude some applications of pesticides – they do not apply to those used in industrial or manufacturing processes, for example. The Regulations do apply when pesticides are used in the workplace other than in the industrial process, such as the use of herbicides on industrial land, or rodenticides in a factory. Pesticides which are not covered by the Regulations are listed in **Regulation 2(2)**, and include those used in paint, water supply systems and swimming pools.

The Regulations prohibit the advertisement, supply, storage, and use of pesticides unless they have been approved (**Regulation 4**). As a result, all manufacturers, importers and suppliers of pesticides must obtain approval for each product they develop or market for use in the UK.

The approval may be given in the form of a provisional approval or a full approval (**Regulation 5**). An experimental approval may be granted for pesticides which are being developed but these cannot be sold or advertised. The Regulations also empower ministers to impose conditions on approval and review, revoke or suspend the approvals that have been given in appropriate circumstances.

In the event of a breach of the Regulations, powers are granted for the seizure or disposal of the pesticide (or anything treated with it), its removal from the UK (in the case of imported pesticides) or any other remedial action as may be deemed necessary. It is also possible under **Regulation 8** for ministers to make available to the public evaluations of study reports submitted on pesticides which have been granted provisional or full approval. Commercial use or unauthorised publication of information made available under this Regulation is prohibited.

More specific conditions are set out in the **four Schedules to the Regulations**. These were established over a period of five years, with all controls being fully implemented by 1 January 1989. The Schedules stipulate that:

Only approved pesticides (including anti-fouling paints and surface coatings) may be supplied, stored or used. Only provisionally or fully approved pesticides may be sold or advertised (and then only in relation to their approved uses). Printed or broadcast advertisements must mention the active ingredients and include the phrase "Read the label before you buy: Use pesticide safely". More specific warning phrases may be considered appropriate during approval and these also have to be included.

Only pesticides specifically approved for aerial application may be applied from the air. Detailed rules are established for this specialised type of application in Schedule 4.

General obligations must be complied with by all sellers, suppliers, storers and users, including the general public. They must take all reasonable precautions to protect the health of humans, animals, plants and the environment. In particular, users must take precautions to avoid the pollution of water.

Sellers, suppliers, storers and commercial users must be competent in their duties, and commercial users must have received adequate instruction and guidance in the safe, efficient and humane use of pesticides. A recognised certificate of competence is required for anyone who sells, supplies, stores or uses pesticides in agriculture, horticulture and forestry. Anyone engaged in these activities who does not hold a certificate must be under the direct supervision of someone who does. Certificates are also required for anyone using pesticides who was born after 31 December 1964, unless they are working under direct supervision.

All users must comply with the conditions of approval relating to use. These will be clearly stated either on the label or in the published approval for the pesticide, and usually will include directions as to the protective clothing to be worn, limitations of use, maximum application rates, minimum harvest intervals, protection of bees, and keeping humans and animals out of treated areas.

No person is allowed to use a pesticide in the course of business unless he/she has received adequate instruction and guidance in the safe, efficient and humane use of pesticides, and is competent for the duties he/she will carry out. Employers have particular responsibility to ensure that every employee required to use a pesticide is provided with the necessary instruction and guidance to enable him/her to comply with the Regulations and achieve the standard of competence recognised by ministers.

Any conditions of approval relating to the sale, supply and storage of pesticides must be complied with. Tank mixing of pesticides is controlled with lists of approval published by ministers.

Pesticides approved for agricultural use (which includes agriculture, horticulture, forestry and animal husbandry) can only be used in a commercial service (applying them to other people's property or premises) by a holder of a certificate of competence or by someone under the direct supervision of a certificate holder.

Revision

The Control of Pesticides Regulations 1986 provide for:
- approval for products
- conditions of supply, storage and use
- competence training and certification for commercial use in agriculture generally

32 The Health and Safety (Information for Employees) Regulations 1989

Introduction
Health and safety information is given to employees in a variety of ways. Ensuring that employees are aware of the contents of the employer's health and safety policy statement is perhaps the most obvious of these. However, earlier enactments had also required that copies of printed abstracts of the major Acts and regulations dealing with workplace health and safety be displayed at or near the workplace. Their ability to inform was doubtful, as they were simply displays of the legal requirements and not specially written for comprehension. These Regulations repealed most of the former specific requirements, and introduced a statutory set of approved written information material in the form of a poster or leaflet which must be made available to all employees regardless of the nature of their work. Further legislation has revoked the remaining requirements to display particular pieces of legislation on the walls of premises.

Objectives of the Regulations
The Regulations require information relating to health, safety and welfare to be furnished to employees by means of posters or leaflets in the form approved and published by the HSE. Details of the appropriate enforcing authority and Employment Medical Advisory Service (EMAS) are also made available to employees by this means. The Regulations came into force on 18 October 1989.

How the objectives are met by the Regulations
Regulation 4 requires employers either to provide each employee with the approved leaflet, or else to ensure that for each employee while at work the approved poster ('Health and safety law – What you should know' – no reference number) is kept displayed in a readable condition at a reasonably accessible place, and so positioned as to be easily seen and read. Any later revisions of the poster or the notice have to replace the previous ones, so out of date material is not in compliance. (The current poster is coloured buff and pink.)

Regulation 5 requires the poster to contain the name and address of the appropriate enforcing authority for the premises, and the address of the EMAS for the area in which the premises lie. Any necessary changes to either addresses or areas have to be made within six months of the change. If the employer has chosen to give the required information by leaflet, he/she must also give the employee a written notice containing the foregoing name and addresses. For the purpose of establishing which name and addresses are relevant for employees working away from a base, they are treated as working at the premises where the work is administered. If the work is administered from more than one base, the employer may choose any one of them as 'the base'.

Exemption certificates can be issued by the HSE under **Regulation 6** to anyone, covering any or all of the requirements, provided that they are satisfied that the health, safety or welfare of persons likely to be affected by the exemption will not be prejudiced as a result. Conditions and time limits can be attached to exemption certificates to ensure this.

Regulation 7 provides a defence of due diligence for the employer who can prove he/she took all reasonable precautions and exercised all due diligence to avoid the commission of an offence against the Regulations.

Revision
The Health and Safety (Information for Employees) Regulations 1989 require that an information poster or leaflet be available for all employees

33 The social security system and benefits

Introduction

The principle that an injured or sick worker unable to earn a full wage should receive financial support has been recognised in the United Kingdom since 1837. In that year, the first case was recorded of a servant suing his employer (Priestley -v- Fowler (1835–1842) AER Rep 449) – he lost, and in so doing established the doctrine of **common employment**. This stated that an employer could not be held liable to his servant for an injury caused by the negligence of a fellow servant with whom he was engaged in common employment. The case was supported by later decisions, and the doctrine was not abolished until 1948 by the Law Reform (Personal Injuries) Act.

In those early days, binding decisions emerged which further restricted the worker's right to claim against his/her employer. There was an inability to sue at common law if the injured party had also made a claim under the Workman's Compensation Act (this was not amended until 1946, by the National Insurance (Industrial Injuries) Act), and if there was any element of contributory negligence by the employee (amended by the Law Reform (Contributory Negligence) Act 1945).

Because of these reforms, financial support for those injured at work by accident or disease can be obtained by using the social security system of benefits, and by claiming damages under the tort of negligence. A **tort** is defined simply as 'a civil wrong'. Other Sections of this Part outline principles of the employer's and occupier's duties at common law, which depend on the establishing of negligence on their part for the success of the claim. The social security system is governed by specific legislation, and pays benefits to those entitled to them according to defined rules and without reference to the question of liability or fault of the parties involved.

Social security

The present social security benefit system began in 1948; funds for the system are derived from taxation, particularly of employers and employees. At first, benefits were paid for injury, disablement and to widows of those killed at work, regardless of the extent of their personal contribution to the benefit scheme.

The legislation which governs the social security system is complex, and changes frequently. As a consequence the benefits themselves change, as do their extent and the qualifying entitlements. A current guide should be consulted.

The most recent major changes to the system were made by the Social Security (Incapacity for Work) Act 1994 and related regulations, which applied to industrial injuries benefits from April 1995. Other relevant statutes include the Social Security (Contributions and Benefits) Act 1992 and the Statutory Sick Pay Act 1994.

The main forms of benefit currently available to those injured at work (in addition to statutory sick pay) are industrial injuries disablement benefit, constant attendance allowance and exceptionally severe disablement benefit, none of which are taxable. Statutory sick pay is taxable under PAYE (Pay As You Earn).

To be eligible for industrial injuries benefit, a claimant generally has to show that he or she is suffering from a prescribed disease, that the disease is prescribed for the particular occupation followed, and that the disease was caused by the occupation. There are different rules which apply to sufferers from asbestosis, asthma and occupational deafness. The Social Security (Industrial Injuries) (Prescribed Diseases) Regulations 1985 have been amended progressively since their introduction, keeping up to date the list of prescribed industrial diseases for which there is an entitlement to benefit for those suffering from them. Review of the list is the remit of the Industrial Injuries Advisory Council.

Self-assessment questions and answers

Questions and answers: Part 1

Section 1

3 Add another layer to the bottom of the 'accident triangle'. How could you classify the incidents that could be included there? How could they be counted?

'Non-injury accidents' are measurable in theory as long as a means of counting and reporting them is available. A definition can be used to capture them, so that employees are asked to record and report 'property damage in excess of £10', for example. Beneath these in the triangle, there is the huge number of incidents that can be classified as 'near miss'. Because these are incidents where neither person nor property damage is caused, it is very difficult to learn about them – other than by anecdotes – but one way is to encourage their reporting in a 'no blame' system. These are cases where the description typically begins 'I almost', or 'I could have been hurt when'.

Section 2

1 What are the advantages of a works safety committee? Are there any disadvantages?

A works safety committee provides a means of consulting with employees, and encourages their involvement. The committee should have a composition of roughly equal numbers of employees and management, with alternating chairmanship; it also needs a written constitution defining its scope, direction and activities, the authority it possesses (if any), and the role of its members. Debate and discussion on application of local or national standards and regulations will be helpful in achieving their acceptance. Solutions to problems may also be generated by the committee, and problems themselves may be discovered by, or brought to the attention of, the committee.

Disadvantages can arise when the committee is not properly guided by expert advice, when members do not have a positive role to play, and when the committee is seen by its members and others as merely a complaints forum with no positive function other than to record problems which do not receive management attention. Another major disadvantage is that the safety committee can be abused by management in attempting to delegate (or pass over entirely) the management role and effective responsibility to the committee.

Section 3

1 What information will be needed before a risk assessment is carried out?

A risk assessment relates the hazard under consideration to the particular circumstances. Therefore, information about the circumstances (including numbers exposed and details of who may be exposed) is especially important. This can be obtained by consultation with employees, studying of records and statistics in the workplace and also nationally reported trends, and studying inspection and audits records. If appropriate, a fresh inspection including necessary measurements should be carried out to ensure the information is up to date.

The legal requirements of the situation must be available, as well as any guidance issued and relevant standards.

2 Explain the difference between qualitative and quantitative risk assessments with examples, showing where it would be appropriate for each to be used.

Qualitative risk assessments are based on personal judgment and general statements about the level of risk. They are appropriate in circumstances where compliance objectives do not require a more rigorous approach, such as in assessments to comply with the Management of Health and Safety at Work Regulations 1999.

Quantitative risk assessments contain probability estimates of failure, based on observed and interpreted data. They will be required where a more precise view of risk needs to be taken, such as where public concern is raised over planning proposals, eg nuclear power stations.

Section 4

1 **Why are safety policies useful in the management of health and safety? Why would one consisting of only one page be of no value?**

Safety policies are useful because they:
* demonstrate management intentions, support and commitment
* set out safety goals and objectives
* delegate responsibility and accountability for safety
* supply organisational details
* supply information about arrangements made for health and safety
* comply with any relevant statutory requirements.

The test of any safety policy will, therefore, be the extent to which it meets the needs of the organisation in these respects. It may be possible, in the very smallest organisations, to contain the safety policy in one page. By cross-referencing to larger documents containing detailed information, larger organisations may be able to use a single page as a basic summary of safety policy. However, if there is no cross-referencing, a single page could not contain the information and detail which is needed.

Section 5

2 **Write out a safe system of work for moving filing cabinets between offices.**

There are five steps involved in devising and implementing a safe system of work: task assessment, hazard identification, definition of the safe system, its implementation and monitoring the results. This task requires the use of the first three of these.

Overall assessment

Moving (25) cabinets down one set of stairs, across a yard, up one set of stairs to new office. No mechanical handling equipment is presently available. The cabinets are all full, each drawer is full, there are four drawers to each cabinet. The maximum weight of each drawer is 13kg, the maximum weight of each cabinet is therefore 52kg plus the weight of the frame –

20kg. Cabinets must be unlocked prior to removal of any drawers. Doorways and stairs are wide and in good condition. There is some light traffic in the yard. Manual handling operation – task cannot be automated or mechanised, handling aids to be used. (See table overleaf.)

Section 6

1 **What topics should be covered in induction safety training to provide necessary knowledge and skills for fire prevention?**

Fire prevention induction training should cover:
* identification of alarms (tones, number of rings, etc)
* action to take when alarms sound
* basic evacuation plan
* how to raise the alarm
* use of available fire appliances
* fire prevention
  ○ no smoking rules
  ○ housekeeping requirements
  ○ special fire hazards of the work or workplace
  ○ correct use of fire doors.

2 **'Managers don't need to know about safety rules – they are not at risk.' Discuss.**

Regardless of the question of risk, managers are also employees, and have duties under common law and statute law to obey safety rules, as well as to manage their observance by employees in their charge. Equally importantly, employees will take a lead from management, so the personal actions of managers reflect the importance which management as a whole gives to safety rules. Management commitment is an essential ingredient of any safety programme – the commitment must be active and this requires the support to be demonstrated.

The risk of injury for managers is not negligible – managers appear in safety statistics, especially in the construction industry. They are as vulnerable as workers to making errors of personal judgement, to forgetfulness and to taking shortcuts in the

Task	Hazard	Control by safe method
Minimise loads before moving	Attempts may be made to lift too much	All cabinets must be unloaded before work begins
Moving heavy cabinets	Manual lifting (back strains, etc)	Selection of fit staff required, with no history of injury and trained
	Time pressure to complete may force errors in handling	Staff must not be limited by time on this task, and allow adequate time
	Weight of cabinets	Do not move full or part-full units. Staff to be instructed to carry each between two, use sack trolleys where possible
Remove drawers	Weight of drawers when full	As above, drawers to be emptied to lower weight as necessary
Carrying drawers	Obstruction in path	Select door props and put in place before lifting, **not** fire extinguishers, or assign staff to open and close them as needed
Crossing yard	Passing traffic	Limit traffic, use cones/barriers, or assign staff to control traffic
Reassemble units	Put in wrong place (doors, passages may be obstructed)	Confirm proper placing with local supervisor before reassembling
	Overtipping	Confirm presence of anti-tilt mechanism on each cabinet. If not, place wooden batten under front edge of each

Other notes: All concerned must wear protective gloves and safety shoes. The task must not be proceeded with if a problem occurs which is not covered by this safe system of work. Alternative solutions to existing problems, and solutions for new problems, are not acceptable unless verified by the safety department.

interests of expediency. Managers are often injured as a result of the actions of others – eye injuries in machine shops can happen to anyone present, not just the operators. Wearing safety glasses in this situation is therefore important for everyone, as risk is no respecter of status.

Section 7

1 **Outline the maintenance system which would lead to the rapid repair of a faulty machine guard.**

1 Notification of a defective guard is made by the operator to the supervisor, and the machine is removed from use. (Variances from this are to be authorised only by the departmental manager, in writing.)

2 A written record of the defect is made by the supervisor, noting the defect, the date and the time reported. A copy of this record goes to the maintenance department, with a priority rating given by the supervisor to ensure appropriate prompt action is taken.

3 An entry is made by the supervisor in the machine's log, or other database. Regular occurrences of the same fault will be reported and referred for management action.

4 A priority system for safety rectifications is set up in the maintenance department.

5 There is a special budget set aside for safety maintenance work.

6 Each supervisor has a follow-up system in place to ensure maintenance is carried out within a given timescale.

7 The supervisor makes contact with the maintenance department if the timescale is exceeded.

8 Ongoing regular inspections are made of machinery and guarding facilities (this may be a statutory requirement).

9 All concerned require training in the system, which should also be outlined in the safety policy or manual.

10 An audit system should be in place to look at the practical operation of the defect rectification system at intervals.

Section 8

2 Develop a prequalification questionnaire suitable for use by contractors such as window cleaners, caterers and repair firms, which, on completion of the contracted work, would help you evaluate their safety performance.

All of the steps given in the text can be shortened as appropriate, depending on the risks of the work. Some contractors, such as window cleaners, engage in a limited range of high-risk activities, so the questionnaire for them should concentrate on perceived problems which may arise from their work.

Prequalification questionnaire

1 In what type(s) of work are you skilled?
2 Who has responsibility at board level in your organisation for health and safety?
3 Who provides you with surveillance and advice on health and safety, and what are his/her qualifications?
4 What health and safety training have you carried out in respect of the management/ supervision you will provide, and of the operatives you will provide?
5 What is your system for investigating and reporting accidents, diseases and dangerous occurrences?
6 Has your organisation won any safety performance awards?
7 Is your organisation a member of a recognised safety organisation?
8 What are your systems for the maintenance of plant, tools and equipment?
9 Have you any written safety procedures or manuals?
10 How do you prequalify your own subcontractors on health and safety?
11 Have you supplied us with a copy of your safety policy and specified risk assessments?
12 Please supply details of all lost-time injuries to your employees recorded during the past three years.
13 Is there any further information which would assist us in assessing your organisation's safety performance?

Section 9

2 Is 'carelessness' an adequate sole conclusion on an accident report?

As a description, 'carelessness' can mean almost everything – and almost nothing. Accidents are caused by human failings and by failures in, or the absence of, control measures. Accident reports should address both of these if appropriate, not merely the human element (which is often easier to identify, especially if blame is to be laid – whose fault was it?).

'Human failings' covers all aspects of personal behaviour which result in mistakes being made; failure to take care is only one aspect. It can be argued that no-one fails to take care, in the sense that a decision (conscious or unconscious) is taken before every action, which is seen by the individual at the time to be the most appropriate response, or at least the best choice between available alternatives. Some of these alternatives should be provided or indicated by the control measures available. The decision taken may be shown afterwards to have been inappropriate or incorrect, but this does not mean that care was not taken in arriving at it – merely that it was not the right or best one. Training and experience have major influences on the quality of decisions taken.

It can also be said that almost no-one deliberately sets out to have or cause an accident. Making an error of judgement is not necessarily due to carelessness; therefore, further investigation is required of those accidents with a 'human failing' component (possibly all of them?), and the listing of 'carelessness' as the sole conclusion of an accident report is never justified. It reveals more about the person making the report than about the accident.

Section 11

2 Explain with examples the difference between a primary and secondary source of information. Mark the following sources as primary or secondary.

Primary sources are original documents, and collections of these separately published are secondary sources. The telephone directory and this written answer are primary sources, as are personal letters and papers given at meetings. Bibliographies and summaries of papers held on databases are secondary sources.

Your company's annual report	Primary
Full proceedings of an annual technical conference	Primary
Your shopping list	Primary
An international standard on eye protectors	Primary
A reference library	Secondary

Section 12

2 **If effective change must start at the top, what do you think are the barriers to successful change at lower levels in your organisation?**

The acceptance of change is one of the features of a positive safety culture. The elements of a good safety culture apply to all levels in the organisation, and lack or inadequacy of any of them can be a barrier to change. The elements are:

- **good communications** between and with employees and management
- ensuring a real and visible commitment to high standards by **everyone**
- maintaining good **training standards** to achieve competence
- achievement of good **working conditions**
- workplace **culture** in relation to the acceptability of all forms of change.

The perception of the safety culture is key to the behaviour that will result. If it is felt that senior management give only lip service to safety, then that is what will be given at lower levels in the organisation. Peer pressure is very strong, with the desire to conform to organisational norms. Any of the above may violate existing norms and values.

Questions and answers: Part 2

Section 1

1 **Define a 'workplace'.**

A workplace is defined in the Workplace (Health, Safety and Welfare) Regulations 1992 as any non-domestic premises or part of premises made available to any person as a place of work. Therefore, the Regulations do not cover homeworkers.

Section 2

2 **In what circumstances would plastic sheeting not be an adequate material to use in machine guarding?**

Material used in machine guarding needs to be strong enough to withstand vibration and wear caused by the normal functioning of the machine, and not be flexible enough to allow the guard to be bypassed by operators or others, or by parts, offcuts, etc, which may deflect the guard from the inside. Plastic sheeting of adequate thickness and stiffness may be able to meet these requirements, but particular circumstances where it would not will depend on the machine's characteristics and the task(s) being done.

The presence of ultraviolet light, heat and excessive vibration would be circumstances where plastic sheeting is not suitable. This is because ultraviolet light attacks plastic over time, heat can deform it, and vibration can cause cracking in the plastic or at the attachment points.

Section 3

1 **What checks should be made before a crane is used in a workplace?**

Checks should be made on the crane itself, the method of operation proposed, and the work environment. The crane must be properly installed, including a strong base, and having regard to the loads likely to be lifted and the capacity of the crane. The crane structure and chassis must be sound and adequately tested, and reliance may be placed on the seller to certify this if it is a new piece of equipment. The ability of the parts to take the loads and lift safely is most important, and this can be verified first in documentation, then by a load test at installation and at intervals afterwards. The safe working load established by a competent person should be marked prominently on the structure. Indicators supplied should give the operator basic information about the load, as a minimum. Visual inspection of the lifting parts should be done frequently.

Important aspects of the crane's method of operation include the selection of and training for operators, a survey of potential hazards such as obstructions in the possible path of the crane, overhead wires, and isolation procedures during maintenance. Accessibility during regular maintenance will have to be established. Lifting equipment to be used in conjunction with the crane will need to be reviewed for capacity and suitability (slings, chains, ropes and so on).

2 **A powered lift truck overturns and the operator is injured. List the potential causes of the overturning.**

Powered trucks can overturn because of overloading, inability of the truck to deal with ground conditions under the circumstances at the time, and speed. The following factors are relevant:

Overloading by exceeding the truck's maximum capacity, travelling with the load elevated or tilted, travelling downhill with the load in front of the truck, and a shifting load.

Ground conditions include striking against obstructions or running across uneven sections of floor, turning on or crossing slopes at an angle, unsuitable floor or road for the use of the truck.

Speed includes sudden braking or changes in speed, and turning while changing speed abruptly.

Section 4

1 **What factors should be considered by management before manual lifting of loads is authorised?**

The manual lifting and handling of loads should be avoided completely where reasonably practicable. Failing this, an assessment must be made of the hazards and risks, and preventive measures considered. These steps can include alterations to the task or mechanisation. Assessment of the hazards involves an examination of the process to note the stages at which manual handling need not be used or can be limited. Before manual handling is authorised, a safe system of work will be required which provides solutions to the remaining hazards. Such a system will include training for those carrying out manual handling and their supervisors, and also adequate health screening to ensure fit and healthy workers are selected for the task.

Section 5

2 **A scaffold collapses as a result of overloading. How could this have been prevented?**

Overloading of scaffolds can be prevented by limiting the loads to be placed on the scaffold, ensuring the correct type of scaffold has been specified and erected, making those working on it aware of its load limitations, and by making arrangements for its competent inspection before use and at intervals to identify any weakness which may develop. Inspection is also required to ensure the scaffold is properly supported and constructed to withstand the expected loads, and complies with any special design requirements. Mobile towers are especially prone to overturning, often because of overloading combined with inadequate base dimensions for the working height. Stability of all types of scaffold is assisted by tying them to adjacent fixed structures or by fitting raker support tubes to increase their base dimensions.

Section 6

1 **A vehicle maintenance fitter insists he is not exposed to danger when carrying out his work. Identify the hazards to which he is exposed.**

Maintenance fitters are exposed to hazards from the vehicles on which they work, from the work environment, from work tools, equipment and procedures, and from hazardous substances they work with or become exposed to.

Vehicle hazards include working under non-propped bodies, where wheels are not chocked, brakes are not applied or the vehicle is not adequately stabilised. Jacks should be used in conjunction with axle stands. Hot work near fuel tanks requires special procedures. Inflation of tyres at high pressure should be carried out in cages because of the potential for bursting, especially with split rim wheels.

Work environment hazards include noise and vibration from engines and tools, and exposure to radiation from welding equipment.

Work tools and equipment hazards include risk of eye injury from use of abrasive wheels and wire brushes, explosion of battery-charging systems, hand injuries from common tools and equipment, and electrical hazards from equipment and circuits.

Hazardous substances include dust from brake drums which may contain asbestos, fumes and gases such as carbon monoxide from exhausts, and welding fumes, solvents and degreasing chemicals.

2 **Almost a quarter of all deaths involving vehicles at work happen while vehicles are reversing, most at low speeds. Apply basic principles of hazard control to prepare a list of ways in which reversing accidents can be prevented.**

The safety precedence sequence can be used to identify basic controls and prioritise them in terms of effectiveness. Several controls may be used together. (See table opposite.)

Section 7

2 **Can you think of other, simple, ways in which substances could be classified in your workplace so as to give an indication of their potential for harm? Are there any drawbacks to your classification?**

Priority	Example control measure
Hazard elimination	Prohibit reversing manoeuvres. Provide through access on a one-way system
Substitution	Examine the issue to see whether an alternative transport method is practicable which eliminates or reduces the need to reverse
Use of barriers	Erect barriers or refuges near reversing areas. Plan traffic routes to separate people and vehicles
Use of procedures	Written safe system of work involving one or more banksmen
Use of warning systems	Training for banksmen, safety signs warning of reversing dangers, instructions to visiting drivers
Use of personal protective equipment	Issue high-visibility jackets to pedestrians likely to be affected

Substances could be classified in-company, using a simple coding system to indicate their assessed potential for harm:

1 life threatening in all circumstances of exposure
2 life threatening in some circumstances, serious injuries are probable results of exposure
3 serious injuries possible in some circumstances, minor injuries are probable results of exposure
4 minor injuries are the only likely outcome of exposure
5 no injuries are likely to result from exposure.

Some drawbacks to this system are:
- it is easily applied only if data exists for each substance about its likely effects in the circumstances of use
- it takes no account of repeated exposures at low doses
- hazardous substances generated in the workplace or in processes are likely to be overlooked when making classifications
- the system does not provide enough positive information to workers to enable them to take sensible decisions about precautions.

Section 8

1 **Is operator training sufficient by itself to prevent chemical accidents?**

Operator training is only one aspect of the safe control of chemicals at work. It is therefore necessary, but not sufficient in itself. What is required is identification of all (chemical) hazards, assessment of the risks in practice, control of these, and then operator training in the selected control techniques. Final parts of the control system are the monitoring of effectiveness and necessary record-keeping.

2 **Give an example of a substance with different levels of risk when used and when stored.**

Dynamite is such a substance. In practice, it is kept in controlled conditions, including at a lowered temperature and away from the means of detonating it. In storage, it is therefore a low risk, which rises as the physical conditions become less controlled when use is near, and when operating procedures are relied on for control of risk thereafter.

3 **What zoonoses might be relevant to work in industrial premises?**

The presence of animals in premises is the most common source of zoonoses. Examples include rats (which can spread Weil's disease) and pigeons, whose droppings can be infected by micro-organisms which can lead to psittacosis. Legionella bacteria are found in natural water sources, but can become trapped and will multiply rapidly within water systems such as wet cooling towers and hot water systems. Showers and other fixed equipment can harbour the bacteria, which can cause a number of pneumonia-like diseases.

Section 9

1 **What preventive measures can be taken against electrical failure?**

Electrical failures and interruptions are caused by poor or damaged insulation, overheating, earth leakage currents, loose connections, inadequate circuit and component ratings, poor maintenance, inadequate systems of work, and human mistakes. These can be minimised by inspecting, testing and improving earthing standards; introducing safe systems of work including standard methods of testing for lack of power to circuits before starting work and not working on live equipment where possible; use of insulators, fuses and circuit breakers; installation of residual current devices; and use of only competent workers on electrical systems.

2 **An electric shock occurs as a result of using a drill outside. What factors might have contributed to this accident?**

- use of higher voltage than necessary for the task
- inadequate earthing
- use of equipment unsuitable for the circumstances (conditions or task) – wrong drill (electric), faulty drill (not subject to maintenance?), unsuitable cable (not waterproof?), damaged cable (jointed wrongly?)
- failure to maintain drill, cable, sockets, plugs
- failure to train or instruct the worker using the drill, the supervisor(s), or the management.

Section 10

1 **How would you handle a fire involving a flammable gas cylinder?**

Foam or dry powder extinguishers can be used on flammable gas fires. Cylinders containing propane or butane come into this category. Water should be used to cool a cylinder if it is leaking because it is necessary to shut off the gas supply at the cylinder as soon as possible. If this is not done, re-ignition can occur.

Questions and answers: Part 3

Section 1

2   **Operatives in your workplace report general discomfort when working with a material. Discuss how you would assess the problem.**

The first step is to find out as much as possible about the material. This information should be provided by the manufacturer or seller, if the material is brought into the workplace. Chemical information can be found in reference books or databases if the material is a production by-product. This enables the hazard to be recognised. Measurement of the hazard is then done by finding out what standards are applicable in relation to normal exposure to the material. The information for this is likely to be available already from the first step.

The third step, evaluation of the risk in practice, is carried out by examining the method of use, handling, storage, transport and disposal of the material. Control of the risk(s) by design, engineering, elimination and substitution can be carried out with this knowledge. Control may also include the introduction of revised systems of work, PPE and other techniques. The final requirement of assessment is to keep in mind the need for review as conditions change, and arrangements should be made to monitor the chosen control methods at intervals to ensure that they are still effective and that the risks have not altered.

3   **Describe how the body can defend and repair itself when the skin is cut.**

As soon as the skin is cut, blood flows and cleans the area of the cut. It rapidly clots, preventing germs from getting into the blood or surrounding tissue through the cut. Any foreign material entering through the cut will be attacked by white blood cells, which can deal with it by chemical actions or by absorbing it. The presence of these cells is encouraged by the release of histamine in the area of the cut, which dilates local blood vessels and increases blood flow to the area.

Repair of the area of the cut involves the removal of dead blood cells and other tissue,

and the creation of scar tissue to repair physical damage.

4   **Give examples of reflexes which take part in the body's response to the presence of foreign substances.**

The coughing reflex is produced by irritating the lining of the respiratory tract – the airway from the outside into the lungs. This reflex stimulates the diaphragm to expel most of the air in the lungs in a violent movement, which forces air at speed through the respiratory tract and helps clear any particles stuck on the walls or floating in the air within the tract.

Vomiting and diarrhoea are reflexes of the gastro-intestinal tract, the pathway for food and drink through the body from entry to exit. These reflexes stimulate violent contractions of the muscular walls of the tract, which then expels substances or quantities which the body is not able to deal with by conventional means.

5   **'Ingestion of toxic chemicals is a rare method of contracting industrial disease.' Discuss this statement, giving examples from your own experience.**

Ingestion of substances happens when substances pass into the digestive system in contrast or in addition to inhalation through the respiratory system. Any toxic substance which can get into the mouth directly, such as when eaten by mistake or by intention, or indirectly as when carried on food, is a hazard and potentially disease-producing. Direct entry can be guarded against by clear use of containers and labelling to prevent errors. Indirect entry is often an outcome of poor hygiene and welfare standards. Ability to wash the hands before eating is important, and to take meals in areas that are not contaminated with airborne toxic substances.

The rarity of ingestion of toxic chemicals is likely to be related to the industry concerned. Those working with lead and asbestos, for example, are known to be prone to ingestion unless scrupulous hygiene standards are observed.

6 **Outline the measures which can be taken to prevent outbreaks of dermatitis.**

Dermatitis is often a result of use of materials which produce an abrasive effect on the skin, or which remove natural skin secretions and allow the skin surface to dry and crack. Examination of materials in use which have the potential to cause dermatitis is a necessary first step to take.

Workers should be subject to regular simple inspection of hands and other exposed skin if they are at risk of dermatitis. Good personal hygiene practised by informed workers is the best defence in the absence of alternative materials. Use of barrier creams and protective clothing are two other methods which are part of the personal hygiene routine, but the choice of steps to take should not be left to individuals by management.

Section 2

1 **What are the sources of occupational exposure limits?**

Occupational exposure limits are recommended or compulsory national and international standards for airborne contaminants, including most gases. In the UK, the HSE publishes limits annually in the Guidance Note EH40, and as necessary. In the USA, the American Conference of Governmental Industrial Hygienists (ACGIH) publishes annual lists of threshold limit values (TLVs) and the Occupational Safety and Health Administration (OSHA) publishes national standards on the recommendation of the Research Section of NIOSH, the National Institute for Occupational Safety and Health.

Section 3

1 **Explain the advantages of monitoring air quality using stain detector tubes.**

Stain detector tubes are cheap and easy to obtain. The pumps in which they are used are also relatively cheap, and are simple measuring devices which are unlikely to go wrong. Detector tubes give fast indication of the presence of a contaminant, and a measurement

can be made easily without any special skills except in the simple method of taking a sample. When used, the tube is easy to dispose of. Only the general atmospheric level of a contaminant can be measured in this way, and the contaminant must be one of those for which stain detector tubes are available.

Section 4

1 **Differentiate between local exhaust ventilation and dilution ventilation.**

Local exhaust ventilation traps contaminants at or close to their origins, and removes them through specially built ventilation systems before they enter the breathing zones of workers. Examples include the use of a small hood, a ventilated enclosure or a booth. General or dilution ventilation uses fresh air to dilute a contaminated atmosphere. Although there is a general need for a regular supply of fresh air to a workplace, this should not be used as a control measure for removing contaminants unless there is only a small amount to be removed, which is evenly distributed in the workplace and is of low toxicity.

Section 5

1 **Explain how noise-induced hearing loss can be caused by noise at work.**

Sound is both rapid pulsations in air pressure produced by a vibrating source, and the auditory sensation produced by the ear. The mechanism of the ear is sensitive to the pulses. These are passed to it by bone conduction, and by the hearing system. This begins with the collecting action of the outer ear which captures sound, then the mechanical vibration of the ear drum produces movements of the three middle-ear connecting bones. They pass the pulses into the fluid contents of the inner ear, the cochlea. The fluid's motion causes hair cells to rub against a membrane, which causes an electrical discharge in the hair cells. This is passed to the auditory nerve and received by the brain.

Any interruption in this process, by the body as a defence mechanism or as a result of wear or injury by excessive sound, will result in hearing defects. There are several ways in which the hearing system can fail. Noise-induced hearing loss is usually associated with the wearing out of the hair cells in the inner ear. This condition is not reversible.

Section 6

1 **An oxygen-deficient atmosphere has been established. Discuss the forms of RPE which should be used for (a) continuous work in the area and (b) short-term work only.**

In oxygen-deficient atmospheres, RPE must provide a continuous supply of air. There are three types of equipment commonly available which will do this. The choice will depend on the circumstances of the work to be done, including its duration, degree of difficulty and accessibility required.

Fresh air hose apparatus is relatively inexpensive, and can supply air from an external source by pump bellows or by breathing action. This would be suitable for continuous work, provided that mobility is not required, that high demand for air is not likely, and that the hose is not longer than 10m.

Compressed airline apparatus involves more equipment than the above and has the same mobility restriction, but may be useful if the outside air is contaminated or if a high air demand is required.

The final type of RPE which can be used here is self-contained breathing apparatus, which is independent of external supply or control. It has a relatively short duration of use before the cylinders need to be refilled, and it can be heavy, so it would be very suitable for short-duration entry where mobility is needed but potentially less so for continuous work.

A more detailed analysis of the circumstances would be needed before a choice can be made of equipment for continuous work.

Section 7

1 **What are the main arguments for and against the precautionary removal of asbestos-containing material from the workplace?**

Materials containing asbestos could release fibres into the atmosphere under certain conditions, including deliberate disturbance through building work or vandalism, and unintentional disturbance such as collisions of passing vehicles with coated structures. As a precaution, removal of the hazard is the best way of controlling the risk, and the size of the risk will determine the decision. Frequency of contact and the type of asbestos are two of the factors to be considered.

Arguments in favour of removal are:
- removal of the hazard.

Arguments against removal are:
- cost
- possible low risk of disturbance and release of fibres
- release of fibres during removal
- risk to employees and third parties.

Section 8

1 **Explain the difference between ionising and non-ionising radiations.**

Ionising radiation is a form of energy which causes the ionisation of matter with which it interacts. The energy of the radiation is sufficient to dislodge electrons from matter to which it is exposed, and in the case of the human body this can produce tissue changes. Examples are alpha, beta, gamma and X-rays.

Non-ionising radiation does not cause ionisation and its effects, although it can have other negative consequences for the body. Non-ionising radiation includes the electromagnetic spectrum between ultraviolet and radio waves, and artificially generated laser beams.

2 **What control measures would be appropriate to ensure the health and safety of employees using or working near a microwave oven?**

Microwaves are non-ionising radiations, produced artificially in ovens, where the heat of the absorbed energy cannot be dispersed so that the temperature rises proportionately with the inability to absorb the heat.

The most important control measure is to ensure, by inspection, testing and original purchase verification, that the microwave oven complies with current national laws and/or standards, particularly those relating to output power restrictions, door seals and warning lights. The warning lights and door seals then require inspection and testing at least annually, depending on use.

Training for users is another control measure which is appropriate. This should include knowledge of the function of the door seals and the need to maintain them properly; also in the need to use a dummy load when testing (water in a suitable container) to absorb unwanted energy. Other necessary knowledge is the need to check the warning light indicating operation is functioning on each use, and not to put metallic or metal-finished objects inside the oven. As a precaution against leaking seals, the oven should not be operated close to a permanent workstation or seating area.

Section 9

2 **How can a knowledge of ergonomics help in assessing manual handling risks?**

Important elements of the assessment of manual handling operations are: task, load, individual characteristics and the environment in which the task will be carried out. A knowledge of ergonomic principles can help in each of these, with the result being to fit the task to the worker and not the other way around.

Understanding the limitations on human performance, designing handling equipment for safe operation, effects of extremes of cold or heat, and preferred load parameters for ease of handling are examples of the practical uses of ergonomic principles.

Questions and answers: Part 4

Section 2

1 Write notes on the structure and functions of magistrates' courts, and explain the extent of the powers of the court in relation to health and safety legislation.

Magistrates' courts hear criminal and civil cases; over 90 per cent of all criminal prosecutions begin and end there. Health and safety cases nearly always start in these courts, and most are completed there as well. Magistrates are selected members of the public, not required to have legal training. Their powers of sentencing include fines of up to £5,000 and/or up to six months' imprisonment for the more serious cases, and fines of up to £20,000 for breaches of the 'general duties' sections of the Health and Safety at Work etc Act 1974. If they feel their powers are insufficient, a case can be referred to the Crown court for trial and/or sentence.

2 How do European directives affect health and safety in the UK?

Directives are documents, created through complex consultation, negotiating and voting systems within the European Union, which set out standards to be achieved within each member state after a date stated in the directive. Therefore, each member state must review existing legal provisions covering the subject, and pass appropriate internal legislation to meet or exceed the standards required by the directive.

Section 3

1 Outline the main differences between common law and statute law.

Common law has evolved over hundreds of years, resulting in a system of precedents or existing cases of record which are binding on future similar cases unless overruled by a higher court or by statute.

Statute law is the written law of the land and consists of Acts of Parliament and the rules, regulations and orders made within the provisions of the Acts. Acts of Parliament dealing with health and safety matters usually set out a framework of objectives and use specific regulations or orders to achieve them.

The penalty for a proven breach of duties at common law is to pay compensation for the injury or damage caused after the event. The penalty for breach of statutory duty is usually a fine or imprisonment for a criminal offence, and a case may be made for such a breach even if no accident has taken place.

2 Explain the meaning of 'reasonably practicable'.

Use of this term in legislation permits an employer, or the duty holder, to balance the cost of taking action (in terms of time and inconvenience as well as financial) against the risk being considered. If the risk is insignificant, such that the balance between the two is 'grossly disproportionate', the steps need not be taken, but lesser steps must then be evaluated in the same way. The meaning of this and other terms in health and safety law may be either defined within the particular law, or derived from legal precedents where binding opinions on meaning and interpretation have been given by a superior court.

Section 4

1 What are the powers of an HSE inspector? What enforcement action can be taken by an inspector to prevent dangerous acts taking place?

HSE inspectors can:
- gain access to any place of work at any time
- seek support from the police if entry is made difficult
- take equipment onto premises to assist them
- carry out inspections and examinations
- insist that areas be left undisturbed
- take measurements, photographs and samples
- remove equipment for test or examination
- take statements, records or documents
- request facilities to be made available to assist enquiries
- do anything else necessary to carry out enforcement duties.

Inspectors can issue prohibition or improvement notices, and seize, render harmless or destroy items considered to be a source of imminent danger, and prosecute offenders.

2 Summarise the employer's duties under the Health and Safety at Work etc Act 1974.

Generally, the employer must safeguard the health, safety and welfare at work of his/her employees, with specific regard to the provision of:

- safe plant and systems of work
- safe handling, storage, transport and maintenance of articles and substances used at work
- necessary information, instruction, training and supervision
- a safe place of work and access and egress
- a safe working environment and adequate welfare facilities.

The employer must provide a safety policy setting out how these objectives will be achieved, in writing if he/she employs five or more. He/she must prevent risks arising from his/her work activities which affect the self-employed, members of the public and employees of other employers. All the employer's duties above are subject to the test of reasonable practicability. He/she must also establish an adequate consultation process with his/her workforce.

Section 6

2 Summarise the main ways in which the Management of Health and Safety at Work Regulations 1999 expand the duties of the employer under the Health and Safety at Work etc Act 1974.

The Regulations have been claimed to be simple extensions and definitions of the general duties of the employer under the Health and Safety at Work etc Act 1974. However, there are several additional duties, including the requirement to record information of various kinds.

Other duties include making assessments, establishing management frameworks,

appointment of two types of competent person, formulating procedures for action in the event of serious and imminent danger and access to danger areas, co-ordination with other employers, training and capability assessment, and health surveillance.

Section 7

2 Write notes on the application of the Workplace (Health, Safety and Welfare) Regulations to a forestry worker in a wood.

Forestry workplaces which are outdoors and away from the undertaking's main buildings are excluded from the requirements of the Regulations, except for those on sanitary conveniences, washing facilities and drinking water (Regulations 20–22), which apply 'so far as is reasonably practicable'.

Section 10

2 Write notes on how the Manual Handling Operations Regulations affect a company supplying and installing double glazing.

A list of all manual handling tasks must be made, from fabrication through to installation. It may be reasonable to automate or mechanise some of these, but there will be occasions when the products have to be handled manually. Written risk assessments will be required to determine the preventive measures, including any requirement for protective clothing or other equipment. Account will be taken of the more general requirements of the MHSWR, especially those covering capability assessment of individuals. The company will have to decide how best to provide information to the handlers about the weight and centre of gravity of the loads, and also the packaging weight and content.

Section 11

2 Explain the difference between a DSE 'user' and an 'operator'.

A user is any employee who habitually uses display screen equipment as a significant part of his or her normal work. An operator is a self-employed user.

Section 12

2 **Write notes on the duties of employers under the Personal Protective Equipment at Work Regulations.**

The employer is required to: provide PPE where risks have not been adequately controlled by other means equally or more effective; ensure suitability through assessment, and compatibility with other PPE where used; maintain and replace lost or defective items; provide accommodation for PPE; and provide instruction, information and training for users.

Section 13

2 **List the contents of a fire certificate. Why are they crucial to planning for fire safety in the workplace?**

The contents of a fire certificate may include the following details: the use of the premises, the means of escape in case of fire and ensuring its maintenance, the means of fighting fire, the means of raising the alarm, and particular requirements relating to explosives or highly flammable liquids kept on the premises. Additional provisions may be written into the certificate, at the discretion of the local fire authority. As a result of the potential for extra provisions, the cost implications must be taken into account and may have a variety of consequences.

Section 14

1 **A new substance is to be introduced into your workplace. What basic steps should be taken to ensure compliance with the COSHH Regulations?**

Obtain full information about the substance, using manufacturer's data sheets, labels, reference works, Guidance Note EH40 and other appropriate sources. Examine how the substance is to be used in the workplace, including all points in the user chain from bulk entry to end disposal, as appropriate. Evaluate the risks of the proposed use, and assess what, if any, control measures are necessary. Introduce

the control measures. The (written) assessment and control measures should be communicated to employees and others who may be affected by means of written information, instruction and training as necessary. Exposure should be monitored together with the effectiveness of control measures where necessary.

Section 15

1 **What are the conditions under which live electrical working is permissible?**

There are three basic conditions to be fulfilled. It must be unreasonable in all the circumstances of the situation under consideration for the equipment to be made dead. It must be similarly reasonable for the work to be carried out live, and adequate precautions must be taken to prevent injury.

2 **What are the qualities a person must have to carry out work with electrical apparatus?**

He or she must possess the necessary knowledge and/or experience to do the work safely, or must be under appropriate supervision given the nature of the work.

Section 17

1 **The noise exposure of some of your workers has been measured and found to fall between the First and Second Action Levels. What steps should be taken to ensure compliance with the Noise at Work Regulations?**

There is a general duty to reduce the risk of hearing damage generally to the lowest level reasonably practicable. The establishment of the given noise level also requires the following steps to be taken:
- make noise assessments (to be made by a competent person)
- record the results of the noise assessments
- provide information, instruction and training to employees about the risks and how to minimise them
- provide hearing protectors to those people who ask for them

- maintain and repair hearing protectors provided
- ensure that any additional noise controls (such as baffles and enclosures) are used and maintained.

2 **What are the main differences between the action required to be taken when the Second or Peak Action Levels are exceeded, and that required when the First Action Level is exceeded?**

Exceeding the Second or Peak Action Levels requires the following steps in addition to those required when the First Action Level is reached:

- reduction of noise exposures as far as is reasonably practicable, by means other than by provision of hearing protectors
- marking out of hearing protection zones, with notices displayed
- provision of hearing protectors to all exposed employees
- ensuring that the protectors are used by all persons exposed.

Section 19

1 **What considerations would have to be made to determine first aid requirements for a construction site?**

Matters to be taken into consideration in determining the adequacy of first aid arrangements include the nature of the industry, the nature of the work and its location, the likely risks involved at the particular location, the size of the site and the numbers employed there. It could be useful to arrange for the provision of central first aid facilities by one contractor and have these made available for the use of all.

Construction is considered to be a high-risk industry, often involving sites where access to medical treatment is difficult and may take time. Under these circumstances, a more extensive first aid facility would be called for, staffed by appropriately qualified personnel.

2 **Explain the difference between a first aider and an appointed person.**

A first aider is a person who has received training and obtained qualifications in first aid which are approved by the HSE, the qualifications being current and renewable at intervals. An 'appointed person' is a person who is capable of taking charge in an emergency and in the absence of a first aider. They should be capable of administering emergency first aid, and are responsible for the facilities provided under their charge.

Section 20

2 **Are the following reportable under the Regulations?**

1 A resident in a nursing home assaults a nurse, causing absence for a week.

Answer: **Yes**

2 Two cars are involved in an accident on a motorway. One of the drivers, who is driving on business during normal working hours, is killed.

Answer: **No**

3 An employee is unloading materials from a lorry parked in a road outside a building site. He is hit by a passing car, admitted to hospital for observation and released two days later.

Answer: **Yes**

4 A child falls in a caravan park while playing football and is taken to hospital by his parents. Examination reveals no damage other than a twisted knee and the child is discharged.

Answer: **Yes**

Index